COMPUTING FOR SCIENTISTS AND ENGINEERS

COMPUTING FOR SCIENTISTS AND ENGINEERS

A Workbook of Analysis, Numerics, and Applications

WILLIAM J. THOMPSON
University of North Carolina
Chapel Hill, North Carolina

A Wiley-Interscience Publication
JOHN WILEY & SONS, INC.
New York / Chichester / Brisbane / Toronto / Singapore

Copyright © 1992 by John Wiley & Sons, Inc.

Library of Congress Cataloging in Publication Data:

Thompson, William J. (William Jackson), 1939 -
Computing for Scientists and Engineers : a workbook of analysis,
numerics, and applications / William J. Thompson.
    p.   cm.
        Rev. ed. of: Computing in applied science. c1984.
        "A Wiley-Interscience publication."
        Includes bibliographical references and indexes.
        ISBN 0-471-54718-2 (cloth)
        1. Numerical analysis–Data processing. 2. Science–Data
processing. 3. Engineering–Data processing. I. Thompson, William
J. (William Jackson), 1939- Computing in applied science.
II. Title.
QA297.T5 1992
519.4–dc20   OCLC: 25873613   92-16744

PREFACE

This preface is addressed to readers who are interested in computing but who seldom either consult manuals or read prefaces. So, I will be brief.

Computing requires an integrated approach, in which scientific and mathematical analysis, numerical algorithms, and programming are developed and used together. The purpose of this book is to provide an introduction to analysis, numerics, and their applications. I believe that a firm grounding in the basic concepts and methods in these areas is necessary if you wish to use numerical recipes effectively. The topics that I develop extensively are drawn mostly from applied mathematics, the physical sciences, and engineering. They are divided almost equally among review of the mathematics, numerical-analysis methods (such as differentiation, integration, and solution of differential equations from the sciences and engineering), and data-analysis applications (such as splines, least-squares fitting, and Fourier expansions).

I call this a workbook, since I think that the best way to learn numerically oriented computing is to work many examples. Therefore, you will notice and, I hope, solve many of the exercises that are strewn throughout the text like rocks in the stream of consciousness. I try to introduce you to a technique, show you some of the steps, then let you work out further steps and developments yourself. I also suggest new and scenic routes rather than overtraveled highways. There are occasional diversions from the mainstream, often to point out how a topic that we are developing fits in with others in computing and its applications.

The programming language in which I present programs is C. This language is now used extensively in systems development, data-acquisition systems, numerical methods, and in many engineering applications. To accommodate readers who prefer Fortran or Pascal, I have used only the numerically oriented parts of C and I have structured the programs so that they can usually be translated line-by-line to these other languages if you insist. The appendix summarizes the correspondences between the three languages. All programs and the functions they use are listed in the index to programs. The fully-developed programs are usually included in the projects near the end of each chapter.

This book may be used either in a regular course or for self-study. In a one-semester course for college students at the junior level or above (including graduate students), more than half the topics in the book could be covered in detail. There is adequate cross-referencing between chapters, in addition to references to sources for preparation and for further exploration of topics. It is therefore possible to be selective of topics within each chapter.

Now that we have an idea of where we're heading, let's get going and compute!

William J. Thompson

Chapel Hill, July 1992

CONTENTS

COMPUTING FOR SCIENTISTS AND ENGINEERS

Chapter 1

INTRODUCTION TO APPLICABLE MATHEMATICS AND COMPUTING

The major goal for you in using this book should be to integrate your understanding, at both conceptual and technical levels, of the interplay among mathematical analysis, numerical methods, and computer programming and its applications. The purpose of this chapter is to introduce you to the major viewpoints that I have about how you can best accomplish this integration.

Beginning in Section 1.1, is my summary of what I mean by applicable mathematics and my opinions on its relations to programming. The nested hierarchy (a rare bird indeed) of computing, programming, and coding is described in Section 1.2. There I also describe why I am using C as the programming language, then how to translate from to Fortran or Pascal from C, if you insist. The text has twelve projects on computing, so I also summarize their purpose in this section. Section 1.3 has remarks about the usefulness of graphics, of which there are many in this book. Various ways in which the book may be used are suggested in Section 1.4, where I also point out the guideposts that are provided for you to navigate through it. Finally, there is a list of general references and many references on learning the C language, especially for those already familiar with Fortran or Pascal.

1.1 WHAT IS APPLICABLE MATHEMATICS ?

Applicable mathematics covers a wide range of topics from diverse fields of mathematics and computing. In this section I summarize the main themes of this book. First I emphasize my distinctions between programming and applications of programs, then I summarize the purpose of the diversion sections, some new paths to familiar destinations are then pointed out, and I conclude with remarks about common topics that I have omitted.

Analysis, numerics, and applications

In computing, the fields of analysis, numerics, and applications interact in complicated ways. I envisage the connections between the areas of mathematical analysis, numerical methods, and computer programming, and their scientific applications as shown schematically in Figure 1.1.

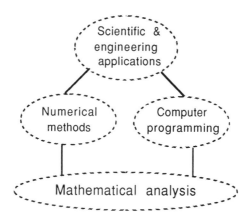

FIGURE 1.1 Mathematical analysis is the foundation upon which numerical methods and computer programming for scientific and engineering applications are built.

You should note that the lines connecting these areas are not arrows flying upward. The demands of scientific and engineering applications have a large impact on numerical methods and computing, and all of these have an impact on topics and progress in mathematics. Therefore, there is also a downward flow of ideas and methods. For example, numerical weather forecasting accelerates the development of supercomputers, while topics such as chaos, string theory in physics, and neural networks have a large influence on diverse areas of mathematics.

Several of the applications topics that I cover in some detail are often found in books on mathematical modeling, such as the interesting books by Dym and Ivey, by Meyer, and by Mesterton-Gibbons. However, such books usually do not emphasize much computation beyond the pencil-and-paper level. This is not enough for scientists, since there are usually many experimental or observational data to be handled and many model parameters to be estimated. Thus, interfacing the mathematical models to the realities of computer use and the experience of program writing is mandatory for the training of scientists and engineers. I hope that by working the materials provided in this book you will become adept at connecting formalism to practice.

By comparison with the computational physics books by Koonin (see also Koonin and Meredith), and by Gould and Tobochnik, I place less emphasis on the physics and more emphasis on the mathematics and general algorithms than these authors provide. There are several textbooks on computational methods in engineering and science, such as that by Nakamura. These books place more emphasis on specific problem-solving techniques and less emphasis on computer use than I provide here.

Data analysis methods, as well as mathematical or numerical techniques that may be useful in data analysis, are given significant attention in this book. Examples are spline fitting (Chapter 5), least-squares analyses (Chapter 6), the fast Fourier transform (Chapter 9), and convolutions (Chapter 10). My observation is that, as part of their training, many scientists and engineers learn to apply data-analysis methods without understanding their assumptions, formulation, and limitations. I hope to provide you the opportunity to avoid these defects of training. That is one reason why I develop the algorithms fairly completely and show their relation to other analysis methods. The statistics background to several of the data-analysis methods is also provided.

Cooking school, then recipes

Those who wish to become good cooks of nutritious and enjoyable food usually go to cooking school to learn and practice the culinary arts. After such training they are able to use and adapt a wide variety of recipes from various culinary traditions for their needs, employment, and pleasure. I believe that computing — including analysis, numerics, and their applications — should be approached in the same way. One should first develop an understanding of analytical techniques and algorithms. After such training, one can usually make profitable and enlightened use of numerical recipes from a variety of sources and in a variety of computer languages.

The approach used in this book is therefore to illustrate the processes through which algorithms and programming are derived from mathematical analysis of scientific problems. The topics that I have chosen to develop in detail are those that I believe contain elements common to many such problems. Thus, after working through this book, when you tackle an unfamiliar computing task you will often recognize parts that relate to topics in this book, and you can then probably master the task effectively.

Therefore I have not attempted an exhaustive (and exhausting) spread of topics. To continue the culinary analogs, once you have learned how to make a good vanilla ice cream you can probably concoct 30 other varieties and flavors. I prefer to examine various facets of each topic, from mathematical analysis, through appropriate numerical methods, to computer programs and their applications. In this book, therefore, you will usually find the presentation of a topic in this order: analysis, numerics, and applications.

The level I have aimed at for mathematical and computational sophistication, as well as for scientific applications, is a middle ground. The applications themselves do not require expert knowledge in the fields from which they derive, although I give appropriate background references. Although I present several topics that are also in my earlier book, *Computing in Applied Science* (Thompson, 1984), the preliminary and review materials in this book are always more condensed, while the developments are carried further and with more rigor. The background level necessary for the mathematics used in this book is available from mathematics texts such as that by Wylie and Barrett, and also from the calculus text by Taylor and Mann.

Diversions and new routes

In order to broaden your scientific horizons and interests, I have included a few diversion sections. These are intended to point out the conceptual connections between the topic under development and the wider world of science, technology, and the useful arts.

The diversions include a discussion of the interpretation of complex numbers (in Section 2.5), the relationships between recursion in mathematics and computing (Section 3.4), the development of computers and the growing use of spline methods in data analysis (Section 5.6), and the Wilbraham-Gibbs overshoot in the Fourier series of discontinuous functions (Section 9.6). Other connections are indicated in subsections of various chapters. Although these diversions may not necessarily be of much help in making you an expert in your field of endeavor, they will help you to appreciate how your field fits into the larger scientific landscape.

This book also uses some seldom-traveled routes to reach known destinations, as well as a few tracks that are quite new. These routes and their results include linearized square-root approximations (Section 3.3), the relations between maximum likelihood, least squares, and Fourier expansion methods (Sections 6.1, 6.2, 9.2), algorithms for least-squares normalization factors (Section 6.4), logarithmic transformations and parameter biases (Section 6.5), generalized logistic growth (Section 7.3), a novel approach to catenaries (Section 8.2), the discrete and integral Fourier transforms of the complex-exponential function (Sections 9.2 and 10.2), and the Wilbraham-Gibbs overshoot (Section 9.6).

Roads not taken

Because this book is directed to a readership of mathematicians, scientists, engineers, and other professionals who may be starting work in fields from which examples in this book are drawn, there are many byways that I have not ventured upon. Rather, I emphasize principles and methods that are of both general validity and general applicability.

One road not taken is the one leading to topics from linear algebra (except incidentally) and matrix manipulation. Mathematics texts are replete with examples of methods for 3×3 matrices, many of which will usually fail for matrices of typically interesting size of 100×100. I believe that matrix computation is best taught and handled numerically by using the powerful and somewhat advanced methods specially developed for computers, rather than methods that are simply extensions of methods suitable for hand calculations on small matrices.

Symbolic calculation, also misleadingly termed "computer algebra," is not discussed here either. I believe that it is dangerous to just "give it to the computer" when mathematical analysis is necessary. Machines should certainly be used to solve long and tedious problems reliably and to display results graphically. However, the human who sets up the problem for the computer should always understand clearly the problem that is being solved. This is not likely to be so if most of one's training has been through the wizardry of the computer. I have the same objection to the overuse of applications programs for numerical work until the principles have been mastered. (Of course, that's one reason to set oneself the task of writing a book such as this.) In spite of my warnings, when you have the appropriate understanding of topics, you should master systems for doing mathematics by computer, such as those described in Wolfram's *Mathematica*.

Finally, I have deliberately omitted descriptions of computational methods that are optimized for specialized computer designs, such as vectorized or parallel architectures. You should learn and use these methods when you need them, but most of them are developed from the simpler principles that I hope you will learn from this book. You can become informed on both parallel computation and matrix methods by reading the book by Modi.

1.2   COMPUTING, PROGRAMMING, CODING

When computing numerically there are three levels that I envision in problem analysis: program design, coding, and testing. They are best described by Figure 1.2.

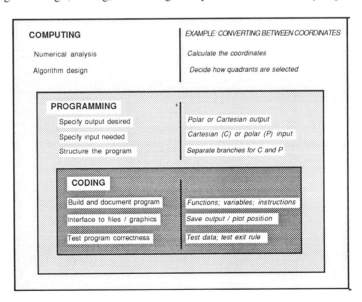

FIGURE 1.2 Computing, programming, and coding form a nested hierarchy. An example of this nesting activity is that of converting between coordinates (Section 2.6).

In Figure 1.2 the activity of computing includes programming, which includes coding. The right side of the figure shows the example of converting between Cartesian and polar coordinates — the programming project described in Section 2.6.

The aspects of *computing* that lie outside programming and coding are numerical analysis and (to some degree) algorithm design. In the example, the formulas for calculating coordinates are part of numerical analysis, while deciding how quadrants are selected is probably best considered as part of algorithm design.

At the *programming* level one first has to decide what one wants to calculate, that is, what output the program should produce. Then one decides what input is needed to achieve this. One can then decide on the overall structure of the program; for example, the conversions for Cartesian and polar coordinates are probably best handled by separate branches in the program. At this level the choices of computing system and programming language usually become important.

Finally, as shown in Figure 1.2, one reaches the *coding* level. Here the program is built up from the language elements. In the C language, for example, functions are written or obtained from function libraries, variable types are declared, and variables are shared between functions. Detailed instructions are coded, the interfaces to files or graphics are written, and the program is tested for correctness of formulas and program control, such as the method used to terminate program execution. If you think of the activity of computing as a nested three-level system, as schematized in Figure 1.2, then you will probably produce better results faster than if you let your thoughts and actions become jumbled together like wet crabs scuttling in a fishing basket.

In the following parts of this section, I make remarks and give suggestions about programming as described in this book. First, I justify my choice of C as the language for preparing the programs, then I give you some pointers for learning to program in C, and for translating the programs in this book to Fortran or Pascal from C (if you insist on doing this). Next, I summarize programming aspects of the projects that occur toward the end of each of Chapters 2 through 10, and I remark on the correctness and portability of these programs. Finally in this section, I draw your attention to a convenience, namely, that all the programs and functions provided in this book are listed by section and topic in the index to computer programs before the main index.

The C language for the programs

I decided to write the sample programs in C language for the following reasons. First, C is a language that encourages clear program structure, which leads to programs that are readable. My experience is that C is easier to write than Pascal because the logic of the program is usually clearer. For example, the use of a top-down structure in the programs is closer to the way scientists and enginners tackle real problems. In this aspect C and Fortran are similar. The C language is more demanding than Fortran, in that what you want to do and the meanings of the variables must all be specified more accurately. Surprisingly, scientists (who pride themselves on precise language) often object to this demand from a computer. I estimate that C is the easiest of the three languages in which to write programs that are numerically oriented. Ease is indicated by the time it takes to produce a correctly executing program in each language.

The second reason for using C in this book is that it is intermediate in complexity between Fortran and Pascal, as illustrated by the comparison chart in the appendix. That is, there are very few elements in the Fortran language (which makes it simple to write but hard to understand), most of the elements of Fortran are in C, and some of the elements of Pascal are also in C. For data handling, C is much more powerful and convenient than Fortran because facilities for handling characters and strings were designed into the original C language.

The third reason for my choice of C is that it is now the language of choice for writing applications programs for workstations and personal computers. Therefore, programs that you write in C for these machines will probably interface easily with such applications programs. This is a major reason why C is used extensively in engineering applications.

A fourth reason for using C is that its developers have tried to make it portable across different computers. Lack of portability has long been a problem with Fortran. Interconnectivity between computers, plus the upward mobility of programs developed on personal computers and workstations to larger machines and supercomputers, demand program portability. Since very few large computer systems have extensive support for Pascal, C is the current language of choice for portability.

One drawback of C lies with input and output. Some of the difficulty arises from the extensive use of pointers in C, and some inconvenience arises from the limited flexibility of the input and output functions in the language. For these reasons, I have written the input and output parts of the sample programs as simply as practicable, without any attempt to produce elegant formats. Since you probably want to modify the programs to send the output to a file for processing by a graphics program, as discussed in Section 1.3, for this reason also such elegance is not worthwhile in the sample programs.

Complex numbers are not part of the C language, although they are used extensively in numerical applications, as discussed in Chapter 2. Our calculations that use complex variables convert the complex numbers to pairs of real numbers, then work with these. Extensive practice with programming using complex numbers in this way is given in Sections 2.1 and 2.6. In *Numerical Recipes in C*, Press et al. also discuss (Chapter 1 and Appendix E) handling complex numbers in C.

Learning to program in C

This book does not claim to be a guide to learning the C programming language. It will, however, provide extensive on-the-job training for the programming of numerical applications in C. If you wish to learn how to program in C, especially for the numerically oriented applications emphasized herein, there are several suitable textbooks. Starting with books which do not assume that you have much familiarity with programming then moving upward, there are Eliason's *C, a Practical Learning Guide* and Schildt's *Teach Yourself C*, then the text by Darnell and Margolis, *C, a*

Software Engineering Approach. Many of C's more subtle and confusing aspects are described by Koenig in *C Traps and Pitfalls.*

If you are familiar with other programming languages and wish to use the C programs in this book, there are several texts that should be of considerable help to you. For general use there is the book by Gehani, which emphasizes the differences between C and other procedural programming languages. Gehani also discusses the advanced aspects of C as implemented on UNIX systems. A second cross-cultural book that will help you with the language barrier is Kerrigan's *From Fortran to C,* which has extensive discussions and examples of how to learn C and to reprogram from Fortran. The book by Müldner and Steele, *C as a Second Language,* and that by Shammas, *Introducing C to Pascal Programmers,* are especially suitable for those who are familiar with Pascal but wish to learn to program effectively in C.

Finally, for detailed references on the C language there are *C: A Reference Manual* by Harbison and Steele, and *The Standard C Library* by Plauger. You should also consult the programming manuals for C provided with the implementation of C for your computing environment, and the manuals that explain the connection between C and your computer's operating system.

The references on learning to program in the C language are listed together in the reference section at the end of this chapter. The appendix provides examples of translating between C, Fortran, and Pascal that are drawn from the C programs in the first chapters.

Translating to Fortran or Pascal from C

By choosing to present the example programs and the project programs in C language, I know that I will have made a few friends but I may have alienated others. Especially for the latter, I have tried to decrease their animosity by avoiding use of some of the useful constructions in C that are sometimes not available or are awkward to implement in other numerically oriented procedural languages. This should make the programs easier to translate on-the-fly into Fortran or Pascal. Among my main concessions are the following.

In arrays the [0] element is usually not used by the program, so that the used elements of the array range from [1] upward. The only confusion this may cause when programming is that the array size must be declared one larger than the maximum element that will ever be used. For example, if you want to be able to use elements 1...100, then the maximum array size (which is always defined as MAX) should be 101. This labeling starting with [1] usually also makes the correspondence between the mathematics and the coding simpler, because most of the summation and iteration indices in formulas (k or j) begin with unity: any zeroth-index value in a summation or iteration has usually to be treated specially. I use the [0] element in an array only if this tightens the connections between the mathematical analysis, the algorithm, and the code.

In summations and indexing, the C construction of $++$ to denote incrementing by one, and similarly $--$ for decrementing, is not used except in *for* loops. Although general avoidance of $++$ and $--$ is less efficient, it is less confusing when translating to Fortran or Pascal, which do not allow such useful constructions.

The *for* loop in C is such a practical and convenient programming device that I use it without concession to Fortran programmers, who are often confined to the much clumsier *DO* loop. However, I use the *for* loop in a consistent style to which you can readily adapt.

I have avoided *go to* statements, so there are no statement labels in the programs. Consequently, you will have to go elsewhere if your favorite computing recipes include spaghetti. (The *come from* statement, which might rescue many a programmer from distress, is also scrupulously avoided.) These omissions do not make C programs difficult to write or to use.

There are a few operators in C, especially the logical operators, that look quite different and may be confusing to Fortran and Pascal programmers. I explain these operators where they appear in programs, especially in the early chapters. They are listed with their Fortran and Pascal counterparts at the end of the appendix on translating between C, Fortran, and Pascal.

In C the `exit` function terminates execution when it is called. (Technically, it terminates the calling process. All open output streams are flushed, all open files are closed, and all temporary files are removed.) There is a conventional distinction, which we follow, between `exit(0)` and `exit(1)`. The first is for successful termination and graceful exit, while the second is to signal an abnormal situation. In some computing environments the process that refers to the terminating program may be able to make use of this distinction.

Within the text, the font and style used to refer to names of `programs`, `functions`, and `variables` is 10-point Monaco (since all programming involves an element of gambling). All the programs and functions are listed in the index to computer programs, which is discussed below.

The computing projects and the programs

Several of the exercises and projects, as described in Section 1.1, require that you modify programs that are in the text. By this means you will practice what is so common in scientific and engineering computing, namely the assembling of tested and documented stand-alone function modules to make a more powerful program tailored for your use. One advantage of this method is that, provided you are careful how you make the modifications, you will usually be able to check the integrity of the program module by comparison with the stand-alone version.

The sample programs, both in the text and in the projects, are written for clarity and efficiency of writing effort. In particular, when there are choices between algorithms, as in the numerical solution of differential equations, the different algorithms are usually coded in-line so that it is easy for you to compare them. Therefore, if you wish to transform one of the chosen sections of in-line code into a function you will need to be careful, especially in the type declarations of variables used.

I have not attempted to make the sample programs efficient in terms of execution speed or use of memory. If you want to use a particular computing technique for production work, after you have understood an algorithm by exploring with the programs provided, you should use a program package specially developed for the purpose and for the computer resources that you have available. At an intermediate level of efficiency of your effort and computer time are the programs available (in C, Fortran, and Pascal) as part of the *Numerical Recipes* books of Press et al.

My view of the connections among materials in this book, the C language, the *Numerical Recipes* books, systems such as *Mathematica*, and their uses in scientific applications is summarized in Figure 1.3.

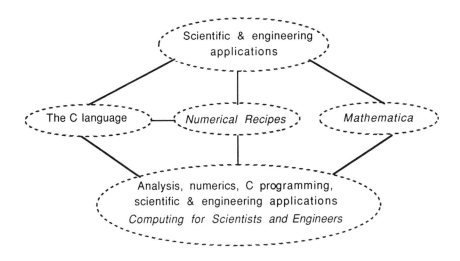

FIGURE 1.3 Connections among topics in this book, C language, the *Numerical Recipes* books, the *Mathematica* system, and scientific applications.

In Figure 1.3 the lines are connectors, not arrows. They indicate the strongest two-way connections between the topics and books (names written in italics). Some significant links have been omitted, mostly for topological reasons. For example, many of the scientific applications examples in this book do not require C programs or use of the *Mathematica* system. Also, much of the latter is programmed in C, and it can convert its symbolic results into C (or Fortran) source code, as described in Wolfram's book on *Mathematica*.

Caveat emptor about the programs

The sample programs included in this book have been written as simply as practical in order that they could readily be understood by the human reader and by the compiler. In order to keep the programs easy to read, I have not included extensive checking of the allowed range of input variables, such as choices that control program options. My rule of thumb has been to put in a range check if I made an input error while testing a program, or if lack of a check is likely to produce confusing results. There are checks for array bounds if they are simple to code and do not interrupt program flow. Errors of use or input that are diagnosed by our C programs always begin with a double exclamation, !!, followed by an explanation of the error. Program execution will often continue after some reasonable fix-up is attempted. A typical fix-up is just to request another input for the troublesome variable.

Because the programs are written to be translated easily to Fortran or Pascal, as described in a previous subsection and shown in the appendix, I have tried to avoid nonstandard parts of C. The compiler that I use claims to follow ANSI standards. I also checked for compatibility with the C language as described in the second edition of Kernighan and Ritchie's book.

In spite of all these precautions, I have two words of advice: *caveat emptor* — let the buyer beware. The programs are supplied as is and are not guaranteed. For each program you use, I suggest that you make at least the checks I have indicated. If you can devise other tests of program correctness, I encourage you to do so.

The index to computer programs

Because this book has many computer programs with associated functions, I have included an annotated index to all the programs and functions. They are listed, by order of appearance, in the index that immediately precedes the regular index. The programs and functions also appear alphabetically by name in the regular index.

1.3   ONE PICTURE IS WORTH 1000 WORDS

In this book graphical output is usually suggested as a way to improve the presentation of results, especially in the projects. Since graphics are so hardware dependent, my suggestions for graphics in the projects are necessarily vague. You should familiarize yourself as much as practicable with techniques of graphical presentation. If you have access to a powerful system that combines graphics and mathematics, such as *Mathematica* as described by Wolfram or by Wagon, you may wish to develop some of the projects by using such a system.

Why and when you should use graphics

Tufte, in his two books on displaying and envisioning quantitative information, has given very interesting discussions and examples of effective (and ineffective) ways of displaying quantitative information from a variety of fields. In numerical applications of mathematics, graphics are especially important because of the enormous number of numbers that come spewing out of the computer in a stream of numerical environmental pollution. If there are many values to be compared, or if you want to show trends and comparisons (as we usually do), it is worth the effort to write a graphical interface for your program. If there are just a few check values to output, it is not worth the extra coding and possible lack of clarity that graphics may produce.

If you have access to *Mathematica* or some other system that combines mathematics, numerics, and graphics, your learning will be enhanced if you combine the three elements when working the exercises and projects. Wagon's book provides many examples of graphics techniques that would be useful in conjunction with this workbook.

Impressive graphics, or practical graphics

In many books that relate to computing you will see elegant and impressive graphics that have been produced by long runs on powerful computers using special-purpose programs. Although these illustrations may improve your comprehension, and perhaps inspire you to become a computer-graphics artist, their production is usually not practicable for most computer users. Therefore, I urge you to find a simple graphics system that interfaces easily to your programming environment, that is readily available to you, and that is inexpensive to use. For example, there are about a hundred line drawings in this book. They were all produced by using only two applications programs (one for graphics and one for drafting). The graphics program used input files that the C programs produced, so the numbers were seldom touched by human hand, and the graphics output was produced on the same laser printer that printed the text.

Many of the programs in this book produce simple output files. I most often used these files for input to graphics, and sometimes for preparing tables. If you make a similar interface and use it often to produce graphics (perhaps through the intermediary of a spreadsheet), I think it will improve your comprehension of the numerical results, without burdening you with much coding effort.

If you have convenient access to a state-of-the-art graphics system, it may be useful for a few of the projects in this book. Just as I believe that an approach to numerical computing that is completely recipe-based is unwise, I believe that using computer-graphics systems without an understanding of their background is similarly unwise. A comprehensive treatment of many aspects of computer graphics is provided in the treatise by Foley et al. Methods for preparing high-resolution graphics, and how to implement them in Pascal, are described in the book by Angell and Griffith.

1.4 SUGGESTIONS FOR USING THIS BOOK

This book may be used for both self-study and for class use. I have some suggestions that should help you to make most effective use of it. First I indicate the conections between the remaining nine chapters, then there are remarks about the exercises and the projects.

Links between the chapters

Because we cover a large amount of territory and a variety of scientific and engineering landscapes in this book, it is useful to have an indication of the connections between its nine other chapters. Table 1.1 summarizes the strength of the links between the chapters.

TABLE 1.1 Cross-reference chart for use of this book.

Key: • chapter above is *necessary* preparation

   • chapter above is *desirable* preparation

   • chapter above is *optional* preparation

	2	3	4	5	6	7	8	9	10
Complex variables	2								
Power series	•	3							
Numerical derivatives and integrals	•	•	4						
Fitting curves through data	•	•	•	5					
Least-squares analysis of data	•	•	•	•	6				
Introduction to differential equations	•	•	•	•	•	7			
Second-order differential equations	•	•	•	•	•	•	8		
Discrete Fourier transforms and series	•	•	•	•	•	•	•	9	
Fourier integral transforms	•	•	•	•	•	•	•	•	10

For example, if you plan to work through Chapter 7 (introduction to differential equations), use of Chapters 2, 5, and 6 is optional, Chapter 3 (power series) is desirable, whereas Chapter 4 (numerical derivatives and integrals) is necessary preparation. Within each chapter you should read not only the text, but also the exercises, which are embedded in the text. Exercises containing equations are especially important to be read, since these equations often form part of the development. Therefore, read over every exercise, even if you don't work it through in detail.

The exercises and projects

Since this book has an overwhelming number of exercises, many of them nontrivial, a guide to use of the exercises is appropriate. It will be clear to you that I always insert an exercise whenever I don't want to show you all the steps of a development. This is not laziness on my part, because I assure you that I have worked through every step. Rather, an exercise provides a checkpoint where you should pause, take stock of what you have been reading, then test your understanding by trying the exercise. If you have difficulty with this exercise, reread all of the subsection containing the exercise, even past the troublesome exercise. Then work the exercise once more. I believe that by doing this you will have a realistic estimate of your progress in comprehension. If you are using this book for self-study, this procedure should help you considerably.

Some exercises are more than checkpoints, they are crossroads where concepts and techniques developed previously are focused on relevant and interesting problems. This type of exercise, which often appears in a project toward the ends of chapters, is always indicated. Such exercises provide very good tests of your overall comprehension of the material in the current and previous chapters.

The projects, of which there is at least one per chapter after this introductory chapter, are designed to bring together many of the aspects of analysis and numerics emphasized within the chapter. They provide you with opportunities to explore the numerics and the science by using the number-crunching and graphical powers of computers. Programming aspects of the projects are discussed in Section 1.2.

REFERENCES FOR THE INTRODUCTION

General references

Angell, I. O., and G. Griffith, *High-Resolution Computer Graphics Using Pascal,* Macmillan Education, Basingstoke, England, 1988.

Dym, C. L., and E. S. Ivey, *Principles of Mathematical Modeling*, Academic Press, New York, 1980.

Foley, J. D., A. van Dam, S. K. Feiner, and J. F. Hughes, *Computer Graphics*, Addison-Wesley, Reading, Massachusetts, third edition, 1990.

Gould, H. and J. Tobochnik, *An Introduction to Computer Simulation Methods*, Addison-Wesley, Reading, Massachusetts, 1988.

Koonin, S. E., *Computational Physics*, Addison-Wesley, Redwood City, California, 1986.

Koonin, S. E., and D. C. Meredith, *Computational Physics: FORTRAN Version*, Addison-Wesley, Redwood City, California, 1990.

Mesterton-Gibbons, M., *A Concrete Approach to Mathematical Modelling*, Addison-Wesley, Reading, Massachusetts, 1989.

Meyer, W. J., *Concepts of Mathematical Modeling*, McGraw-Hill, New York, 1984.

Modi, J. J., *Parallel Algorithms and Matrix Computation*, Clarendon Press, Oxford, England, 1988.

Nakamura, S., *Computational Methods in Engineering and Science*, Wiley-Interscience, New York, 1986.

Taylor, A. E., and W. R. Mann, *Advanced Calculus*, Wiley, New York, third edition, 1983.

Thompson, W. J., *Computing in Applied Science*, Wiley, New York, 1984.

Tufte, E. R., *The Visual Display of Quantitative Information*, Graphics Press, Cheshire, Connecticut, 1983.

Tufte, E. R., *Envisioning Information*, Graphics Press, Cheshire, Connecticut, 1990.

Wagon, S., *Mathematica in Action*, W. H. Freeman, New York, 1991.

Wolfram, S., *Mathematica: A System for Doing Mathematics by Computer*, Addison-Wesley, Redwood City, California, second edition, 1991.

Wylie, C. R., and L. C. Barrett, *Advanced Engineering Mathematics*, McGraw-Hill, New York, fifth edition, 1982.

References on learning and using C

Darnell, P. A., and P. E. Margolis, *C, a Software Engineering Approach*, Springer-Verlag, New York, 1991.

Eliason, A. L., *C, a Practical Learning Guide*, Macmillan, New York, 1988.

Gehani, N., *C: An Advanced Introduction*, Computer Science Press, Rockville, Maryland, 1985.

Harbison, S. P., and G. L Steele, Jr., *C: A Reference Manual*, Prentice Hall, Englewood Cliffs, New Jersey, third edition, 1991.

Kernighan, B. W., and D. M. Ritchie, *The C Programming Language*, Prentice Hall, Englewood Cliffs, New Jersey, second edition, 1988.

Kerrigan, J. F., *From Fortran to C*, Windcrest Books, Blue Ridge Summit, Pennsylvania, 1991.

Koenig, A., *C Traps and Pitfalls*, Addison-Wesley, Reading, Massachusetts, 1989.

Müldner, T., and P. W. Steele, *C as a Second Language*, Addison-Wesley, Reading, Massachusetts, 1988.

Plauger, P. J., *The Standard C Library*, Prentice Hall, Englewood Cliffs, New Jersey, 1992.

Press, W. H., B. P. Flannery, S. A. Teukolsky, and W. T. Vetterling, *Numerical Recipes in C,* Cambridge University Press, New York, 1988.

Schildt, H., *Teach Yourself C*, Osborne McGraw-Hill, Berkeley, California, 1990.

Shammas, N., *Introducing C to Pascal Programmers*, Wiley, New York, 1988.

Chapter 2

A REVIEW OF COMPLEX VARIABLES

The purpose of this chapter is to review your understanding of complex numbers and complex variables, and to summarize results that are used extensively in subsequent chapters. Complex variables are treated elegantly and completely in many mathematics texts, but a treatment in which the aim is to develop intuition in the scientific applications of complex numbers may have much more modest goals, and is best done from a geometrical perspective, rather than from an analytic or algebraic viewpoint.

We start with the algebra and arithmetic of complex numbers (in Section 2.1) including a simple program, then turn in Section 2.2 to the complex-plane representation because of its similarities to plane-polar coordinates and to planar vectors. The simplest (and most common) functions of complex variables — complex exponentials and hyperbolic functions — are reviewed in Section 2.3. Phase angles, vibrations, and complex-number representation of waves, which are all of great interest to scientists and engineers, are summarized in Section 2.4 before we take a diversion in Section 2.5 to discuss the interpretation of complex numbers. Project 2, which includes program `Cartesian & Polar Coordinate Interconversion` for converting between plane-polar and Cartesian coordinates, concludes the text of this chapter. This program serves to emphasize the ambiguities in calculating polar coordinates from Cartesian coordinates, and it will be useful in the later programming applications. References on complex numbers complete the chapter.

The discussion of complex variables is limited to the above topics, and does not develop extensively or with any rigor the topics of analytic functions in the complex plane, their differentiation, or their integration. Although several derivations later in this book, especially those involving integrals, would be simplified if the methods of contour integration were used, the methods of derivation used here are usually direct and do not require the extra formalism of contours and residues. Readers who have experience with functions of a complex variable will often be able to substitute their own methods of proof, which may be more direct than those provided here.

Many of the examples and exercises in this chapter anticipate steps in our later developments that use complex variables, especially the material on Fourier expansions (Chapters 9 and 10). Since we always have a reference to the later material, you may wish to look ahead to see how the complex variables are used.

2.1   ALGEBRA AND COMPUTING WITH COMPLEX NUMBERS

In this section we summarize the algebraic properties of complex numbers, their properties for numerical computation, the relation between complex numbers and plane geometry, and the special operations on complex numbers — complex conjugation, modulus, and argument.

When you have reviewed this section, you will have the conceptual and technical skills for the more extensive developments of complex numbers that are presented in the remainder of this chapter. In particular, Project 2 — the program for converting between coordinates (Section 2.6) — requires most of the ideas from this section. If you are experienced with complex variables, you may try the project before working this section. If you have difficulties with the mathematics in the project (as distinguished from the programming involved), return and rework this section.

The algebra of complex numbers

We indicate a complex number, z, symbolically by

$$z \equiv x + iy \tag{2.1}$$

in which x and y are understood to be both real numbers. The sign $+$ in this formula does not mean arithmetic addition, although it has many rules that are similar to those for addition. You will have already encountered yet another meaning of $+$ as a sign used in the addition of vectors, which is also distinct from arithmetic addition.

In (2.1), the symbol i has the property that

$$i^2 = -1 \tag{2.2}$$

with a unique value being assumed for i itself. In engineering contexts it is more usual to find the symbol i replaced by the symbol j, thus avoiding possible confusion when complex numbers are used to describe currents (i) in electrical circuits, as in our Section 8.1. We will use i, recalling its origins in the initial letter of the historical term "imaginary."

Complex numbers may also be thought of as pairs of numbers, in which the order in the pair is significant. Thus we might write

$$z \equiv (x, y) \tag{2.3}$$

analogously to vectors in a plane. Just as the coordinates (x, y) and (y, x) are usually distinct, so are the analogous complex numbers. The notation in (2.3) avoids ambiguities in using the $+$ sign and in the necessity of inventing a symbol satisfying (2.2). Further, many of the rules for manipulating complex numbers have a strong similarity to those for vectors in a plane if the notation (2.3) is used. Although we will write our results for complex-number algebra and arithmetic in the notation (2.1), you are invited to try the number-pair notation in Exercise 2.1 (c). This notation is also used in some programming languages that allow complex-arithmetic operations.

The rules for manipulating complex numbers must be consistent with those for purely real numbers ($y = 0$) and for purely imaginary numbers ($x = 0$). In the following, let $z = x + i\,y$ generically, and let $z_1 = x_1 + i\,y_1$, $z_2 = x_2 + i\,y_2$ represent two particular complex numbers. Then the following properties hold:

$$z_1 = z_2 \iff x_1 = x_2 \text{ and } y_1 = y_2 \tag{2.4}$$

Negation of a complex number is defined by

$$-z \equiv (-x) + i\,(-y) \tag{2.5}$$

which is often written casually as

$$-z \equiv -x - i\,y \tag{2.6}$$

A complex number is zero only if both its real and imaginary parts are zero, which is consistent with zero being the only solution of the equation $z = -z$.

Addition or *subtraction* of two complex numbers is accomplished by

$$z_1 \pm z_2 \equiv \left(x_1 \pm x_2\right) + i\left(y_1 \pm y_2\right) \tag{2.7}$$

Multiplication of complex numbers is performed by

$$z_1 z_2 \equiv (x_1 x_2 - y_1 y_2) + i\,(x_1 y_2 + y_1 x_2) \tag{2.8}$$

Reciprocal of a complex number is defined by

$$\frac{1}{z} \equiv \left(\frac{x}{x^2 + y^2}\right) + i\left(\frac{-y}{x^2 + y^2}\right) \quad z \neq 0 \tag{2.9}$$

which has the property that $z\,(1/z) = 1$, as for the arithmetic of real numbers.

Division of one complex number into another is based on the reciprocal of the divisor, and is therefore undefined if the divisor is zero:

$$\frac{z_1}{z_2} \equiv z_1\left(\frac{1}{z_2}\right) \quad z_2 \neq 0 \tag{2.10}$$

In order to check your comprehension of these rules for complex arithmetic, try the following exercise.

Exercise 2.1

(*a*) Verify that the rules (2.7) through (2.10) are consistent with those for real arithmetic by checking them for $y_1 = y_2 = 0$.

(*b*) Check the consistency of (2.7) through (2.10) for purely imaginary numbers by setting $x_1 = x_2 = 0$ and noting the condition on i, (2.2).

(*c*) Use the notation for complex numbers as ordered-number pairs, as indicated by (2.3), to write down the preceding complex-arithmetic rules, (2.4) through (2.10). ■

Now that we have summarized the formal basis of complex-variable algebra, it is time to consider complex arithmetic, especially for computer applications.

Programming with complex numbers

Few computer languages are designed to include complex-variable types in their standard definition. They are available in Fortran, but not in C or Pascal. In Wolfram's *Mathematica* system for doing mathematics by computer, which has both symbolic and numeric capabilities, complex numbers can be handled readily. An introduction to their use is provided in Section 1.1 of Wolfram's book.

To appreciate why computer hardware is not built and computer software is not designed to assume that numbers they handle are complex, consider the following exercise.

Exercise 2.2

Show that the total number of real-arithmetic operations needed for complex-number addition and subtraction is 2, the number for multiplication is 6, and the number for division is 11 or 14, depending on whether or not the divisor in (2.9) is stored. ■

We now show a simple program in C language for performing complex arithmetic by the rules given in the preceding subsection. The purpose of this program is twofold: if you are unfamiliar with the C language the program will provide a simple introduction, while it will also develop your understanding of complex arithmetic.

The program `Complex-Arithmetic Functions` takes as input `x1, y1, x2, y2` for the components of two complex numbers `z1` and `z2`. After checking that both numbers are nonzero, it calls the functions for addition, subtraction, multiplication, and division, namely `CAdd`, `CSub`, `CMult`, and `CDiv`, then prints the results before returning for more input. Here is the program.

PROGRAM 2.1 Functions for performing complex arithmetic; addition, subtraction, multiplication, and division.

```c
#include <stdio.h>
#include <math.h>

main ()
{
/* Complex-Arithmetic Functions */
double  x1,x2,y1,y2,x1a2,y1a2,x1s2,y1s2,x1m2,y1m2,x1d2,y1d2;
void CAdd(),CSub(),CMult(),CDiv();

printf("Complex-Arithmetic Functions\n");
x1 = 1;  y1 = 1;  x2 = 1;  y2 = 1;
/* Check that at least one complex numbers is not zero */
while ( x1 !=0  |  y1 != 0  |  x2 != 0  |  y2 != 0 )
  {
  printf("\nInput x1,y1,x2,y2 (all zero to end):\n");
  scanf("%lf%lf%lf%lf",&x1,&y1,&x2,&y2);
  if ( x1 == 0  &&  y1 == 0  &&  x2 == 0  &&  y2 == 0 )
    {
    printf("\nEnd Complex-Arithmetic Functions");
    exit(0);
    }

  CAdd(x1,y1,x2,y2,&x1a2,&y1a2); /* complex Add */
                                 /* returns x1a2, y1a2 */

  CSub(x1,y1,x2,y2,&x1s2,&y1s2); /* complex Subtract */
                                 /* returns x1s2, y1s2 */

  CMult(x1,y1,x2,y2,&x1m2,&y1m2); /* complex Multiply */


                                 /* returns x1m2, y1m2 */
  CDiv(x1,y1,x2,y2,&x1d2,&y1d2); /* complex Divide */
                                 /* returns x1d2, y1d2 */
  printf("\nz1+z2=(%lf) + i(%lf)",x1a2,y1a2);
  printf("\nz1-z2=(%lf) + i(%lf)",x1s2,y1s2);
  printf("\nz1*z2=(%lf) + i(%lf)",x1m2,y1m2);
  printf("\nz1/z2=(%lf) + i(%lf)\n",x1d2,y1d2);
  }
}
```

```
void CAdd(x1,y1,x2,y2,x1a2,y1a2)
/* Complex Addition function */
double x1,y1,x2,y2,*x1a2,*y1a2;
{
*x1a2 = x1+x2;   *y1a2 = y1+y2;
}

void CSub(x1,y1,x2,y2,x1s2,y1s2)
/* Complex Subtraction function */
double x1,y1,x2,y2,*x1s2,*y1s2;
{
*x1s2 = x1-x2;   *y1s2 = y1-y2;
}

void CMult(x1,y1,x2,y2,x1m2,y1m2)
/* Complex Multiplication function */
double x1,y1,x2,y2,*x1m2,*y1m2;
{
*x1m2 = x1*x2-y1*y2;
*y1m2 = x1*y2+y1*x2;
}

void CDiv(x1,y1,x2,y2,x1d2,y1d2)
/* Complex Division function */
double x1,y1,x2,y2,*x1d2,*y1d2;
{
double den;

den = x2*x2+y2*y2;
if ( den == 0 )
   {
   printf("!! CDiv denominator = 0; dividend set to zero");
   *x1d2 = 0;   *y1d2 = 0;
   }
else
   {
   *x1d2 = (x1*x2+y1*y2)/den;
   *y1d2 = (y1*x2-x1*y2)/den;
   }
}
```

The program reveals an immediate difficulty with modifying a language to include complex variables, in that two values must be returned by a complex-variable function. In C this cannot be done simply by a conventional function (which returns just one value, at most). One can get around the problem by using the indirection (dereferencing) operator, written as an asterisk (*) preceding a variable name, as used for each of the two real-variable values returned by the program functions.

Here are some suggestions for exploring complex numbers by using the program Complex-Arithmetic Functions.

Exercise 2.3

(*a*) Use several pairs of real numbers for both inputs ($y1 = 0, y2 = 0$) in order to verify that the complex numbers contain real numbers as special cases.

(*b*) Input purely imaginary numbers ($x1 = 0, x2 = 0$) to the program and verify the correctness of the arithmetic.

(*c*) Show by a careful analytical proof that if the product of two complex numbers is zero, then at least one of the complex numbers is identically zero (both real and imaginary parts zero). Prove that if one of a pair of complex numbers is zero, their product is zero. Verify this by using the program. ∎

With this background of algebraic and arithmetic properties of complex numbers, we are prepared to review some more formal definitions and properties.

Complex conjugation, modulus, argument

In complex-variable algebra and arithmetic one often needs complex quantities that are related by reversal of the sign of just their imaginary parts. We therefore have the operation called *complex conjugation*. In mathematics texts the notation for this operation is often denoted by a bar, $^-$, while other scientists often use an asterisk, as $*$. In the latter notation the complex-conjugate value of z is

$$z* = x - i\,y \tag{2.11}$$

if and only if

$$z = x + i\,y \tag{2.12}$$

From the definition of complex conjugation we can readily derive several interesting results.

Exercise 2.4

(*a*) Prove that

$$z + z* = 2x = 2\,\mathrm{Re}\,z \tag{2.13}$$

where the notation Re stands for "real part of."

(*b*) Similarly, prove that

$$\frac{z - z^*}{i} = 2y = 2 \operatorname{Im} z \qquad (2.14)$$

where Im denotes "imaginary part of."

(*c*) Derive the following properties of complex conjugation:

$$(z_1 \pm z_2)^* = z_1^* \pm z_2^* \qquad (2.15)$$

$$(z_1 z_2)^* = z_1^* \, z_2^* \qquad (2.16)$$

$$(z_1/z_2)^* = z_1^*/z_2^* \qquad z_2 \neq 0 \qquad (2.17)$$

which show that complex conjugation is distributive over addition, subtraction, multiplication, and division. ■

The identity

$$z z^* = x^2 + y^2 \qquad (2.18)$$

shows that $z z^*$ is zero only if z is identically zero, which is an example of the condition from Exercise 2.3 (*c*) for vanishing of the product of two complex numbers.

The frequent occurrence of $z z^*$ and its connection with vectors in two dimensions lead to the notation of the *modulus* of a complex number z, denoted by

$$\operatorname{mod} z = |z| = +\sqrt{x^2 + y^2} \qquad (2.19)$$

Thus mod z indicates the magnitude of z if we picture it as a vector in the x-y plane. Another name for modulus is *absolute value*. For example, the modulus of a real number is just its value without regard to sign, that is, its absolute value. The modulus of a pure imaginary number is just the value of y without regard to sign. The modulus of a complex number is zero if and only if both its real and its imaginary part are zero.

The *argument* of a complex number is introduced similarly to polar coordinates for two-dimensional vectors. One defines the arg function for a complex variable by the requirement that

$$\cos[\arg z] = \frac{x}{\operatorname{mod} z} = \frac{x}{\sqrt{x^2 + y^2}} \qquad (2.20)$$

and the requirement that

$$\sin[\arg z] = \frac{y}{\mathrm{mod}\ z} = \frac{y}{\sqrt{x^2 + y^2}} \tag{2.21}$$

which are necessary and sufficient conditions for definition of $\arg z$ to within a multiple of 2π.

Exercise 2.5

Explain why the commonly given formula

$$\arg z = \mathrm{atan}\left(\frac{y}{x}\right) \tag{2.22}$$

is not sufficient to specify $\arg z$ uniquely, even to within a multiple of 2π. ∎

In Section 2.6, in the program to convert between coordinates, we return to this problem of ambiguous angles. The argument of a complex number is sometimes called the *phase angle*, or (confusingly) the *amplitude*.

One aspect of the argument relates to the analogy between complex variables and planar vectors. If the pair (x, y) formed the components of a vector, then $\arg z$ would be the angle that the vector makes with the positive x axis. For example, $\arg(\mathrm{Re}\ z) = \pm\pi$, $\arg i = \pi/2$, and $\arg(-i) = -\pi/2$.

A program for complex conjugate and modulus

For applications with complex variables it is worthwhile to have available programs for complex functions. We provide here a program that invokes functions returning complex conjugate and modulus values. The more-involved coding for the argument function is provided in the programming project, Section 2.6. Here is the program Conjugate & Modulus Functions.

PROGRAM 2.2 Conjugate and modulus functions for complex numbers.

```
#include <stdio.h>
#include <math.h>

main ()
{
/* Conjugate & Modulus Functions */
double  x,y,xc,yc,zmod;
void CConjugate();
double CModulus();
```

```
printf("Complex Conjugate & Modulus Functions\n");
x = 1; y = 1;
  /* Check for x not zero or y not zero */
while ( x != 0  |  y != 0 )
  {
  printf("\nInput x,y (both zero to end):\n");
  scanf("%lf%lf",&x,&y);
  if ( x == 0  &&  y == 0 )
    {
    printf("\nEnd Conjugate & Modulus Functions");
    exit(0);
    }
  CConjugate(x,y,&xc,&yc); /* conjugate x+iy */
  zmod = CModulus(x,y); /* modulus of x+iy */
  printf("z* = (%lf) + i(%lf)",xc,yc);
  printf("\nmod [(%lf) + i(%lf)] = %lf\n",x,y,zmod);
  }
}


void CConjugate(x,y,xc,yc)
/* Complex Conjugation function */
double x,y,*xc,*yc; /*  xc & yc are returned */
{
*xc = x;   *yc = -y;
}


double CModulus(x,y)
/* Complex Modulus function */
double x,y;
{
double mod;
mod = sqrt(x*x+y*y);
return mod;
}
```

Some remarks on programming the functions CConjugate and CModulus in the C language are in order:

1. Note that complex conjugation performs an operation on a complex number, albeit a simple one. So it does not return a value in the sense of a C function value. Therefore, the "function" CConjugate is declared to be "void." The value of the

complex conjugate is returned in the argument list of `CConjugate` as `xc` and `yc`, which are dereferenced variables (preceded by a `*`, which should not be confused with complex conjugation, or even with `/*` and `*/` used as comment terminators).

2. On the other hand, `CModulus` is declared as "double" because it is a function which returns a value, namely `mod` (inside `CModulus`), which is assigned to `zmod` within the main program. Note that `CModulus` might also be used within an arithmetic statement on the right-hand side of the `=` sign for `zmod`.

3. The program continues to process complex-number pairs while the input number pair is nonzero. If the zero-valued complex number (`x` = 0 and `y` = 0) is entered, the program exits gracefully by `exit(0)` rather than with the signal of an ungraceful termination, `exit(1)`.

With this background to programming complex conjugation and modulus, plus the program for arguments in Section 2.6, you are ready to compute with complex variables.

Exercise 2.6
Run several complex-number pairs through the program `Conjugate & Modulus Functions`. For example, check that the complex conjugate of a complex-conjugate number produces the original number. Also verify that the modulus values of (x, y) and (y, x) are the same. ∎

2.2   THE COMPLEX PLANE AND PLANE GEOMETRY

In the preceding section on algebra and computing with complex numbers we had several hints that there is a strong connection between complex-number pairs (x, y) and the coordinates of points in a plane. This connection, formally called a "mapping," is reinforced when we consider successive multiplications of $z = x + iy$ by i itself.

Exercise 2.7
(*a*) Show that if z is represented by (x, y), then iz is $(-y, x)$, then $i^2 z$ is $(-x, -y)$, $i^3 z$ is $(y, -x)$, and $i^4 z$ regains (x, y).

(*b*) Verify that in plane-polar geometry these coordinates are displaced from each other by successive rotations through $\pi/2$, as shown in Figure 2.1. ∎

The geometric representation of complex numbers shown in Figure 2.1 is variously known as the *complex plane*, the *Argand diagram*, or the *Gauss plane*. The relations between rotations and multiplication with complex numbers are summarized in the following subsections.

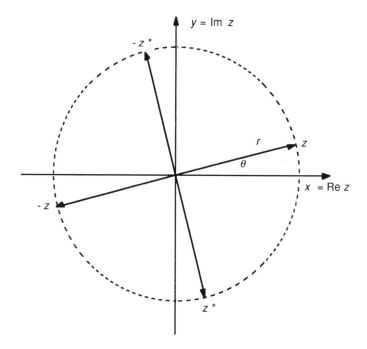

FIGURE 2.1 Rotations of complex numbers by $\pi/2$ in the complex plane. Note that rotations do not change the length of a complex number, as indicated by the dashed circle.

Cartesian and plane-polar coordinates

Before reviewing the connections between the complex plane and plane geometry, let us recall some elementary relations between Cartesian and plane-polar coordinates. If the polar coordinates are r, the (positive) distance from the origin along a line to a point in the plane and θ the angle (positive in the anticlockwise direction) that this line makes with the x axis, then the Cartesian coordinates are given as

$$x = r \cos \theta \qquad y = r \sin \theta \tag{2.23}$$

in a right-handed coordinate system. The polar coordinates are indicated in Figure 2.1. Inverting these equations to determine r and θ, which is not as trivial as it may look, is discussed in the first part of Section 2.6.

In the complex plane we may therefore write z as

$$z = r \left[\cos \theta + i \sin \theta \right] \tag{2.24}$$

so that the modulus, which is also the radius, is given by

$$r = \text{mod } z \geq 0 \tag{2.25}$$

and the polar angle with respect to the x axis is given by

$$\theta = \arg z \tag{2.26}$$

The *principal value* of the polar angle θ is the smallest angle lying between $-\pi$ and $+\pi$. Such a specification of the principal value allows unique location of a point in the complex plane. Other choices that limit θ may also be encountered, for example, the range 0 to 2π.

Complex conjugation is readily accomplished by reflecting from θ to $-\theta$, since

$$z* = r\left(\cos\theta - i\sin\theta\right) = r\left[\cos\left(-\theta\right) - i\sin\left(-\theta\right)\right] \tag{2.27}$$

In the language of physics, z and its complex conjugate are related through a parity symmetry in a two-dimensional space.

With this angular representation of complex variables, we can derive several interesting results.

De Moivre's theorem and its uses

A theorem on multiplication of complex numbers in terms of their polar-coordinate representations in the complex plane was enunciated by Abraham De Moivre (1667–1754). We derive his theorem as follows. Suppose that we have two complex numbers, the first as

$$z_1 = r_1\left(\cos\theta_1 + i\sin\theta_1\right) \tag{2.28}$$

and the second as

$$z_2 = r_2\left(\cos\theta_2 + i\sin\theta_2\right) \tag{2.29}$$

Their product can be obtained by using the trigonometric identities for expanding cosines and sines of sums of angles, to obtain

$$z_1 z_2 = r_1 r_2\left[\cos\left(\theta_1 + \theta_2\right) + i\sin\left(\theta_1 + \theta_2\right)\right] \tag{2.30}$$

From this result we see that multiplication of complex numbers involves conventional multiplication of their moduli, the $r_1 r_2$ part of (2.30), and addition of their angles.

Therefore, multiplication in the complex plane, as well as addition, can readily be shown, as in Figure 2.2.

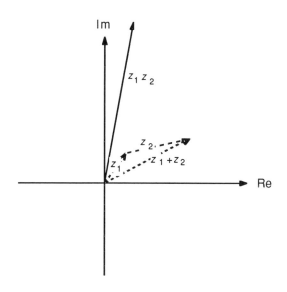

FIGURE 2.2 Combination of complex numbers in the complex plane. The complex numbers and their sum are indicated by the dashed lines, while their product is shown by the solid line.

Equation (2.30) can be generalized to the product of n complex numbers, as the following exercise suggests.

Exercise 2.8

(a) Prove by the method of mathematical induction that for the product of n complex numbers, $z_1, z_2, ..., z_n$, one has in polar-coordinate form

$$z_1...z_n = r_1...r_n \left[\cos\left(\theta_1 + ... + \theta_n\right) + i \sin\left(\theta_1 + ... + \theta_n\right) \right] \qquad (2.31)$$

(b) From this result, setting all the complex numbers equal to each other, prove that for the (positive-integer) nth power of a complex number

$$z^n = r^n \left[\cos\left(n\theta\right) + i \sin\left(n\theta\right) \right] \qquad (2.32)$$

which is called *De Moivre's theorem.* ∎

This remarkable theorem can also be proved directly by using induction on n.

Reciprocation of a complex number is readily performed in polar-coordinate form, and therefore so is division, as you may wish to show.

Exercise 2.9

(a) Show that for a nonzero complex number, z, its reciprocal in polar-coordinate form is given by

$$\frac{1}{z} = \frac{1}{r}\left(\cos\theta - i\sin\theta\right) = \frac{1}{r}\left[\cos(-\theta) + i\sin(-\theta)\right] \tag{2.33}$$

(b) From the result in (a) show that the quotient of two complex numbers can be written in polar-coordinate form as

$$\frac{z_1}{z_2} = \frac{r_1}{r_2}\left[\cos(\theta_1 - \theta_2) + i\sin(\theta_1 - \theta_2)\right] \tag{2.34}$$

where it is assumed that r_2 is not zero, that is, z_2 is not zero. ∎

Thus the polar-coordinate expressions for multiplication and division are much simpler than the Cartesian-coordinate forms, (2.8) and (2.10).

Although we emphasized in Exercise 2.2 that complex-number multiplication and division in Cartesian form are much slower than such operations with real numbers, these operations may be somewhat speedier in polar form, especially if several numbers are to be multiplied. An overhead is imposed by the need to calculate cosines and sines. Note that such advantages and disadvantages also occur when using logarithms to multiply real numbers.

2.3   FUNCTIONS OF COMPLEX VARIABLES

In the preceding two sections we reviewed complex numbers from algebraic, computational, and geometric viewpoints. The goal of this section is to summarize how complex variables appear in the most common functions, particularly the exponential, the cosine and sine, and the hyperbolic functions. We also introduce the idea of trajectories of functions in the complex plane.

Complex exponentials: Euler's theorem

In discussing De Moivre's theorem at the end of the preceding section we noticed that multiplication of complex numbers may be done by adding their angles, a procedure analogous to multiplying exponentials by adding their exponents, just the procedure used in multiplication using logarithms. Therefore, there is probably a connection between complex numbers in polar-coordinate form and exponentials. This is the subject of Euler's theorem.

A nice way to derive Euler's theorem is to write

$$E(\theta) \equiv \cos\theta + i\sin\theta \tag{2.35}$$

then to note the derivative relation

$$\frac{d\,E(\theta)}{d\theta} = -\sin\theta + i\cos\theta = i\left(\cos\theta + i\sin\theta\right) = iE\,(\theta) \qquad (2.36)$$

But the general solution of an equation of the form

$$\frac{d\,E\,(\theta)}{d\theta} = \beta E\,(\theta) \qquad (2.37)$$

is given by

$$E\,(\theta) = \alpha e^{\beta\theta} \qquad (2.38)$$

Exercise 2.10
Show, by identifying (2.36) and (2.37) with the result in (2.38), that $\beta = i$. Then choose a special angle, say $\theta = 0$, to show that $\alpha = 1$. Thus, you have proved *Euler's theorem*,

$$\cos\theta + i\sin\theta = e^{i\theta} \qquad (2.39)$$

which is a remarkable theorem showing a profound connection between the geometry and algebra of complex variables. ∎

It is now clear from Euler's theorem why multiplication of complex numbers involves addition of angles, because the angles are added when they appear in the exponents. Now that we have the formal derivation of Euler's theorem out of the way, it is time to apply it to interesting functions.

Applications of Euler's theorem

There are several interesting and practical results that follow from Euler's theorem and the algebra of complex numbers that we reviewed in Section 2.1. The trigonometric and complex-exponential functions can be related by noting that, for real angles,

$$\cos\theta - i\sin\theta = e^{-i\theta} \qquad (2.40)$$

which follows by taking complex conjugates on both sides of (2.39). On solving this for the cosine we have

$$\cos\theta = \frac{e^{i\theta} + e^{-i\theta}}{2} = \operatorname{Re} e^{i\theta} \qquad (2.41)$$

while solving for the sine function gives

$$\sin \theta \; = \; \frac{e^{i\theta} - e^{-i\theta}}{2i} \; = \; \operatorname{Im} e^{i\theta} \tag{2.42}$$

Both formulas are of considerable usefulness for simplifying expressions involving complex exponentials.

Exercise 2.11
Use the above complex-exponential forms of the cosine and sine functions to prove the familiar trigonometric identity

$$\cos^2 \theta + \sin^2 \theta \; = \; 1 \tag{2.43}$$

the familiar *theorem of Pythagoras.* ■

Although our derivation of Euler's theorem does not justify it, since θ is assumed to be real in the differentiation (2.36), the theorem holds even for θ being a complex variable. Thus the Pythagoras theorem also holds for complex θ.

Some remarkable results, which are also often useful in later chapters, are found if multiples of $\pi/2$ are inserted in Euler's identity, (2.39). Derive them yourself.

Exercise 2.12
Use Euler's theorem to show that

$$e^{i\pi/2} = i \quad e^{i\pi} = -1 \quad e^{3i\pi/2} = -i \quad e^{2i\pi} = 1 \tag{2.44}$$

which gives the values for successive rotations in the complex plane by $\pi/2$. Compare these results with Figure 2.1. ■

The exponential form $e^{i\theta}$ is generally much more symmetric and therefore is easier to handle analytically than are the cosine and sine functions, with their awkward function changes and sign changes upon differentiation compared with the simplicity of differentiating the complex exponential. This simplicity is very powerful when used in discussing the solution of differential equations in Chapter 8, and also in deriving the fast Fourier transform (FFT) algorithm in Chapter 9.3.

An interesting application of the complex-exponential function that anticipates its use in the FFT algorithm is made in the following exercise.

Exercise 2.13
Consider the distributive and recurrence properties of the complex-exponential function defined by

$$E(N) \equiv e^{-2i\pi/N} \tag{2.45}$$

(*a*) Prove the following properties of powers of $E(N)$:

$$E^0(N) = E^1(N) = 1 \qquad E^{-1}(N) = E^*(N) \qquad (2.46)$$

$$E^{a+b}(N) = E^a(N) E^b(N) \qquad E^{pa}(N) = E^a(N/p) \qquad (2.47)$$

for any a, b, and for any $p \neq 0$.

(*b*) Using these results, show that if $N = 2^v$, where v is a positive integer, then, no matter how many integer powers of $E(N)$ are required, only one evaluation of this complex-exponential function is required. ■

As a further topic in our review of functions of complex variables, let us consider the hyperbolic and circular functions.

Hyperbolic functions and their circular analogs

Exponential functions with complex arguments are required when studying the solutions of differential equations in Chapters 7 and 8. A frequently occurring combination is made from exponentially damped and exponentially increasing functions. This leads to the definition of hyperbolic functions, as follows.

The *hyperbolic cosine*, called "cosh," is defined by

$$\cosh(u) \equiv \frac{e^u + e^{-u}}{2} \qquad (2.48)$$

while the hyperbolic sine, pronounced "sinsh," is defined by

$$\sinh(u) \equiv \frac{e^u - e^{-u}}{2} \qquad (2.49)$$

If u is real, then the hyperbolic functions are real-valued. The name "hyperbolic" comes from noting the identity

$$\cosh^2(u) - \sinh^2(u = 1 \qquad (2.50)$$

in which, if u describes the x and y coordinates parametrically by

$$x = \cosh(u) \qquad y = \sinh(u) \qquad (2.51)$$

then an x-y plot is a rectangular hyperbola with lines at $\pi/4$ to the x and y axes as asymptotes.

Exercise 2.14

(*a*) Noting the theorem of Pythagoras,

$$\cos^2(u) + \sin^2(u) = 1 \qquad (2.52)$$

for any (complex) u, as proved in Exercise 2.11, explain why the cosine and sine functions are called "circular" functions.

(b) Derive the following relations between hyperbolic and circular functions

$$\cosh(u) = \cos(iu) \qquad (2.53)$$

and

$$\sinh(u) = -i\sin(iu) \qquad (2.54)$$

valid for any complex-valued u. ∎

These two equations may be used to provide a general rule relating signs in identities for hyperbolic functions to identities for circular functions:

An algebraic identity for hyperbolic functions is the same as that for circular functions, except that in the former the product (or implied product) of two *sinh* functions has the *opposite sign* to that for two *sin* functions.

For example, given the identity for the circular functions

$$\cos(2u) = \cos^2(u) - \sin^2(u) \qquad (2.55)$$

we immediately have the identity for the hyperbolic functions

$$\cosh(2u) = \cosh^2(u) + \sinh^2(u) \qquad (2.56)$$

Exercise 2.15
Provide a brief general proof of the hyperbolic-circular rule stated above. ∎

Note that derivatives, and therefore integrals, of hyperbolic and circular functions do not satisfy the above general rule. The derivatives of the hyperbolic functions are given by

$$\frac{d\cosh(u)}{du} = \sinh(u) \qquad (2.57)$$

and by

$$\frac{d\sinh(u)}{du} = \cosh(u) \qquad (2.58)$$

in both of which the real argument, u, is in radians. There is no sign change on differentiating the hyperbolic cosine, unlike the analogous result for the circular cosine.

The differential equations satisfied by the circular and hyperbolic functions also differ by signs, since the cosine and sine are solutions of

$$\frac{d^2 y(u)}{du^2} = -y(u) \qquad (2.59)$$

which has oscillatory solutions, whereas the cosh and sinh are solutions of

$$\frac{d^2 y(u)}{du^2} = +y(u) \qquad (2.60)$$

which has solutions exponentially increasing or exponentially decreasing.

Exercise 2.16

(a) Prove the two derivative relations (2.57) and (2.58) by starting with the defining equations for the cosh and sinh.

(b) Use the relations between hyperbolic and circular functions, (2.53) and (2.54), to compute the derivatives of the hyperbolic functions in terms of those for the circular functions.

(c) Verify the appropriateness of the circular and hyperbolic functions as solutions of the differential equations (2.59) and (2.60), respectively. ∎

To complete the analogy with the circular functions, one also defines the *hyperbolic tangent*, called "tansh," by

$$\tanh(u) \equiv \frac{\sinh(u)}{\cosh(u)} \qquad (2.61)$$

which is analogous to the circular function, the tangent, defined by

$$\tan(u) \equiv \frac{\sin(u)}{\cos(u)} \qquad \cos(u) \neq 0 \qquad (2.62)$$

Among these six hyperbolic and circular functions, for real arguments there are three that are bounded by ± 1 (sin, cos, tanh) and three that are unbounded (sinh, cosh, tan). Therefore we show them in a pair of figures, Figures 2.3 and 2.4, with appropriate scales.

By displaying the bounded hyperbolic tangent on the same scale as the sine function in Figure 2.3, we notice an interesting fact — these two functions are equal to within 10% for $|x| < 2$, so may often be used nearly interchangeably. The explanation for their agreement is given in Section 3.2, where their Maclaurin series are presented. Figure 2.4 shows a similar near-coincidence of the cosh and sinh functions for $x > 1.5$, where they agree to better than 10% and the agreement improves as x increases because they both tend to the exponential function.

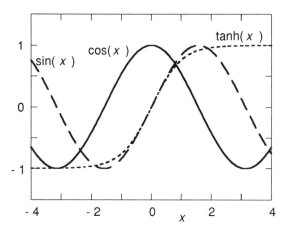

FIGURE 2.3 Bounded circular and hyperbolic functions, sine, cosine, and hyperbolic tangent.

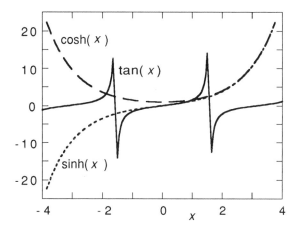

FIGURE 2.4 The unbounded circular and hyperbolic functions, tangent, hyperbolic cosine, and hyperbolic sine. For x greater than about 1.5, the latter two functions are indistinguishable on the scale of this figure.

The tangent function is undefined in the limit that the cosine function in the denominator of its definition (2.62) tends to zero. For example, in Figure 2.4 values of the argument of the tangent function within about 0.1 of $x = \pm \pi/2$ have been omitted.

Trajectories in the complex plane

Another interesting concept and visualization method for complex quantities is that of the trajectory in the complex plane. It is best introduced by analogy with particle trajectories in two space dimensions, as we now summarize.

When studying motion in a real plane one often displays the path of the motion, called the trajectory, by plotting the coordinates $x(t)$ and $y(t)$, with time t being the parameter labeling points on the trajectory. For example, suppose that $x(t) = A \cos(\omega t)$ and $y(t) = B \sin(\omega t)$, with A and B positive, then the trajectory is an ellipse with axes A and B, and it is symmetric about the origin of the x-y coordinates. As t increases from zero, x initially decreases and y initially increases. One may indicate this by labeling the trajectory to indicate the direction of increasing t. The intricate Lissajous figures in mechanics, obtained by superposition of harmonic motions, provide a more-involved example of trajectories.

Analogously to kinematic trajectories, in the complex plane real and imaginary parts of a complex-valued function of a parameter may be displayed. For example, in Section 8.1 we discuss the motion of damped harmonic oscillators in terms of a real dimensionless damping parameter β. Expressed in polar-coordinate form, the amplitude of oscillation is

$$y(x) = e^{V_{\pm} x} \tag{2.63}$$

where the complex "frequency" (if x represents time) is given by

$$V_{\pm} = -\beta \pm i \sqrt{1 - \beta^2} \tag{2.64}$$

The trajectory of v depends on the range of β and on the sign associated with the square root in (2.64).

Exercise 2.17

(a) Show that if $|\beta| < 1$, which gives rise to damped oscillatory motion, then

$$\left(\text{Re}\, v_{\pm}\right)^2 + \left(\text{Im}\, v_{\pm}\right)^2 = 1 \tag{2.65}$$

and that the trajectory of v_+ is a counterclockwise semicircle in the upper half plane, while the trajectory of v_- is a clockwise semicircle in the lower half plane. In both trajectories β is given as increasing from -1 to $+1$.

(b) Suppose that $\beta > 1$, which produces exponential decay called overdamped motion. Show that $v_{\pm}$ is then purely real and negative, so the trajectory lies along the real axis. Show that v_+ increases from -1 toward the origin as β increases, while v_- decreases toward $-\infty$ as β increases. ∎

The complex-plane trajectory, with β as parameter, expressed by (2.64) is therefore as displayed in Figure 2.5.

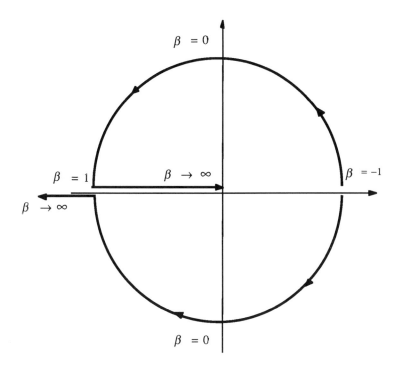

FIGURE 2.5 Frequency trajectory in the complex plane according to (2.64) as a function of the damping parameter β.

As a final note on this example, there is no acceptable solution for v_- if $\beta < -1$ and if $x > 0$ is considered in (2.63), since $y(x)$ is then divergent.

Another interesting example of a trajectory in the complex plane arises in the problem of forced oscillations (Section 10.2) in the approximation that the energy dependence is given by the Lorentzian

$$L(\omega) = c|L_\pm(\omega)|^2 \qquad (2.66)$$

where c is a proportionality constant and the complex Lorentzian amplitudes $L_\pm$ are given by

$$L_\pm(\omega) \equiv \frac{\gamma_0}{\omega - 1 \pm i\gamma_0} \qquad (2.67)$$

where ω and γ_0 are dimensionless frequency and damping parameters. The analysis of this example is similar to the first one.

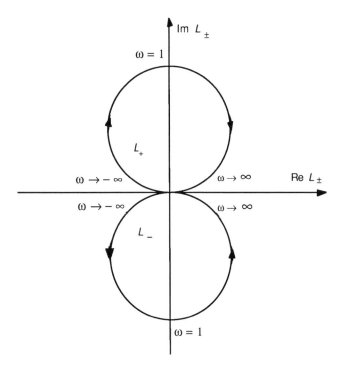

FIGURE 2.6 Frequency trajectory in the complex plane for the Lorentzian amplitude described by (2.67).

Exercise 2.18

(*a*) Show that the Lorentzian amplitudes in (2.67) satisfy

$$\left(\operatorname{Re}L_{\pm}\right)^{2} + \left(\operatorname{Im}L_{\pm} \pm \tfrac{1}{2}\right)^{2} = \left(\tfrac{1}{2}\right)^{2} \tag{2.68}$$

so that the trajectories of $L_{\pm}$ lie on circles of radius 1/2 in the complex plane.

(*b*) Investigate the details of the trajectory by showing that L_{+} describes the anticlockwise semicircle in the lower half plane, while L_{-} describes the clockwise semicircle in the upper half complex plane. For both of these trajectories the directions are for ω going from $-\infty$ to $+\infty$. ∎

The trajectories of the Lorentzian amplitudes are shown in Figure 2.6. They are discussed more completely in Chapter 3 of the text by Pippard in the context of Cole-Cole plots of complex-valued dielectric constants as a function of frequency.

In both of our examples the trajectories in the complex plane lie on circles. This is neither a mere coincidence nor is it uncommon, as the following exercise should convince you.

Exercise 2.19

Consider the following complex function z, called a linear fractional transformation of the real variable p according to

$$z(p) \equiv \frac{\alpha + \beta p}{\gamma + \delta p} \tag{2.69}$$

in which α, β, γ, and δ are complex constants, with $\delta \neq 0$. Now consider the function z' that is obtained from z by the shift and scaling transformation

$$z'(p) \equiv \frac{(z\delta - \beta)\gamma}{\alpha\delta - \beta\gamma} = \frac{\gamma/\delta}{p + \gamma/\delta} \tag{2.70}$$

By analogy with the result in Exercise 2.18, argue that z' lies on a circular trajectory and therefore that the original z in (2.69) lies on a circular trajectory. ∎

Thus, the functions (2.67) and (2.69) are both special cases of the more general circular trajectory given by (2.70). From these examples we see that the notion of a trajectory in the complex plane is useful for visualizing the properties of complex-valued functions.

2.4   PHASE ANGLES, VIBRATIONS, AND WAVES

The angle θ in the complex plane often has interesting interpretations in scientific applications, particularly in the context of vibrations and waves. In this section we summarize some of the main results. An encyclopedic treatment is provided in Pippard's book on the physics of vibration.

The topics introduced here are developed and applied throughout this book. In particular, Section 8.1 discusses free-motion and resonant vibrations in mechanical and electrical systems, then the quantum oscillator is considered briefly in Section 8.5. In Chapters 9 and 10 we develop Fourier expansions, emphasizing the complex-exponential treatment for the discrete, series, and integral expansions.

Phase angles and phasors

Suppose that we have a (real) angle $\theta = \omega t$, where ω is a constant angular frequency, $\omega = 2\pi f$ (with f the frequency) and t denotes time. Then

$$z_1(t) = e^{i\theta} = e^{i\omega t} \tag{2.71}$$

describes in the complex plane uniform circular motion of the point z_1, while the projections onto the real and imaginary axes (x and y) describe simple harmonic motions.

Exercise 2.20

Prove that the motion of z_1 is periodic by showing that

$$z_1(t+T) = z_1(t) \tag{2.72}$$

where the period $T = 2\pi/\omega$. ∎

If a second uniform circular motion in the complex plane is described by

$$z_2(t) = e^{i(\theta+\phi)} = e^{i(\omega t+\phi)} \tag{2.73}$$

then this motion has the same period as that described by z_1, but at a given time z_2 has its phase advanced by ϕ over the phase of z_1.

Whether one refers to a positive value of ϕ as a lag or a lead depends on the scientific field in which one is working. If $\phi > 0$, in mechanics the phase of z_2 is said to *lag* that of z_1, whereas in electrical-circuit applications z_2 is said to *lead* z_1.

A complex-plane diagram showing the magnitude of z and a relative phase (with ωt usually suppressed) is called a *vibration diagram* or *phasor diagram*. Its use gives a visualization of complex-variable relationships which often improves comprehension and interpretation.

Vibrations and waves

We can broaden the discussion of phases to include both spatial as well as temporal variation in the amplitude of a complex vibration. For example, a wave that has constant amplitude of unity at all points along the x direction and at all times t can be described by

$$z(x,t) = e^{i(kx-\omega t)} \tag{2.74}$$

in which the wavenumber, k, is given in terms of wavelength, λ, by

$$k = \frac{2\pi}{\lambda} \tag{2.75}$$

Although the wavenumber is physically a less intuitive quantity than is the wavelength, computationally and in most applications k is a much simpler quantity to deal with. Note that k has the dimensions of an inverse length, just as the angular frequency, ω, has the dimensions of inverse time. Thus, the argument of the exponential function in (2.74) is dimension-free, as should be the argument of any function in mathematics.

Recall also that the wave described by (2.74) is monochromatic (unique values of ω and k) and that points of constant phase have an associated *phase velocity*, v_p, given by

$$v_p = \omega / k = f \lambda \tag{2.76}$$

Exercise 2.21
Discuss from the viewpoint of wave motion why v_p in (2.76) is called the phase velocity. ■

The superposition of such waves of constant phase to build up a dispersive wave in which components with different frequencies transport energy at different speeds is an extension of the Fourier expansions in Chapters 9 and 10. A comprehensive and lucid discussion is given by Baldock and Bridgeman in their book on wave motion.

In Chapters 9 and 10 on Fourier expansions we make detailed study of phenomena described in terms of x or in terms of the complementary variable k, or in terms of t and its complementary variable ω. Clear expositions of the relations between complex exponentials and vibrations are given in detail with many applications in Pippard's omnibus book. Vibrations and waves are described very completely at an introductory level in the book by Ingard.

2.5 DIVERSION: INTERPRETING COMPLEX NUMBERS

The development of the interpretation of complex numbers provides an example of the consequences of education and of the dominance of scientific thought by mathematical representations. Since many scientists claim that a phenomenon is not understood until it can be described mathematically, it is interesting to discuss the relation between "the queen and servant of science" and the natural sciences.

Are complex numbers real?

Before the quantum physics revolution of the 1920s, scientists usually apologized for using complex numbers, since they provided only mathematically convenient shortcuts and shorthand for problem solving. Indeed, Leonard Euler of Euler's theorem in Section 2.3 coined the Latin "imaginarius" for the quantity $i = \sqrt{-1}$. The first major use of complex numbers was made by C. P. Steinmetz (1865–1923), a research electrical engineer who used them extensively (as we do in Section 8.1) to simplify the analysis of alternating-current circuits.

In quantum mechanics, for example in the Schrödinger equation that we use in Section 8.5, the wave function is fundamentally a complex variable that is not a shorthand for two real quantities such as magnitude and phase. Many quantities derived from wave functions, such as scattering amplitudes, are also intrinsically complex-valued. This leads to the scientific use of "analytic continuation," a concept and technique familiar in mathematics but of more recent use in the natural sciences.

Analytic continuation

We have displayed in Figures 2.3 and 2.4 the circular functions and the hyperbolic functions, respectively. In terms of complex variables, however, these hyperbolic functions are essentially just the circular functions evaluated for purely imaginary arguments, or vice versa. It is therefore interesting, and sometimes useful, to think of there being just a single set of functions, say the circular functions, which may be evaluated along the real axis (then they are the conventional trigonometric functions) or they may be evaluated along the imaginary axis (then they are the hyperbolic functions, within factors of i), or they may be evaluated for the argument which takes on a value anywhere in the complex plane.

When we make this last bold step off either the real or the imaginary axis and into the complex plane we are making an *analytic continuation* of the functions. The concept of analytic continuation and some understanding of the techniques applied to it are best appreciated by working the following exercise.

Exercise 2.22

Consider the behavior of the complex-valued function of complex-variable argument, z, defined as follows:

$$A(z) \equiv \frac{e^z + e^{-z}}{2} \tag{2.77}$$

(*a*) Show that for a real argument A is just the hyperbolic cosine function discussed in Section 2.3, while for purely imaginary z it is the circular cosine function.

(*b*) Sketch the graph of $\cos x$ along the real axis of the complex-z plane and the graph of $\cosh y$ along the imaginary axis of the same plane. They look quite different, don't they?

(*c*) Devise a graphical representation of $A(z)$ that is suitable for arbitrary complex z, and make some representative sketches of the function thus graphed. One possible form of representation is to sketch contours of constant $\text{Re}\,A$ and of constant $\text{Im}\,A$. ■

In scientific research analytic continuation is often a useful technique. As an example, experiments on wave scattering (such as in acoustics, optics, electromagnetism, and subatomic physics) are, at best, obtained in the range of scattering angles from zero to π. How would the data look if they could be analytically continued into the complex-angle plane? Similarly, data obtained at real energies or frequencies may be interesting to extrapolate to complex energies or complex frequencies. Indeed, we explore this possibility in discussing the Lorentzian resonances in Section 10.2.

2.6 PROJECT 2: PROGRAM TO CONVERT BETWEEN COORDINATES

The program Cartesian & Polar Coordinate Interconversion developed in this project serves both to develop your understanding of the relations between these two coordinate systems and to give you practice with writing programs in C.

The conversion from plane-polar to Cartesian coordinates is straightforward and unambiguous. Given r and θ, one has immediately (as discussed in Section 2.2)

$$x = r\cos(\theta) \qquad y = r\sin(\theta) \qquad (2.78)$$

which can be programmed directly.

Stepping into the correct quadrant

The transformation from Cartesian coordinates to polar coordinates is less direct than the inverse transformation just considered. The required formulas are

$$r = \sqrt{x^2 + y^2} \geq 0 \qquad (2.79)$$

which is straightforward to compute, and

$$\theta = \operatorname{atan}\left(\frac{y}{x}\right) \qquad (2.80)$$

which is ambiguous. This formula does not uniquely determine the quadrant in which θ lies because only the sign of the quotient in (2.80) is available after the division has been made. The relative signs of the circular functions in the four quadrants indicated in Figure 2.7 may be used to determine the angle uniquely from the signs of x and y.

In some programming languages, including C, two functions are available for the inverse tangent. In one, such as the atan(t) in C language (with t the argument of the function), the angle is usually returned in the range $-\pi/2$ to $\pi/2$, and the user of this function has to determine the appropriate quadrant by other means.

In the second function for the inverse tangent, such as atan2(y,x) in the C language, the angle is located in the correct quadrant by the function itself. If we were to use atan2 in the program, the conversion from Cartesian to polar representation would be very direct. For practice in C and to reinforce your understanding of plane-polar coordinates we use the simpler function atan.

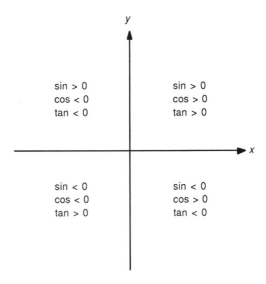

FIGURE 2.7 Signs of the circular functions in each quadrant of the Cartesian plane.

If we begin with an angle calculated into the first quadrant by using the absolute value of y/x, then multiples of $\pi/2$ have to be added or subtracted to get the angle into the correct quadrant. Computationally, the best way to get π at the accuracy of your computer system is to use the identity

$$\pi = 4 \operatorname{atan}(1) \tag{2.81}$$

This relation is used in this program and in many others throughout this book.

Coding, testing, and using the program

The structure of program `Cartesian & Polar Coordinate Interconversion` is typical of the interactive programs listed in the index to computer programs at the end of the book. The program is controlled by a `while` loop that terminates when both x and y (going from Cartesian to polar) and also r (going from polar to Cartesian) have been input with values of zero. Within each execution of the loop there is a choice of input of either Cartesian coordinates (chosen by `c` or `C`) or polar (`p` or `P`) coordinates.

The Cartesian-input option uses the function `MakePolar` to locate an angle in the first quadrant and to use the signs of x and y to adjust the quadrant, as discussed above. Some care is required for angles that coincide with the axes. The polar-input option uses (2.78) directly in the function `MakeCartesian` and returns the values of the x and y variables. The program listing is as follows.

PROGRAM 2.3 Interconversion between Cartesian and polar coordinates.

```c
#include <stdio.h>
#include <math.h>

main ()
{
/* Cartesian & Polar Coordinate Interconversion */
double pi,x,y,r,theta;
char isCP;
void MakePolar(),MakeCartesian();

pi = 4*atan(1); /* pi to machine accuracy */
printf("Cartesian & Polar Coordinates\n");
x = 1;  y = 1;  r = 1;

while ( (( x != 0 ) && ( y != 0 )) || ( r != 0 ) )
  {
  printf("\nCartesian (C) or Polar (P) input?\n");
  scanf("%s",&isCP);
  if ( isCP == 'c' || isCP == 'C' )
    {
    printf("Input Cartesian values as  x y\n");
    scanf("%le %le",&x,&y);
    if ( ( x != 0 ) || ( y != 0 ) )
      {
      MakePolar(pi,x,y,&r,&theta);
      printf("Polar values: r=%le, theta=%le (rad)\n",r,theta);
      }
    }
  else
    {
    if ( isCP == 'p' || isCP == 'P' )
      {
      printf("Input Polar values as  r theta (rad)\n");
      scanf("%le %le",&r,&theta);
      if ( r != 0 )
        {
        MakeCartesian(r,theta,&x,&y);
        printf("Cartesian values: x=%le, y=%le\n",x,y);
        }
      }
    }
  }
printf("\nEnd  Cartesian & Polar Coordinates");  exit(0);
}
```

```
void MakePolar(pi,x,y,r,theta)
/* Make polar coordinates from Cartesian coordinates */
double pi,x,y; /* are input variables */
double *r,*theta; /* are calculated variables */
{
double angle;

*r = sqrt(x*x+y*y);
if ( x == 0 )
  angle = pi/2;
else   angle = atan(fabs(y/x));
if (x>=0 && y>=0)   *theta = angle; /* first quadrant */
if (x<=0 && y>=0)   *theta = pi-angle; /* second quadrant */
if (x<=0 && y<=0)   *theta = pi+angle; /* third quadrant */
if (x>=0 && y<=0)   *theta = -angle; /* fourth quadrant */
}


void MakeCartesian(r,theta,x,y)
/* Make Cartesian coordinates from polar coordinates */
double r,theta; /* are input variables */
double *x,*y; /* are calculated variables */
{
*x = r*cos(theta);
*y = r*sin(theta);
}
```

Now that we have a source program for converting between coordinates, it is appropriate to test and use it.

Exercise 2.23

Test program Cartesian & Polar Coordinate Interconversion for the two options by using as input values combinations of x and y or of r and θ that correspond to points in each of the four quadrants and along the boundaries between quadrants, a total of 16 combinations. Also verify that the program terminates if all of x, y, and r are input as zero. ∎

You now have two functions that are useful in converting coordinates. Extensions of these programs may be used to interconvert spherical-polar and three-dimensional Cartesian coordinates.

REFERENCES ON COMPLEX NUMBERS

Baldock, G. R., and T. Bridgeman, *Mathematical Theory of Wave Motion*, Ellis Horwood, Chichester, England, 1981.

Ingard, K. U., *Fundamentals of Waves and Oscillations*, Cambridge University Press, Cambridge, England, 1988.

Pippard, A. B., *The Physics of Vibration*, Cambridge University Press, Cambridge, England, 1988.

Wolfram, S., *Mathematica: A System for Doing Mathematics by Computer*, Addison-Wesley, Redwood City, California, second edition, 1991.

Chapter 3

POWER SERIES AND THEIR APPLICATIONS

The goal of this chapter is to understand the analysis, numerics, and applications of power series, which form the basis for many of the powerful methods used in computing. The chapter begins with a discussion of the motivation for using series, summarizing some properties and applications of the geometric series, then enunciating and proving Taylor's theorem on power series. This general theorem is then used in Section 3.2 to develop expansions of useful functions — the exponential, cosine, sine, arcsine, logarithm, and the hyperbolic cosine and sine. In this section there are preliminary numerical and computing studies in preparation for the project at the end of the chapter.

The binomial approximation and various of its applications are described in Section 3.3, followed by a diversion (Section 3.4) in which we discuss iteration, recurrence, and recursion in mathematics and computing. The computing project for this chapter is testing the numerical convergence of the series for the functions considered in Section 3.2. References on power series then complete the chapter.

3.1 MOTIVATION FOR USING SERIES: TAYLOR'S THEOREM

In this section we review the properties of power series and we explore their numerical properties, especially their convergence. Since the geometric series is easy to handle algebraically but also exhibits many of the problems that power series may exhibit, we consider it first. Then we consider the very powerful Taylor's theorem, which allows differentiable functions to be characterized by power series. A discussion of the interpretation of Taylor series then prepares us for the several examples in Section 3.2 of expansions of interesting and useful functions.

We emphasize the numerical convergence of series, resting assured by such mathematics texts as those by Taylor and Mann (Chapters 19 – 21) and by Protter and Morrey (Chapter 3) that all the series that we consider are analytically convergent for the range of variables that we consider. After you have understood the materials in this chapter you will be ready to use powerful algebraic and numerical computing systems, such as the *Mathematica* system described in Chapter 3.6 of Wolfram's book, to perform accurately and efficiently much of the work involved in making expansions in power series.

The geometric series

The geometric series provides a good example of a power series that can be readily investigated and that illustrates many properties of series. We quickly review its formal properties, then discuss its numerical convergence properties and its relation to other functions.

The usual definition of the geometric series to n terms is the function

$$G_n(r) \equiv \sum_{k=0}^{n-1} r^k = 1 + r + r^2 + \dots + r^{n-1} \tag{3.1}$$

in which the parameter r determines the convergence of the series. Although the results that we derive generally hold for complex values of r, we will discuss examples only for r a real variable. As is well known, the sum can be expressed in closed form as

$$G_n(r) \equiv \frac{1 - r^n}{1 - r} \qquad r \neq 1 \tag{3.2}$$

As one way of deriving this very useful result, try the following exercise.

Exercise 3.1
Use the method of proof by induction to derive (3.2). That is, show that if the formula is assumed to be true for G_n, then it must be true for G_{n+1}. In order to get the recurrence started show that the result is evidently true for $n = 1$. ■

Note that the geometric series is sometimes written in forms other than (3.2). For example, there may be a common multiplier for each term of the series, which just multiplies the sum by this multiplier. For clarity and effect, in Figure 3.1 we have assumed a common multiplier of r. Therefore, the series in this figure starts with r rather than with unity.

The convergence properties of the geometric series (3.1) are quite straightforward. If $|r| < 1$ the series converges as $n \to \infty$, otherwise the series diverges. Note that if $r = -1$ the value of G_n alternates between 0 and +1, so there is no definite limit for large n.

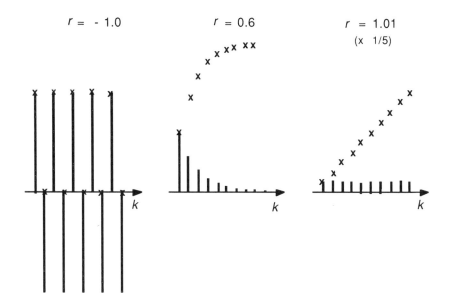

FIGURE 3.1   Terms (shown by bars) and partial sums (crosses, omitting the leading term of unity) for the geometric series. Only for multiplier $r = 0.6$ is the series convergent.

Thus we have the geometric series convergence property

$$G\,(r) \equiv \lim_{n \to \infty} \, G_n\,(r) \equiv \frac{1}{1-r} \qquad |r| < 1 \qquad\qquad (3.3)$$

The convergence conditions are quite evident in Figure 3.1. The series is therefore declared to be divergent, even though its value is always finite. A way around this anomaly is discussed in Exercise 3.3 below.

Programming geometric series

If you want practice in writing C programs involving series, you should code up and run the program `Geometric Series`. This simple program has a function to sum the geometric series directly, using (3.1), and it also computes the closed form (3.2). The program does both of these calculations for a range of r values specified as input with a fixed value of n.

PROGRAM 3.1 Geometric series by direct summing and by the closed form.

```c
#include <stdio.h>
#include <math.h>

main()
{
/* Geometric Series */
double rmin,dr,rmax,r,Sum,ClosedForm,error;
int nterms,kmax;
double GeoSum();

printf("Geometric Series by Summing & Closed Form\n");
nterms = 2;
while ( nterms > 1 )
   {
   printf("\n\nInput rmin,dr,rmax,nterms (nterms<2 to end):\n");
   scanf("%lf%lf%lf%i",&rmin,&dr,&rmax,&nterms);
   if ( nterms < 2 )
     {
     printf("\nEnd Geometric Series");  exit(0);
     }
   kmax = nterms-1;
   for ( r = rmin; r <= rmax; r = r+dr )
     {
     if ( fabs(r) >= 1 )
     {
     printf("\n!! Warning: Divergent series |r|=%g >= 1",fabs(r));
     }
     Sum = GeoSum(r,kmax);   /* direct sum of series */
      /* Formula for sum to  nterms  terms;
           uses  r  to the  nterms  power */
     ClosedForm = (1-pow(r,nterms))/(1-r);
     error = ClosedForm-Sum;
     printf("\n%g  %g  %g  %g",r,ClosedForm,Sum,error);
     }
   }
}

double GeoSum(r,kmax)
/*  Direct sum of geometric series */
double r;
int kmax;
{
double term,sum;
int k;  term = 1;  sum = 1; /* initialize terms & sum */
```

```
for ( k = 1; k <= kmax; k++ )
  {
  term = r*term;
  sum = sum+term;
  }
return sum;
}
```

For the benefit of novice C programmers, we now summarize how the program Geometric Series is organized. Program execution is controlled by the value of n, which is denoted by nterms in the program. If there are fewer than two terms in the geometic series (3.1), then one has only the leading term of unity, which is scarcely an interesting series. Therefore, if the input value of nterms is less than 2 this is used to signal program termination by a graceful exit (parameter zero). If the series is to be summed, then there are kmax = $n-1$ terms to be added to the starting value of unity.

For each value of the multiplier r in the input range, starting at rmin and going up to rmax by steps of dr, the program first checks whether r is beyond the range that leads to convergent series. Although the series has a definite value for any finite n, it is instructive to see how a warning about nonconvergence can be output. Next, the function GeoSum (about which more is told in the next paragraph) is used to sum the series directly.

The formula (3.2) is then coded in-line to produce the value ClosedForm, which requires the C library function pow to compute the power r^n. Finally, any difference between the direct summation and closed-form values is indicated by their difference, called error, which is printed following the r value and the two series values.

The function GeoSum, coded after the main program of Geometric Series, involves a simple for loop over the powers of r. The succesive terms are obtained by the recurrence relation indicated, then they are added to the partial sum. There will be at least one iteration of the loop, since kmax is at least 1 because nmax is at least 2. The value of sum is returned to the function GeoSum for assignment in the main program.

Given this program description, and some nimble fingers, you should be ready to code, test, and use Geometric Series.

Exercise 3.2

(a) Code the program Geometric Series, listed as Program 3.1, modifying the input and output functions for your computing environment. Then test the program termination control (nterms < 2) and warning message for $|r| \geq 1$.

(*b*) Run the program for a range of values of the multiplier r and the number of terms nterms, to convince yourself that for a reasonable range of these parameters of the geometric series the direct-summation and closed-form methods give close numerical agreement.

(*c*) Examine the convergence of the series for values of r that are close to (but less than) unity in magnitude. For example, use $r = -1 + 10^{-3}$, then find out how large nterms must be in order to get agreement with the infinite series value of $1/(1 - r)$ to about one part per million. It would be most convenient to modify the program to calculate this value and compare it with the direct-summation value. ■

Alternating series

As you will have discovered by working part (*c*) of Exercise 3.2, series in which successive terms are of similar magnitude but opposite sign, so-called "alternating series," are particularly troublesome numerically. The geometric series is particularly convenient for investigating and understanding the problems that can arise with alternating series.

For example, suppose that you had to sum the geometric series term by term for $r = -1 + \varepsilon$, where $\varepsilon \sim 10^{-6}$ and is positive. By looking at the leftmost panel of Figure 3.1 you would know that the series is convergent but it would clearly take many terms before the remaining terms could be ignored. Numerically, by that number of terms the roundoff error of your computer may have already overwhelmed the significant digits in your partial sum.

Exercise 3.3

(*a*) Suppose that, for a multiplier an amount ε larger than -1, we wanted to estimate how many terms, k, it takes before a geometric-series term has become $1/2$ of the first term. By using (natural) logarithms and the approximation that $\ln(1 - \varepsilon) \approx \varepsilon$ for small ε, show that $k \approx \ln(2)/\varepsilon \approx 0.7/\varepsilon$. Thus if one had $\varepsilon = 10^{-6}$, more than one million terms would be needed for even reasonable convergence.

(*b*) Show that for $r = -1 + \varepsilon$ as the multiplier in the geometric series the algebraic sum of the series is $1/(2 - \varepsilon)$. Thus show that as $\varepsilon \to 0$ the sum tends to $1/2$, which is just the mean value of the successive partial sums shown on the left side of Figure 3.1, namely 1 and 0.

(*c*) Write down the geometric series for negative r term by term, but expressed in terms of $|r|$, so that the alternating signs are explicit in the sum. Next factor out the $(1 - |r|)$, sum the remaining series in the geometric form, then divide away the first factor. Show that the alternating series sums to $1/(1 + |r|)$, which is $1/2$ at $r = -1$, just as in (*b*). Explain why the two methods agree, and explain whether a limit process similar to that in (*b*) is necessary in the second method. ■

Notice in part (*b*) of Exercise 3.3 that if *r* were allowed to approach the limit of −1 before the summation were attempted, then the result would be undefined, as Figure 3.1 shows. Part (*b*) shows that if the limit is taken after an expression for the sum has been obtained, then the result is well defined. This peculiarity of the strong dependence of the series values on the order of taking limits appears again in our investigation in Section 9.6 of the Wilbraham-Gibbs overshoot phenomenon for Fourier series.

From the example in Exercise 3.3 we see particularly troublesome aspects of alternating series. Specialized techniques have been developed for alleviating some of their troubles, as described in the text by Taylor and Mann. An interesting result on combining geometric series with multipliers of opposite sign is obtained in the following exercise.

Exercise 3.4
(*a*) Show analytically that for a convergent geometric series

$$G(r) + G(-r) = 2G(r^2) \qquad |r| < 1 \qquad (3.4)$$

(*b*) Make a sketch, similar to Figure 3.1, of the terms on the left and right sides of this equation. Thereby explain the result (3.4). ∎

Another property of the geometric series that leads to a very interesting result is the integral of the infinite series (3.1). In order to produce the result in standard form, and assuming that $|r| < 1$, replace *r* by −*r* in (3.1) and (3.2), then integrate both $1/(1 + r)$ and each term of the convergent series in order to produce the series expansion

$$\int dr \, \frac{1}{1 + r} = 1 + \sum_{k=1}^{\infty} \frac{(-r)^k}{k} \qquad |r| < 1 \qquad (3.5)$$

If we now identify the indefinite integral on the left-hand side as $\ln(1 + r)$, one has produced a power series expansion for the natural logarithm. This result agrees with the Taylor-expansion method of deriving logarithms in series expansion that is presented in Section 3.2.

Exercise 3.5
Show the steps leading to the result (3.5) for the power-series expansion of the integral. ∎

Now that we have some experience with the well-behaved geometric series, it is time to expand our horizons to power series that are more general. With this as our goal, we should review Taylor's theorem on power-series expansions and function derivatives.

Taylor's theorem and its proof

In later chapters, especially in Chapters 4, 5, 7, and 8, we often require polynomial approximations to functions. If the function is smooth enough that as many derivatives as needed can be calculated analytically, then Taylor's theorem gives a unique prescription for the polynomial coefficients. We now review the conditions for applicability of the theorem and we review a convenient proof of it.

Suppose that a function $y(x)$ and its first n derivatives $y^{(k)}(a)$ for $k = 1, 2, .., n$ are continuous on an interval containing the points a and x. Then *Taylor's theorem* states that

$$y(x) = y_n(x) = y(a) + \sum_{k=1}^{n} \frac{y^{(k)}}{k!}(a)(x-a)^k + R_n \qquad (3.6)$$

in which the remainder after n terms, R_n, is given by

$$R_n = \int_a^x dt\,(x-t)^n \frac{y^{(n+1)}(t)}{n!} \qquad (3.7)$$

In practice, one tries to choose n and a so that it is a fair approximation that R_n can be ignored for the range of x of current interest, typically x values near a. If R_n may be ignored, then $y(x)$ is aproximated by an n th-order polynomial in the variable $(x-a)$. In order to make the statement of Taylor's theorem plausible, consider the following examples.

Exercise 3.6
(a) Suppose that $y(x)$ is a polynomial of order n in the variable $(x-a)$, so that

$$y(x) = p_n(x) \equiv \sum_{k=0}^{n} a_k (x-a)^k \qquad (3.8)$$

Show that Taylor's theorem, (3.6), is exactly satisfied and that the remainder is identically zero because the derivative therein is zero.

(b) Suppose now that $y(x)$ is a polynomial of order $(n+1)$ in the variable $(x-a)$. Show that the remainder after n terms is given by

$$R_n = a_{n+1}(x-a)^{n+1} \qquad (3.9)$$

which is just the $(n+1)$ th term of the polynomial. ∎

Therefore, we have two examples in which the theorem is true, but (just as two swallows do not a summer make) this is not sufficient for a mathematical proof, especially because the result should not depend on the use of a polynomial which has a particularly simple dependence on the variable a. Notice that the statement of the theorem contains a, but $y(x)$ does not depend on a.

The most direct proof of Taylor's theorem, although not an obvious proof, is by the method of induction, a simple example of which is given in Exercise 3.1. We will show that if the theorem is true for n, then it must be true for $n + 1$. By showing that the theorem indeed holds for $n = 0$, we will then know that it is true for all values of n. The proof proceeds as follows.

Since the statement of the theorem, (3.6), claims that $y(x)$ is independent of n, we should be able to prove that the difference between y_{n+1} and y_n is zero.

Exercise 3.7

(a) Use (3.6) applied for n and then for $n + 1$ to show that

$$y_{n+1}(x) - y_n(x) = \sum_{k=1}^{n} \frac{y^{(n+1)}(a)}{(n+1)!}(x-a)^{n+1} + R_{n+1} - R_n \quad (3.10)$$

Now use integration by parts to show that the difference between the two remainders exactly cancels the first term on the right side of this equation, so that the result is zero and the value of the expansion is indeed independent of n.

(b) Show that for $n = 0$ the expression (3.10) becomes, including the calculation of the remainder,

$$y_0(x) = y(x) \quad (3.11)$$

so that the theorem holds for $n = 0$. Therefore, by the above induction, it holds for $n = 0+1$, and therefore for $n = 1+1$, ... ■

Thus we have proved Taylor's theorem. In order to be mathematically punctilious, we should enquire more carefully into the conditions for the required derivatives to exist at the point $x = a$. Rather than delve into these points here, you may follow up on them in texts such as Protter and Morrey or Taylor and Mann. In the next subsection we consider some general aspects of interpreting Taylor series before we derive Taylor expansions of some common and useful functions.

Interpreting Taylor series

Taylor's theorem may well be true, but what's the point of it? The first puzzle may be to understand the meaning and use of the quantity a in a Taylor series. Formally, this quantity is arbitrary, apart from the requirement of being able to compute the first n derivatives at $x = a$. Practically, we want the power series in the variable $(x - a)$ that appears in (3.1) to converge rapidly for the range of x that is of interest for a particular problem. Therefore, a should preferably be chosen in the midrange of x values concerned.

Your second question may well be, what is to be done with the remainder after n terms, R_n? If it could be calculated exactly, one might as well have computed $y(x)$ itself. One therefore usually wants to choose the parameters a and n in the Taylor series so that, according to some estimate of R_n, its effect on the value of $y(x)$ is suitably negligible.

A common choice of a is that $a = 0$, in which case the Taylor series is called a *Maclaurin series*. An example will clarify the relations between n and the remainder term R_n.

Exercise 3.8
Suppose that $y(x)$ is a polynomial of degree N in the variable x

$$p_N(x) = \sum_{k=0}^{N} a_k x^k \tag{3.12}$$

Show that the coefficients a_k and the derivatives of the polynomial are related by

$$a_k = \frac{p^{(k)}(0)}{k!} \tag{3.13}$$

in accordance with the general formula (3.1). Show also that the remainder R_n in the Taylor expansion of this polynomial is, in general, nonzero for $n < N$ and is identically zero for $n \geq N$. ∎

With this introduction to the interpretation of Taylor's theorem, we are ready to develop specific Taylor expansions. From these examples we will further clarify our interpretation of Taylor and Maclaurin series. Further, they serve as the basis for many other numerical developments that we make.

3.2   TAYLOR EXPANSIONS OF USEFUL FUNCTIONS

In this section we discuss the Taylor expansions of useful functions, starting with the exponential, $\exp(x)$, which will be our paradigm for the other functions considered. These are: the circular functions $\cos x$ and $\sin x$, the inverse circular function $\arcsin x$, the natural logarithm $\ln(1+x)$, the function $x \ln(x)$, and the hyperbolic functions $\cosh x$ and $\sinh x$. The exponential will be developed in enough detail that the techniques of Taylor expansions will become clear to you, then the other functions will be considered in a more cursory way.

The exponential function also serves as the model for distinguishing between analytical and numerical convergence properties of power series. Testing the numerical convergence of series for the other functions that we are about to discuss is the topic of Project 3 in Section 3.5.

Expansion of exponentials

The ingredients of the analytical recipes for expanding functions by using Taylor's theorem, (3.1), are the choice of a and the values of the derivatives of the function evaluated at the point $x = a$. For the exponential function, $y(x) = \exp(x)$, these derivatives are especially simple because they are all equal

$$y^{(k)}(a) = e^a \tag{3.14}$$

Therefore in (3.6) there is a common factor of e^a in every term of the series. The power-series expansion of the exponential can therefore be written

$$e^x = e^a \left[1 + \frac{(x-a)}{1!} + \frac{(x-a)^2}{2!} + \dots + \frac{(x-a)^n}{n!} + R_n \right] \tag{3.15}$$

in which the remainder after n terms of the expansion, R_n, is given by substituting the derivative (3.14) into (3.6):

$$R_n = \int_a^x dt \, (x-t)^n \, e^t/n! \tag{3.16}$$

Exercise 3.9

Show that an upper bound on this remainder can be obtained by substituting the upper bound on the exponential in the range x to a, to obtain

$$R_n \leq (x-a)^{n+1} \, e^{\max(x,a)}/(n+1)! \tag{3.17}$$

which shows that the series will gradually converge because of the factorial term. ∎

Notice in (3.15) that division throughout by the factor $\exp(a)$ produces a power series for $\exp(x-a)$ in terms of the variable $(x-a)$. Thus, apart from the relabeling of the variable (a shift of origin from zero to a), the parameter a is of no significance here. Henceforth, we discuss only the Maclaurin expansion, which has $a = 0$. In compact form the Maclaurin expansion of the exponential function is

$$e^x = 1 + \sum_{k=1}^{\infty} \frac{x^k}{k!} \tag{3.18}$$

By comparison with the geometric series in Section 3.1, whose range of convergence is strictly limited (as shown in Figure 3.1), the exponential function Maclaurin series converges (at least analytically) for all values of the argument x. We show examples of this convergence in Figure 3.2 for the same numerical values of the arguments as used in the geometric series in Figure 3.1.

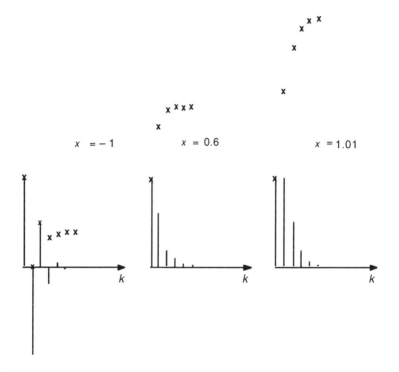

FIGURE 3.2 Convergence of the Maclaurin series for the exponential function. Compare with Figure 3.1 for the geometric series with the same values of the arguments.

For all three arguments of the exponential, $x = -1$, 0.6, and 1.01, the power series converges to within 0.1% of the exponential function value by $n = 5$ terms. Note the key difference between the geometric series (3.1) and the exponential series (3.18), namely that the latter has factorials in the denominator that inexorably diminish the contributions of successive terms.

We should not be completely sanguine about this rapid convergence, because we have restricted the three examples in Figure 3.2 to small values of the argument x, small enough that the powers of x in (3.18) do not initially overwhelm the factorials. It is therefore worthwhile to investigate the series numerically by using a computer program that we now devise.

Computing the exponential series

The C-language function that we write for the power-series expansion of the exponential according to (3.18) also serves as the paradigm for the other power-series functions in Project 3 in Section 3.5.

The power-series expansion of the exponential is given in Program 3.2.

PROGRAM 3.2 Power-series expansion for the exponential function.

```
#include <stdio.h>
#include <math.h>

main()
{
/* Power Series for Exponential */
double xmin,dx,xmax,x,series,error;
int kmax;
double PSexp();

printf("Power Series for Exponential\n");
xmax = 2;
while ( xmax != 0 )
  {
  printf("\n\nInput xmin,dx,xmax (xmax=0 to end),kmax:\n");
  scanf("%lf%lf%lf%i",&xmin,&dx,&xmax,&kmax);
  if ( xmax == 0 )
    { printf("\nEnd Power Series for Exponential");   exit(0); }
  for ( x = xmin; x <= xmax; x = x+dx )
    {
    series = PSexp(x,kmax);
    error = exp(x)-series;
    printf("\n%g  %g  %g",x,series,error);
    }
  }
}

double PSexp(x,kmax)
/*  Power Series function for exponential */
double x;
int kmax;
{
double term,sum;
int k;

term = 1;   sum = 1; /* initialize terms & sum */
for ( k = 1; k <= kmax; k++ )
  {
  term = x*term/k;
  sum = sum+term;
  }
return sum;
}
```

The program `Power Series for Exponential` has a very straightforward structure. The input variables `xmin`, `dx`, and `xmax` control the range and step sizes of the x values. For each x value the input variable `kmax` controls the range of summation for the exponential series (3.18). If the maximum x value, `xmax`, is input as zero, then program execution terminates gracefully. Otherwise, a `for` loop over the x values is run. For each x the series value is computed as `series`, then it is compared with the mathematical library function value `exp(x)` to calculate variable `error`.

The power-series expansion for the exponential is in function `PSexp`. This has a loop over the k values in the sum (3.18), in which each term is computed by recurrence from the immediately preceding term before being added into the partial sum.

Program `Power Series for Exponential` can now be used to explore the numerical convergence of the exponential series. Since the exponential function also serves as a test function for numerical derivatives in Section 4.5, it is worthwhile to understand its numerical properties at this stage.

Exercise 3.10

(*a*) Code and test `Power Series for Exponential` as given in Program 3.2. Tests may be made against tabulated values of the exponential function. The `kmax` you choose for a given test x value should be large enough that convergence to significant accuracy is achieved. Therefore, start with small x values (say 0.1) and small `kmax` values (say 5) before testing with larger values.

(*b*) For input x values in the range zero to 2, check the convergence of the exponential series as a function of varying `kmax`. Compare your results with those in Figure 3.2. What do you conclude about the rate of convergence of this series, and what is the reason for this convergence property?

(*c*) Modify the program so that for a given *x* the power-series expansions of both $\exp(-x)$ and $\exp(x)$ are computed. The program should then multiply together the series from the two expansions and subtract the result from unity. For positive *x* which series do you expect to converge faster? Is this what you observe by changing `kmax`? (Recall the discussion of alternating series in Section 3.1.) ■

Figure 3.3 shows representative values of the series expansion results compared with the exponential function calculated with the mathematical library function `exp`. By `kmax = 10` agreement at parts per million is achieved even for x = 2. The factorials of k in the denominators in (3.18) help greatly in producing convergence, because it is they which distinguish the exponential series (which converges everywhere) from the geometric series (which converges only within the unit circle).

Once we have analytical and numerical experience with power series for the exponential function, other often-used circular and hyperbolic functions can be handled quite directly. They have nice convergence behavior similar to that of the exponential because, as we showed in Section 2.3, they are all functions belonging to the same family.

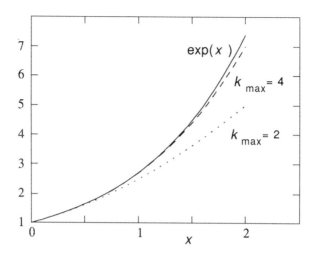

FIGURE 3.3 Convergence of the exponential-function power series. The solid line shows the exponential function, while the dashed lines show the series summed to 2 or 4 terms.

Series for circular functions

The circular functions, cosine and sine, are used frequently, especially when solving problems in oscillatory motion, vibrations, and waves. We discussed this in Section 2.4 in the context of complex-variable representations of these phenomena. The analysis and numerics of such problems can often be simplified if Taylor expansions of the circular functions can be used.

The direct way of deriving the Taylor expansions using (3.6) would be to evaluate the indicated derivatives. This method is tedious and error-prone because the successive derivatives switch between the cosine and sine functions, and they also change signs. A more insightful method of deriving their series is to use the connection between complex exponentials and the circular functions as stated in Euler's theorem, (2.39) in Section 2.3. For simplicity, let us assume that x is a real variable. We write (2.39) as

$$\cos x + i \sin x = e^{ix} = 1 + \sum_{k=1}^{\infty} \frac{(ix)^k}{k!} \tag{3.19}$$

It is now straightforward to derive the expansions for the cosine and sine.

Exercise 3.11
In (3.19) equate real and imaginary parts of the left- and rightmost expressions in order to derive the Maclaurin expansions for the cosine and sine functions, namely

$$\cos x = 1 + \sum_{k=1}^{\infty} \frac{(-x^2)^k}{(2k)!} \tag{3.20}$$

$$\sin x = x \sum_{k=0}^{\infty} \frac{(-x^2)^k}{(2k+1)!} \tag{3.21}$$

which are convergent series, since they arise from the convergent series for the exponential function. ■

Note that in both these formulas the arguments of the circular functions, x, must be in radians, not in degrees. One can see this from the fact that the Euler theorem refers to angles in radians, or alternatively, that the derivatives of cosines and sines are for arguments in radians.

The conversion factors between the two angular measures are given by *radians = π degrees / 180*, thus we have that *radians* = 0.0174533 *degrees* and *degrees* = 57.2958 *radians*. Roughly (in the approximation that $\pi = 3$), we have *radians ≈ degrees / 60*.

The cosine series can often be put to very practical use by considering only the first two terms of its Maclaurin expansion (3.20). This makes the approximation for the cosine a quadratic expression in x, which is usually easy to handle. If we note that cosine or sine need not be computed with $x > \pi/4$ because of the trigonometric formulas for complementary angles, we can see that the error in the use of a quadratic approximation is surprisingly small, and is therefore adequate for many purposes.

Exercise 3.12
(a) Write out the Maclaurin series for the cosine explicitly to show that

$$\cos x = 1 - \frac{x^2}{2} + \frac{x^4}{24} - \dots \tag{3.22}$$

(b) Show that if $x = \pi/4$, then by neglecting the third term one makes an error of about 2% in the value of the cosine, which is $1/\sqrt{2} \approx 0.707$.

(c) In measuring lengths along nearly parallel lines you will have noticed that a small error of nonparallelism does not produce significant error in a length measurement. For example, a good carpenter, surveyor, or drafter can gauge parallelism to within 2°. Show for this angle that the fractional error in a length measurement is then about 1 part in 2000. ■

The sine series given by (3.21) also converges rapidly. It is convenient to write it in the form

$$\frac{\sin x}{x} = 1 - \frac{x^2}{6} + \frac{x^4}{120} - \cdots \tag{3.23}$$

Strictly speaking, this form holds only for $x \neq 0$. It is usual, however, to let the right-hand side of (3.23) be defined as the value of the left-hand-side quotient even for $x = 0$, so that the ratio $(\sin x)/x$ is defined to be unity at $x = 0$. The angles, x, in (3.23) are in radians, just as for the cosine power series (3.20).

By comparison with the cosine series (3.20) for given x, the sine expansion (3.21) converges more rapidly, as the following exercise will show you.

Exercise 3.13

(a) Show that if $x = \pi/4$, then by neglecting the third and higher terms in (3.21) one makes an error of less than 0.4% in the value of $(\sin x)/x$.

(b) Verify that the theorem of Pythagoras, (2.43), is satisfied through the terms of order x^4 in the cosine and sine as given in (3.20) and (3.21). This result provides another justification for the result in Exercise 3.12 (c). ∎

Rapid and accurate numerical approximations for the trigonometric functions are of great importance in scientific and engineering applications. If one is allowed to use both the cosine and sine series for computing either, one can reduce the value of x appearing in the power series (3.20) and (3.21) so that it does not exceed $\pi/8 = 0.392699$. To do this one uses the identities

$$\cos x = 2\cos^2\left(\frac{x}{2}\right) - 1 \qquad \sin x = 2\sin\left(\frac{x}{2}\right)\cos\left(\frac{x}{2}\right) \tag{3.24}$$

The power series (3.20) and (3.21) through terms in x^6 may be written

$$\cos x \approx 1 - y\left(1 - \frac{y}{6}\left(1 - \frac{y}{15}\right)\right) \qquad y = \frac{x^2}{2} \tag{3.25}$$

and in Horner polynomial form

$$\sin x \approx x\left(1 - \frac{y}{3}\left(1 - \frac{y}{10}\left(1 - \frac{y}{21}\right)\right)\right) \qquad y = \frac{x^2}{2} \tag{3.26}$$

For $x < \pi/8$ the neglected terms in the cosine are less than 10^{-9} and in the sine they are less than 10^{-10}. You can test all these numerics by using Program 3.3.

The program `Cosine & Sine` has a straightforward structure. The main program is controlled by a `while` loop over the values of x that are used in the calculation of the cosine and sine polynomial approximations (3.25) and (3.26). Each run of the program allows a range of x values to be evaluated, with `xmax` input as zero to exit gracefully from the program.

PROGRAM 3.3 Cosine and sine in the compact form of polynomials.

```c
#include <stdio.h>
#include <math.h>

main()
{
/* Cosine & Sine in Compact Form */
double xmin,dx,xmax,x,CosVal,CosErr,SinVal,SinErr;
double CosPoly(),SinPoly();

printf("Cosine & Sine in compact form\n");
xmax = 2;
while ( xmax != 0 )
   {
   printf("\n\nInput xmin,dx,xmax (xmax=0 to end):\n");
   scanf("%lf%lf%lf",&xmin,&dx,&xmax);
   if ( xmax == 0 )
     {
     printf("\nEnd  Cosine & Sine in Compact Form");
     exit(0);
     }
   for ( x = xmin; x <= xmax; x = x+dx )
     {
     CosVal = 2*pow(CosPoly(x/2),2)-1;
     CosErr = cos(x)-CosVal; /* compare with computer's cosine */
     SinVal = 2*SinPoly(x/2)*CosPoly(x/2);
     SinErr = sin(x)-SinVal; /* compare with computer's sine */
     printf("\n%g  %g  %g  %g  %g",x,CosVal,CosErr,SinVal,SinErr);
     }
   }
}


double CosPoly(x)
/*  Cosine Polynomial through x to 6-th power */
double x;
{
double y,poly;

y = x*x/2;
poly = 1 - y*(1 - (y/6)*(1 - y/15));
return poly;
}
```

```
double SinPoly(x)
/*  Sine Polynomial through x to 7-th power */
double x;
{
double y,poly;

y = x*x/2;
poly = x*(1 - (y/3)*(1 - (y/10)*(1 - y/21)));
return poly;
}
```

For each x value the program uses the identities (3.24) to halve the argument used in the circular functions. The quantities CosVal and SinVal are obtained in terms of the polynomial-approximation C functions used with the half angles. These quantities are compared with the values produced by the computer mathematical library functions, which are presumed to be exact to within computer roundoff error.

The C functions CosPoly and SinPoly are simple implementations of the polynomial formulas (3.25) and (3.26). By precomputing $x^2/2$ we gain some efficiency, as well as improving the clarity of the coding. The nested form of the polynomial evaluation, the so-called Horner's method that is further discussed in Section 4.2, also is efficient of computer time.

Given the program it is educational to run it and explore numerically the cosine and sine polynomial approximations.

Exercise 3.14

Code and run the program Cosine & Sine for an interesting range of the arguments x. First check that the polynomials maintain the reflection (parity) symmetry of the exact functions, namely that the cosine is an even function of x and that the sine is an odd function of x. Then verify that for $x < \pi/8$ (angles less than 22.5°) that the accuracy is as claimed below (3.26). Finally, expand the range of x values to find out when an unacceptably large error results, say greater than 10^{-6}. ∎

Although the polynomial approximations suggested here are remarkably accurate, one can compute the circular functions more accurately in a comparable time by using various other approximations, such as discussed extensively for the sine function in Chapter 3 of Miller's book on the engineering of numerical software and in the handbook by Cody and Waite. Project 3 in our Section 3.5 again examines the convergence of power series for circular functions.

Having studied Taylor expansions of the circular functions rather exhaustively (and maybe exhaustingly), we are well equipped to consider power series for related functions.

Inverse circular functions

The inverse circular functions are of interest and importance in analysis, numerics, and applications because problem solutions often are given in terms of arccosine or arcsine functions. These are the two inverse functions that we consider, although the arctangent function often also occurs in scientific and engineering problems, but it is considerably more difficult both analytically and numerically.

Because arccosine and arcsine functions with the same argument are related by being complementary angles (summing to $\pi/2$), the value of one function implies the other. If we wish to have a Maclaurin series (an expansion with $a = 0$), this will be difficult for the arccosine, because we know that $\arccos(0) = \pi/2$. So we choose to expand the arcsine function, since $\arcsin(0) = 0$, which promises rapid convergence of its Maclaurin series.

The general prescription for a Taylor series is given by (3.6). With $a = 0$ to produce a Maclaurin series, we need the successive derivatives of the function arcsin x, then these derivatives are to be evaluated at $x = a = 0$. This will provide good mental exercise for you.

Exercise 3.15
(a) The first derivative needed is that of the arcsine function itself. Show that

$$\frac{d \arcsin x}{dx} = \frac{1}{\sqrt{1 - x^2}} \qquad |x| \neq 1 \qquad (3.27)$$

(b) Evaluate the second and higher derivatives of the arcsine function, that is, the first and higher derivatives of the right-hand side of (3.27), then set $x = 0$ in the derivatives for the Taylor expansion (3.6) in order to obtain the power series

$$\arcsin x = x \left(1 + \frac{x^2}{6} + \frac{3 x^4}{40} + \frac{5 x^6}{112} + \ldots \right) \qquad (3.28)$$

which is consistent with the expansion for the sine function, (3.21), through the first two terms. ∎

The general pattern for the coefficients of the powers occurring in the series is not so obvious; it is clarified when we develop the program for the arcsine function in Section 3.5. For the moment we note the close similarity of the Maclaurin series for the sine and arcsine functions. We therefore expect similar rapid convergence of the series (3.28). Since for real angles only values of x with magnitude less than unity are allowed in this expansion, the expansion of the arcsine function is particularly rapid. By continuity, one has $\arcsin(1) = \pi/2$ for angles restricted to the first quadrant.

Hyperbolic function expansions

We introduced the hyperbolic functions cosh and sinh in Section 2.3 in the context of functions of complex variables. On the basis of analysis methods that we have developed, there are three ways to obtain the power series expansions of these hyperbolic functions: we may use Taylor's theorem directly, we may use their definitions in terms of the exponential functions, or we may use the relations between hyperbolic and circular functions that were obtained in Exercise 2.14. The first way is by now rather boring, so consider the other two.

Exercise 3.16

(*a*) Consider the definitions of cosh x and sinh x in terms of the exponentials, as given by (2.48) and (2.49), then note the Maclaurin series for exp (x) given by (3.18). Thus show, by combining terms from the two convergent series, that the cosh gathers the even terms of the exponential and the sinh gathers its odd terms, producing for the hyperbolic cosine

$$\cosh x = 1 + \frac{x^2}{2} + \frac{x^4}{24} + \frac{x^6}{720} + \dots \tag{3.29}$$

and for the hyperbolic sine

$$\sinh x = x\left(1 + \frac{x^2}{6} + \frac{x^4}{120} + \frac{x^6}{5040} + \dots\right) \tag{3.30}$$

(*b*) As an alternative derivation, use relations (2.53) and (2.54) between hyperbolic and circular functions, and assume that the power-series expansions (3.20) and (3.21) can be extended to complex variables, from x to ix. Thus again derive the expansions (3.29) and (3.30). ■

The second method shows, in terms of complex variables, where the sign changes between the power series for circular and hyperbolic functions originate, since the alternating signs for the former series are canceled by the factors of $i^2 = -1$ which arise in forming the latter series.

The convergence properties of the cosh and sinh are very much better than those of the cos and sin, because the sums in both (3.29) and (3.30) have only positive terms. Therefore, given the very rapid convergence of the series for circular functions, the convergence of the series for hyperbolic functions will be even more rapid for small x. Note, however, as shown in Figure 2.4, that these hyperbolic functions are unbounded rather than periodic, so the restrictions to small arguments that we invoked for the circular functions cannot be made so realistically for them.

The numerical properties of the Maclaurin expansions for cosh and sinh are explored in Project 3 in Section 3.5. Their coding is essentially that for the cos and sin expansions, apart from the lack of sign reversals between successive terms.

Logarithms in series expansions

There are many applications for power series expansions of the natural logarithm, ln. Maclaurin expansions of ln (x) cannot be made because the logarithm is divergent as x tends to zero, and its first derivative diverges as $1/x$. Instead, one usually considers the expansion of ln $(1 + x)$ about $x = 0$.

Exercise 3.17
Show that if one wants an expansion of ln $(a + bx)$ about $x = 0$, then an expansion in terms of $(1 + u)$, where $u = bx/a$, is sufficient, except that the region of convergence for u differs by the factor b/a relative to that for x, and is therefore smaller if this factor is of magnitude greater than unity. ■

A power series expansion of ln $(1 + x)$ is easiest made directly from Taylor's theorem, (3.6). The successive derivatives after the first, which is just $1/(1 + x)$, are easy to evaluate. They alternate in sign and grow in a factorial manner, but they are one step behind the Taylor-theorem factorial. Thus you can readily obtain the series expansion

$$\ln (1+x) = x - \frac{x^2}{2} + \frac{x^3}{3} - \frac{x^4}{4} + \ldots \tag{3.31}$$

This series expansion converges only if $|x| < 1$, as can be seen by comparison with the divergent harmonic series.

The result (3.31) and its consequences are of considerable interest.

Exercise 3.18
(a) Show in detail the steps leading to (3.31).

(b) Write down the series expansion for ln $(1 - x)$ by using (3.31) with the sign of x changed, then subtract the series expansion from the expressions in (3.31). Thereby show that

$$\ln\left(\frac{1+x}{1-x}\right) = 2x\left(1 + \frac{x^2}{3} + \frac{x^4}{5} + \frac{x^6}{7} + \ldots\right) \tag{3.32}$$

(c) Check the parity symmetry of the logarithm on the left side of (3.32) by showing that it is an odd function of x, and that this result agrees with the symmetry of the right-side expression. Show also that the simpler expresssion in (3.31) does not have a definite symmetry under sign reversal of x. ■

As you can see from the coefficients of the powers of x in (3.31), the logarithmic series has quite poor convergence, mainly relying for its convergence on the decreasing values of the powers of x, rather than on the presence of factorials in the denominator. We investigate the numerical convergence in more detail as part of the computing project in Section 3.5.

The power series expansion of the logarithm may be adequate in limited ranges of the variable, especially if one is willing to interpolate within a table of logarithms. Suppose we have a table in computer memory with entries at $t - h, t, t + h, \ldots$, and that we want the interpolated logarithm at some point a distance d from t. Clearly, it is sufficient to consider values of d such that $|d/h| < 1/2$. How closely spaced do the table entries have to be in order to obtain a given accuracy, ε, of the interpolated values?

Exercise 3.19

(a) Use the power-series expansion of the logarithm, (3.31), to show that

$$\ln (t + d) = \ln (t) + \frac{d}{t}\left(1 - \frac{d}{2t}\right) + \varepsilon \qquad (3.33)$$

in which the error, ε, can be estimated as the first neglected term, which is

$$\varepsilon \approx \frac{1}{3}\left(\frac{d}{t}\right)^3 < \frac{1}{24}\left(\frac{h}{t}\right)^3 \qquad (3.34)$$

(b) Show that if the tabulated increments satisfy $h/t \approx 0.01$, then the error will be less than 10^{-7}. ∎

Table lookup therefore provides an efficient way to obtain values of logarithms over a small range of its argument. In this example 200 entries would give better than part-per-million accuracy. For applications such as Monte Carlo simulations, where there is intrinsically much approximating, this accuracy is probably sufficient.

Now that we have some experience with handling power series for logarithms, it is interesting to try a more difficult function involving logarithms.

Series expansion of $x \ln(x)$

The function $x \ln(x)$ is of interest because, although it has a Maclaurin expansion, this cannot be obtained directly by taking successive derivatives of this function and evaluating them at $x = 0$, since the derivatives of the logarithm diverge as $x \to 0$. You should first be convinced that $x \ln(x)$ is indeed well-behaved near the origin. To do this, try the following exercise.

Exercise 3.20

(a) Set $x = e^{-y}$, with $y > 0$, so that $x \ln(x) = -y/e^y$. Thence argue that $x \to 0$ as $y \to \infty$, and therefore that $x \ln(x) \to 0$ from below.

(b) Make a graph of $x \ln(x)$ for x in the range 0.05 to 2.00 by steps of 0.05, and check that it agrees with the solid curve in Figure 3.4. Thus this function is well-behaved over the given range of x, and is therefore likely to have a series expansion about the origin, at least over a limited range of x. ∎

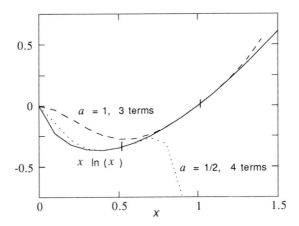

FIGURE 3.4 The function $x \ln (x)$, shown by the solid curve, and its approximations, shown by the dashed curves.

The derivation of the series expansion of $x \ln(x)$, examples of which are shown in Figure 3.4, is easiest made by considering the Taylor expansion about $x = 1$ of the logarithm, written in the form

$$\ln \left(\frac{1}{1-x}\right) = \sum_{n=1}^{\infty} \frac{x^n}{n} \qquad |x| < 1 \qquad (3.35)$$

We also have the geometric series

$$\frac{1}{1-x} = \sum_{m=1}^{\infty} x^m \qquad |x| < 1 \qquad (3.36)$$

If we now multiply the two left-side expressions in (3.35) and (3.36), and their corresponding series on the right sides, then we have a double series in which we can group like powers of x as

$$\frac{1}{1-x} \ln \left(\frac{1}{1-x}\right) = \sum_{m=1, n=1}^{\infty, \infty} \frac{x^{m+n}}{n} \qquad |x| < 1 \qquad (3.37)$$

This is not the conventional form of a series expansion, which usually has a single power, k, of x. You can easily remedy this by working the following exercise.

Exercise 3.21
(a) To understand what the series (3.37) looks like, write out the first few powers of x on the right-hand side, in order to see that the coefficients are 1, $1 + 1/2$, $1 + 1/2 + 1/3$, etc.

(b) Set $k = m + n$ in (3.37), then rearrange the summation variables in order to show that

$$\frac{1}{1-x} \ln\left(\frac{1}{1-x}\right) = \sum_{k=1}^{\infty} h_k x^k \qquad |x| < 1 \qquad (3.38)$$

where the harmonic coefficients, h_k, given by

$$h_k = \sum_{p=1}^{k} \frac{1}{p} \qquad (3.39)$$

are just the sums of the reciprocals of the first k integers. ∎

Thus we have an expansion about $x = 1$ of a series that does not seem to be particularly relevant to our purposes. But we can soon change all that, as follows.
   If in (3.38) we set $y = 1 - x$, then write

$$y \ln(y) = -y^2 \frac{1}{y} \ln\left(\frac{1}{y}\right) = -y^2 \sum_{k=1}^{\infty} h_k (1-y)^k \qquad 0 < y < 2 \qquad (3.40)$$

By relabeling, $y \to x$, we have a power-series expansion. For any number of terms in this expansion it coincides with the function at both $x = 0$ and at $x = 1$, at both of which points it is zero. The expansion to the first three terms (as the dashed curve) is compared with the function (the solid curve) in Figure 3.4. Note that the agreement is fairly poor between the two anchor points, because of the slow convergence of the series. The harmonic series is divergent, so the h_k coefficients grow steadily without limit. Only the decreasing powers of x (less than unity) make the series convergent.
   To reduce the poor convergence properties, we can play tricks with the logarithm series. Suppose that we want an expansion that converges rapidly near $x = a$, where $a \neq 1$. By writing

$$1 - x = (1-a)\left[1 - \frac{x-a}{1-a}\right] \qquad (3.41)$$

and taking logs of this product, we can repeat the above analysis to obtain

$$x \ln(x) = x \ln(1-a) - \frac{x^2}{1-a} \sum_{k=1}^{\infty} h_k \left[1 - \frac{x}{1-a}\right]^k \qquad 0 < x < 2 \qquad (3.42)$$

Exercise 3.22
To verify the improved convergence of this expansion, which is about $x = a$ rather than about $x = 1$, choose $a = 0.5$ and write a small program to calculate the series to, say, 4 terms. For x in the range 0 to 1, compare the series values with the function values, as shown by the dotted curve in Figure 3.4. Why does the series expansion rapidly become a poor approximation as x increases from $x = a$? ∎

3.3   THE BINOMIAL APPROXIMATION

In analysis and numerics there are many problems that can be greatly simplified if they can be approximated by a linear dependence on the variable of interest. One way to produce such a linear dependence is to use the binomial approximation, which we now discuss. Skillful application of this approximation will often produce an orderly solution from an otherwise chaotic problem.

We first derive the binomial approximation, then we give several applications of it and show its connection to exponentials and to problems of compounding interest.

Deriving the binomial approximation

Suppose that we need to estimate $(a + b)^D$, where $|b/a| << 1$ and a^D is known. If D is an integer, this is no problem because for positive D we can use the binomial expansion in terms of combinatorials and powers of a and b. When D is a negative integer it may be sufficient to use the binomial expansion with $|D|$, then to take the reciprocal of the result. But what if D is an arbitrary quantity: how can one reasonably proceed?

The binomial approximation is designed to answer such a question. First, we cast the problem in a simplified form by noting that

$$(a + b)^D = a^D\left(1 + \frac{b}{a}\right)^D = a^D (1 + x)^D \qquad x = \frac{b}{a} \qquad (3.43)$$

If a and b represent quantities with units, these units must be the same, otherwise one is adding apples to bananas. Given this condition, x is dimensionless, so its addition to a pure number (unity) is allowed. Knowing a^D in (3.43), it is sufficient to consider approximating the function $(1 + x)^D$. We do this by devolving the Maclaurin series for this function.

Exercise 3.23
Calculate the successive derivatives of the function, then evaluate them at $x = 0$ in order to show that

$$(1 + x)^D = 1 + Dx\left(1 + \frac{D - 1}{2}x^2\right) + \ldots \qquad (3.44)$$

in which the series terminates after D terms if D is a positive integer. ∎

The binomial approximation is that

$$(1 + x)^D \approx 1 + Dx \qquad \left|\frac{D - 1}{2}x^2\right| << 1 \qquad (3.45)$$

Thus we have approximated a result that usually depends nonlinearly on x by one that has a linear dependence, which often leads to much simpler solutions.

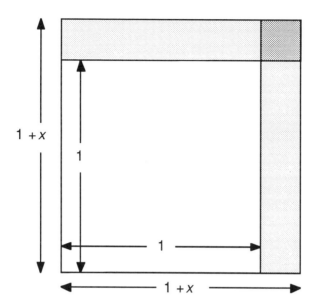

FIGURE 3.5 Geometric representation of the binomial approximation, shown for $D = 2$.

A geometrical viewpoint is appropriate to clarify the interpretation of the binomial approximation. Suppose that in the function to be expanded we have $D = 2$ and $|x| \ll 1$, thus satisfying the condition in (3.45). We can represent the function in a plane (such as a page of this book) by drawing two squares, one with sides of length unity and the other with side length $1 + x$, as shown in Figure 3.5.

The two lightly shaded rectangles in Figure 3.5 are additional contributions proportional to x, and are therefore included in the binomial approximation that $(1 + x)^2 \approx 1 + 2x$ for $D = 2$. The heavily-shaded square is the ignored part, just equal to x^2. Clearly this geometric correspondence can readily be generalized.

Exercise 3.24

(*a*) For practice, sketch the diagrams for the geometric interpretation of the binomial approximation when $D = 0$, 1, and 3. Indicate which parts are included and ignored in the approximation.

(*b*) (Challenge part) Since the interpretation is clear for the above D values, make a sketch for intermediate values of D, such as $D = 1.5$ and $D = 2.5$, as well as for the larger value $D = 4$. ∎

Because this is a geometric representation, we may think of D as the dimension of the binomial approximation exponent. The example of $D = 4$, the fourth dimension, may be explored by the interested reader in the introductory-level book by Rucker.

Applications of the binomial approximation

To illustrate the power of the binomial approximation for numerical work, so much power that you can amaze your friends with your arithmetic skills, try the following numerical examples.

Exercise 3.25

(a) Show that the error in estimating 1.01^{10} by the binomial approximation is only of order 5 parts in 10,000.

(b) Approximate 2.01^{-3} by first removing the known factor of $2^{-3} = 0.125$, then using (3.45) for the remaining factor. Show that the fractional error from using the binomial approximation is only about 10^{-4}.

(c) Estimate by the binomial approximation, then compare with the exact values the ratio of 2001^D to 2000^D for $D = 1, 5, 137, 1/137, 622,$ and 1066. (For the choice of the last three values, consult books on quantum mechanics, on Islamic history, and on English history, respectively.) ∎

With this introduction to the binomial approximation, you should be convinced of its numerical power. It is also used analytically in Chapters 4, 7, and 8 when we develop numerical approximation methods for derivatives, integrals, and the solution of differential equations.

Linearized square-root approximations

Many scientific calculations involve computation of square roots, for example in distance calculations in two and three dimensions. Such calculations are often part of a quite approximate calculation, for example, the distance between two molecules in a Monte Carlo simulation of random motions in a gas or a distance in a computer-graphics display. So one can save computer time by making a linearized approximation to the square root. We first discuss the binomial approximation for the square root, then we explore other linear estimates.

Suppose that we have a base distance of unit length and that we want distances, $D(x)$, given by

$$D(x) = \sqrt{1 + x} \qquad x \le x_m \qquad (3.46)$$

where x_m is the maximum displacement from unity. A linearized approximation to D that is exact at $x = 0$ is of the form

$$D(x) \approx 1 + ax \qquad x \le x_m \qquad (3.47)$$

in which the value of a depends on the linear approximation devised. The binomial approximation uses, according to (3.45),

$$a = a_b = \frac{1}{2} \qquad (3.48)$$

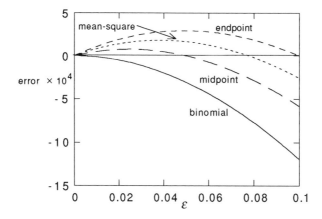

FIGURE 3.6   The errors in various linear approximations to the square root.

Other linear approximations can be devised. One might choose to force the approximation to be exact at the midpoint of the interval $[0, x_m]$ or at the upper end of this interval.

Exercise 3.26

(*a*) Prove that if the square-root approximation is to be exact at the midpoint $x_m/2$, then the coefficient in (3.47) must be chosen as

$$a = a_m \equiv \frac{\sqrt{4 + 2 x_m} - 2}{x_m} \tag{3.49}$$

(*b*) Show that if the square-root approximation is to be exact at the endpoint x_m, then the coefficient in (3.47) must be chosen as

$$a = a_e \equiv \frac{\sqrt{1 + x_m} - 1}{x_m} \tag{3.50}$$

(*c*) Choose $x_m = 0.1$ (about a 5% variation of distances) and calculate the exact distance, then its binomial, midpoint, and endpoint linear approximations. Do this for x values between 0 and 0.1. Calculate the errors as exact distance minus approximate distance. Compare your results with Figure 3.6. ∎

Figure 3.6 shows the errors in the various approximations for $x_m = 0.1$. Notice that, among these three approximations, the endpoint approximation probably does the best job of minimizing the error, except that the error is always such that the approximation systematically underestimates the distance, which may not be desirable. The midpoint approximation does not have this latter defect, but it deviates strongly (rather like the much poorer binomial approximation) toward the endpoint.

Can one make a optimum linear approximation in the least-squares sense of minimizing the averaged squared deviation between the exact and approximated values? Although we justify the least-squares criterion more completely in Chapter 6, the following analysis shows its effectiveness in constraining parameters.

Exercise 3.27
Consider the mean-square error, *MSE*, in fitting a linear approximation to a square root, defined as

$$MSE \equiv \int_0^{x_m} dx \left[\sqrt{1 + x} - (1 + ax) \right]^2 \qquad (3.51)$$

Differentiate *MSE* with respect to the parameter *a* and set this derivative to zero for the optimum value $a = a_{MS}$. Next, approximate the square root under the integrand by its first three terms in powers of *x*. Evaluate the indicated integrals in order to show that the value of *a* is then given by

$$a = a_{MS} \approx \frac{1}{2} - \frac{3}{32} x_m \qquad (3.52)$$

which is shown for $x_m = 0.1$ in Figure 3.6. ■

Notice how the least-squares optimization of the linear approximation provides tighter bounds on the errors than do the other methods. For example, with $x_m = 0.1$ (as in Figure 3.6) the error in this approximation for the distance is never more than about 2 parts in 10^4. In Monte Carlo simulations one frequently needs the distance between two points in a calculation involving square roots. Linearized approximations such as those derived here may be sufficiently accurate for such a simulation, while greatly speeding up program execution. If larger inaccuracies are acceptable, then faster approximations can be devised, as explained in Paeth's article and implementation in C code.

From this example of optimizing a linear approximation to calculation of square roots and distances we have seen that a variety of realistic approximations is possible. We have also gained valuable practice in analyzing numerical methods.

Financial interest schemes

The crass economics of various schemes for calculating interest due in financing might seem to be of little relevance to binomial approximations, and probably not worth the attention of scientists. When that scientist *par excellence*, Isaac Newton, was Keeper of the Mint of England, he was the first to pose and solve the problem of exponential interest, and he used his newly-invented method of the differential calculus to do so. Therefore, let us look at financial interest schemes and their connection to series and the binomial approximation.

Consider an interest rate *r*, usually quoted as an annual percentage rate (APR) paid on the principal. Thus, if the time for interest repayment, *t*, is in years, then after this time the fraction of the principal that is due is $1 + rt$, and the interest fraction is this minus unity. Such a form of interest payment is denoted *simple interest*. Thus, the simple-interest payment after time t, $I_1(t)$, is given by

$$I_1(t) = (1+rt) - 1 = rt \tag{3.53}$$

Such an interest payment is applied at the end of each year, rather than continuously.

Suppose that the interest is paid n times a year on the amount held during the previous n th of one year. For $n > 1$ this is called *compound interest*. For example, quarterly compound interest has $n = 4$. The interest payment after t years is $I_n(t)$, given by

$$I_n(t) = \left(1 + r\frac{t}{n}\right)^n - 1 \tag{3.54}$$

Figure 3.7 shows simple interest and quarterly-compounded interest for a 20% interest rate, $r = 0.2$. The difference is insignificant after one year, but at the end of two years about 15% more has been paid in compound interest.

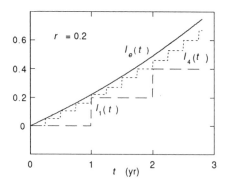

FIGURE 3.7 Simple interest (long dashes), quarterly interest (short dashes), and exponential interest (solid curve) for an interest rate of 20%.

Exercise 3.28
(a) Expand (3.54) in a Maclaurin series in the variable rt/n. Show that

$$I_n(t) = rt + \frac{(n-1)(rt)^2}{2n} + \ldots \tag{3.55}$$

Thus show that the extra fraction by which interest paid n times annually has accumulated to after t years compared with simple interest is

$$\frac{I_n(t) - I_1(t)}{I_1(t)} = \frac{(n-1)rt}{2n} + \ldots \tag{3.56}$$

(b) Use the binomial approximation to show that to this order of approximation there is no difference between compound interest and simple interest. ∎

What happens as one increases the frequency of payments until at any instant one has interest payment proportional to the amount currently held? That is, in (3.54) we allow n to become very large.

Exercise 3.29

(a) Continue the Maclaurin expansion begun in Exercise 3.28 (a). Consider its limit for very large n, so large that $n(n-1)/n^2 \to 1$, and similarly for higher powers of rt. Thus show that

$$\lim_{n \to \infty} I_n(t) = 1 + \frac{rt}{1!} + \frac{(rt)^2}{2!} + \ldots = e^{rt} \tag{3.57}$$

in which the second equality comes from the Maclaurin expansion of the exponential, which is given by (3.18). Thus the interest scheme becomes *exponential interest*.

(b) As a more abstract way of getting to the same result, consider the function

$$E_n(x) \equiv \left(1 + \frac{x}{n}\right)^n \tag{3.58}$$

Show that this function has the derivative property

$$\frac{dE_n(x)}{dx} = E_n(x) \tag{3.59}$$

for any positive integer n. This is a property of the exponential function. Also, the solution of a first-order linear differential equation is unique once the value at a single point (here at $x = 0$) is specified. Argue that we must therefore have

$$\lim_{n \to \infty} E_n(x) = e^x \tag{3.60}$$

which again produces the result (3.57).

(c) From the results of either of the last two parts of this exercise show that exponential interest, $I_e(t)$, is given by

$$I_e(t) = e^{rt} - 1 \tag{3.61}$$

(d) By using the first three terms of the expansion of the exponential function with variable rt, show that exponential interest payments exceed those of simple interest by a fraction of about $rt/2$. ∎

In Figure 3.7 we compare exponential interest, which accrues continuously with time, for an interest rate $r = 0.2$. Although it accumulates to nearly 20% more than simple interest, consistently with the result in Exercise 3.29 (*d*), after two years it is only about 6% greater than interest compounded quarterly.

The mathematics of interest schemes is developed in an interesting book by Kellison and in a module for teaching mathematics by Lindstrom. From our excursion into schemes for calculating financial interest, you should be richer in your understanding of power series, the exponential function, and finite step sizes in approximating functions. This last topic is resumed in Chapter 4.

3.4 DIVERSION: REPETITION IN MATHEMATICS AND COMPUTING

It is interesting and useful to make a distinction between three different kinds of repetition patterns in mathematics and (especially) in computing. We distinguish iteration, recurrence, and recursion. Although I have been careful with the usage of these three terms in the preceding sections of this chapter, I have not yet explained what I understand by each of them. In the following their commonalities and differences will be discussed. Figure 3.8 illustrates schematically the three types of repetition.

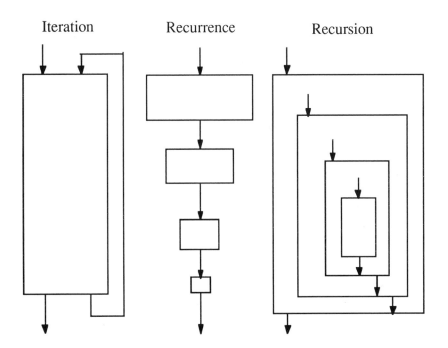

FIGURE 3.8 Three kinds of repetition in mathematics and computing.

We now describe, sequentially, how I intend these three kinds of repetition to be interpreted.

Iteration

By *iteration* is understood repetition of some parts of an analytical or numerical calculation, reusing the same algorithm and parameters but repeating the calculation until some criterion, such as a convergence limit, is satisfied. Summation of series is usually done by iteration over the partial sums. Examples from programming are the C language `for` and `while` statements, the Pascal `FOR` statement, and the Fortran `DO` loop.

The process of iteration can be indicated pictorially by the left-hand diagram in Figure 3.8. In the iteration diagram in Figure 3.8 the elongated rectangle indicates a sequence of program steps. The sequence is entered (top, left arrow), it is executed, then it is repeated (broken arrow re-entering at the top). Eventually, some criterion is satisfied, so the iteration terminates (bottom arrow).

Recurrence

By *recurrence* is understood a formula to be used repeatedly to related successive values of a function when one of its parameters changes by uniform steps. For example, the power-series terms in Section 3.3 and 3.5 are obtained by recurrence with respect to the parameter k. The recurrence process is indicated pictorially in Figure 3.8. In each rectangle there is a procedure for producing this value of a function from the preceding values. For example, the first box may have $k = 1$, the second box $k = 2$, and so on. The procedure within each box is often the same, only the value of the control parameter (such as k) would be changing. As with iteration, recurrence requires a stopping criterion. Indeed, iteration and recurrence are very similar and so they are often not distinguished.

In mathematical analysis recurrence is logically related to the method of proof by induction, as used for the geometric series and the proof of Taylor's theorem in Section 3.1. In his book on induction and analogy in mathematics, Polya discusses clearly the inductive method.

Recursion

In *recursion* a formula refers to itself. This is indicated in Figure 3.8 by the nesting of the boxes, which each describe the (same) function. In mathematical analysis an expansion in continued fractions is an example of recursion. Among programming languages both C and Pascal allow recursive definition of functions, but Fortran usually does not allow recursion. In Fortran if recursive use of a function or subroutine is hidden from the compiler, which can be done because subroutines may be compiled independently, such an error (which is not usually diagnosed by the computer) will usually produce incorrect program execution.

In recursive program functions, a copy of the function has to be saved for each recursion (each box in Figure 3.8). The number of recursions must therefore be finite, and there must be a termination path out of each function copy, as indicated by the arrows at the bottom of each box in Figure 3.8. Recursive functions are not used in this book. Extensive discussions of the pleasures and perils of recursion in C programming are provided in the book by Kernighan and Ritchie, the authors of the C language. The self-referencing nature of recursive functions gives rise to interesting logical and philosophical problems, as discussed by Hofstadter in his book.

3.5 PROJECT 3: TESTING THE CONVERGENCE OF SERIES

In this project we program the functions for the Maclaurin expansions of several of the series that are derived in Section 3.2, then we use these functions to explore the numerical convergence of series, particularly to compare and contrast analytical and numerical convergence properties. We made a beginning on the latter topic in Section 3.2, where we programmed and used the Maclaurin series for the exponential function to examine its numerical convergence.

One of our main goals in this project is to practice developing computer programs by first coding and testing separate modules, then to combine these into a complete and usable program. From this example you will readily see the advantages of such a "divide and conquer" strategy for program construction and verification.

Project 3 begins with coding and checking of five Maclaurin series expansions, for the exponential (initiated in Section 3.2), the cosine and sine, the arcsine, and the logarithm. Then there are suggestions for including the series expansions of the hyperbolic functions, as well as for file output and graphics options. After you are convinced of the correctness of each function (by experience rather than by faith), you may combine them into a composite program Power Series Convergence. This program includes one more function, the inverse cosine, which is efficiently computed from the inverse-sine Maclaurin series. The programming of the arcosine function also illustrates how to pass functions as parameters of other functions in C. Suggestions for using the program to explore the convergence of power series are then given.

Coding and checking each series expansion

In this subsection we summarize the coding and checking of the formulas for six functions of interest. For each of them a main program, similar to that made for the exponential function in Section 3.2, may be used for checking against values obtained from tables of functions or from calculations with an electronic calculator.

1. *Exponential function.* The first function whose Maclaurin expansion we computed numerically was the exponential, discussed in Section 3.2. It is called PSexp in the complete program below. The methods of programming and testing the series

expansions for the three functions cosine, sine, arcsine, and the natural logarithm are similar. Numerical comparisons for the power series of the exponential function are provided in Section 3.2 and Figure 3.3.

2. *Cosine function.* The cosine Maclaurin expansion may be written, by (3.20), as

$$\cos x \approx \sum_{k=1}^{k_{max}} t_k \tag{3.62}$$

where the terms, t_k, are obtained by the recurrence relation

$$t_k = \frac{-\left[x^2\right]}{2k\,(2k-1)} t_{k-1} \qquad k > 0 \tag{3.63}$$

$$t_0 = 1 \tag{3.64}$$

In (3.63) the presence of the brackets around x^2 indicates that this quantity may be computed and saved before the series is begun. In the coding shown in the next subsection for the function PScos this argument is denoted by variable xs.

Exercise 3.30
Show that recurrence formula (3.63) for the expansion coefficients is consistent with the Maclaurin expansion of the cosine function (3.20). ∎

Checking the cosine series may be done by using input values of x less than unity. Values of kmax about 20 should give accuracy better than 1%. Also check that your programmed cosine series is an even function of x, for any value of kmax. If you use values of x more than unity, then convergence will be slow. For a better computation you could use x modulo $\pi/2$ and appropriate sign changes, so that the x value for which the series was evaluated was always less than this number.

Figure 3.9 compares the Maclaurin series expansions of the cosine function for $k_{max} = 1$ and 2. Notice the very rapid convergence of the series that arises from the approximate $1/k^2$ dependence of the successive terms in the series.

3. *Sine function.* The Maclaurin expansion is obtained from (3.21), written as

$$\sin x \approx x \sum_{k=1}^{k_{max}} t_k \tag{3.65}$$

in which the successive terms are obtained by using the recurrence relation

$$t_k = \frac{-\left[x^2\right]}{2k\,(2k+1)} t_{k-1} \qquad k > 0 \tag{3.66}$$

$$t_0 = 1 \tag{3.67}$$

Just as for the cosine series, the x^2 in the brackets is to be precomputed and saved.

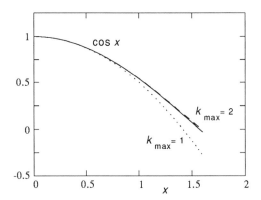

FIGURE 3.9 Power series (dashed curves) for the Maclaurin expansion of the cosine function (solid curve).

Exercise 3.31
Verify that formula (3.66) for the expansion coefficients of the sine functions is consistent with the Maclaurin expansion of the sine function, (3.21). ∎

The sine function, PSsin, can easily be checked against values from tables or from a pocket calculator, with a similar suggestion to that for the cosine series that multiples of $\pi/2$ (with appropriate sign changes) can be removed. The convergence of the sine series as a function of kmax should be similar to that for the cosine series. Also, for any kmax the sine series should be an odd function of x.

In Figure 3.10 we have the power series compared against the sine function for k_{max} values of 1 and 5. Just as for the cosine series, the convergence is very rapid because of the decrease as roughly $1/k^2$ of successive terms in the series.

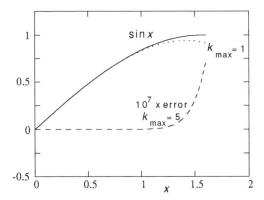

FIGURE 3.10 Maclaurin series expansions for $k_{max} = 1$ (short dashed curve) and the scaled error for $k_{max} = 5$ (long dashed curve) for the sine function (solid curve).

In Figure 3.10, rather than showing the series approximation for $k_{max} = 5$, which could not be visually resolved from the converged series, we show the error (exact minus series) scaled up by a factor of 10^7. Clearly such a truncated series expansion is very accurate, especially for $x < 1$. This leads to the following improvement in accuracy of the power-series expansions.

Given both the cosine and the sine series, these two functions need be used only with arguments that are less in magnitude than $\pi/4$, since if x lies between this value and $\pi/2$ one may use the identity

$$\cos(x) = \sin\left(\frac{\pi}{2} - x\right) \tag{3.68}$$

and vice versa for the sine function. This is a great advantage even over the removal of multiples of $\pi/2$ for the cosine and sine function arguments. Since in the power series for the cosine or sine, successive terms have x^2 factors as large as $(\pi/2)^2 = 2.47 > 1$, or only as large as $(\pi/4)^2 = 0.62 < 1$, the latter (being less than unity) is clearly preferable for producing rapid convergence.

Exercise 3.32

(*a*) Explain, in terms of the recurrence formulas (3.63) and (3.66), why it takes only about half as many terms (k_{max} is halved) to obtain the same accuracy in the powers-series expansions of the cosine and sine if the relation (3.68) is used to limit the function arguments to less than $\pi/4$ than if arguments up to $\pi/2$ are allowed.

(*b*) Verify the analytical estimate in (*a*) by numerical examples, using the functions PScos and PSsin that you have programmed. ∎

4. *Arccosine function.* The arccosine function converges rapidly only for x values where the cosine function is small in magnitude. Therefore, it has much better convergence when it produces angles near $\pi/2$ than angles near zero. We may write

$$\arccos x = \frac{\pi}{2} - \arcsin x \tag{3.69}$$

which is suitable for $x < 1/\sqrt{2} = 0.707$, that is, for obtaining angles less than $\pi/4$. For $x > 1/\sqrt{2}$ one may use

$$\arccos x = \arcsin \sqrt{1 - x^2} \tag{3.70}$$

Exercise 3.33

In order to verify (3.69) and (3.70), sketch a right-angled triangle with hypotenuse of length unity and base of length x. Use geometry and the theorem of Pythagoras to derive these two relations. ∎

Thus, in the function PSarccos we do not evaluate a power series for the arccosine (in spite of the initial letters), but use (3.69) or (3.70) instead. As with any inverse trigonometric function, there are quadrant ambiguities in the arccosine function, as shown in Figure 2.7. Equations (3.69) and (3.70) place the angles in the first quadrant for positive x. Note that the magnitude of x must be less than unity, else the arcsine power series does not converge. Such a check on x is made in both program functions.

Testing of PSarccos is best done after PSarcsin has been coded and tested. For large kmax the two functions should return complementary angles (summing to $\pi/2$) if they are given the same argument value x.

5. *Arcsine function.* The Maclaurin series for this inverse circular function was derived in Section 3.2. We may write the power series as

$$\arcsin x \approx x \sum_{k=1}^{k_{\max}} t_k \qquad (3.71)$$

with successive terms being obtained from the recurrence relation

$$t_k = \frac{\left[x^2\right](2k-1)^2}{2k\,(2k+1)}t_{k-1} \qquad k>0 \qquad (3.72)$$

$$t_0 = 1 \qquad (3.73)$$

Exercise 3.34
Show that formulas (3.72) and (3.73) give the expansion coefficients for the Maclaurin expansion of the arcsine function (3.71). ∎

Figure 3.11 shows the arcsin function and power series approximations to it.

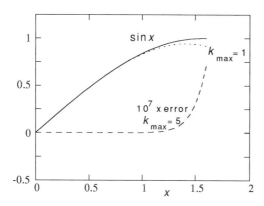

FIGURE 3.11 Arcsine function (solid curve) and its Maclaurin expansions for $k_{\max} = 1$ (dotted curve) and the scaled error for $k_{\max} = 20$.

In the function `PSarcsin` the power series is evaluated after checking that the absolute value of x is less than unity. The function can be tested by using arguments with simple angles, such as $x = 0$ for zero radians, $x = 1/2$ for angle $\pi/6$, $x = 1/\sqrt{2}$ for angle $\pi/4$, and $x = \sqrt{3/2}$ for $\pi/3$. Note that the angles are obtained in radians. The values of `kmax` needed to obtain a close approximation to these angles are large, because the series is not quickly damped out by the k dependence in (3.72).

Notice in Figure 3.11 that the convergence is rapid until x is near unity. Thus, in the recurrence relation for the arcsine series, (3.72), it is the x^2 powers that produce the convergence, rather than the decrease with k (which is rather weak). Even for $k_{max} = 20$ the error for the largest x values is of order 1%.

6. *Natural logarithm.* For this function the power series that is expanded is for the natural logarithm $\ln(1 + x)$, with convergence requiring an absolute value of x that is less than unity. The power series given by (3.31) can be written for the numerical algorithm as

$$\ln(1 + x) \approx x \sum_{k=1}^{k_{max}} \frac{t_k}{k+1} \tag{3.74}$$

with the recurrence relation being

$$t_k = -x\, t_{k-1} \qquad k > 0 \tag{3.75}$$

$$t_0 = 1 \tag{3.76}$$

Exercise 3.35
Show that formulas (3.75) and (3.76), with the denominator in (3.74), give the expansion coefficients for the series expansion of the natural logarithm function (3.31). ∎

Program function `PSln` implements the formulas (3.74) through (3.76), after checking that $|x| < 1$. The function may be tested by comparing its values with those from tables or pocket calculators. Figure 3.12 shows the logarithm function and its power-series approximations for three values of k_{max}.

Notice from Figure 3.12 that convergence of the logarithm series is very slow, as discussed in Section 3.2 for the function $x \ln(x)$. For x larger than about 0.8 there are power-series truncation errors of at least 10^{-4} even after 50 terms have been included in the sum. A similar difficulty was encountered with the power-series expansion of the function $x \ln(x)$ in Section 3.2.

Having completed developing the algorithms, the coding, and the testing of each of the six functions, they can be assembled into a program complete with file-handling or graphics output options, as we suggest in the next subsections.

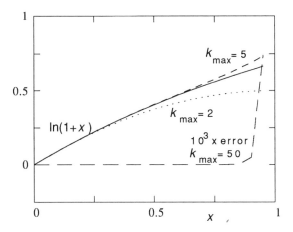

FIGURE 3.12 Logarithm function (solid curve) expanded in Taylor series about unity for 2, 5, and 50 terms in the expansion.

Including the hyperbolic functions

The hyperbolic cosine and sine, cosh and sinh, whose analytical properties are covered extensively in Sections 2.3 and 3.2, are also interesting to study in terms of their power-series expansions. Their Maclaurin series are almost the same (apart from signs) as those for the circular functions, cosine and sine. Deriving the recurrence relations for their series terms, then coding the formulas, is therefore straightforward after you have completed the circular functions.

Exercise 3.36

(*a*) By analogy with the cosine, or by the connection (2.53), show that the Maclaurin series for the hyperbolic cosine can be written as

$$\cosh x \approx \sum_{k=1}^{k_{max}} t_k \tag{3.77}$$

where the successive terms are related by the recurrence relation

$$t_k = \frac{\left[x^2\right]}{2k\,(2k-1)}\,t_{k-1} \qquad k > 0 \tag{3.78}$$

$$t_0 = 1 \tag{3.79}$$

(*b*) Similarly to (*a*), but for the hyperbolic sinh, show that

$$\sinh x \approx \sum_{k=1}^{k_{max}} t_k \tag{3.80}$$

with the recurrence relation

$$t_k = \frac{\left[x^2\right]}{2k\,(2k+1)}\, t_{k-1} \qquad k > 0 \tag{3.81}$$

$$t_0 = 1 \tag{3.82}$$

(c) Code and test functions PScosh and PSsinh for the hyperbolic cosine and hyperbolic sine, respectively. These hyperbolic functions can be obtained from the exponential functions through the definitions (2.48) and (2.49). ∎

These two functions can then be incorporated in the complete program that is suggested below.

File output and graphics options

As you have seen from the preceding four figures, it is usually much easier to interpret numerical results, especially comparison of values, if the comparison is displayed graphically as well as numerically. At the debugging stage numerical output was probably sufficient for you. In order to see both the dependence of the power series approximations on x and k_{max}, graphics are essential. To appreciate this, realize that about 100 numerical values are used to construct each of the figures.

In the composite program in the next subsection, I have included a simple section for file output . One file is provided for each of the six functions, and this file may either be written over or added to at the start of program execution. By using six files you will not be constrained to calculate the series in any particular order.

After you have programmed the file output sections, perhaps testing your ideas beforehand on a short program, you may either include a graphics section in this program or interface the output in the file to a separate graphics program. The latter was my choice, so there are no graphics commands in the following program.

The composite program for the functions

After the considerable preliminaries in the two preceding subsections, you should be ready to join the function modules together to form the program Power Series Convergence. In C the most straightforward, readable, and efficient way to choose which of the functions is to be executed in a given run of the program is to use the switch statement. This is used twice, once near the top of the program to select the output file name, and once for each value of x to select the function to be evaluated.

In the following program the functions for the hyperbolic power series have not been included. There are therefore six choices of function rather than eight.

PROGRAM 3.4 Composite program for power-series expansions.

```c
#include <stdio.h>
#include <math.h>

main()
{
/*   Power Series Convergence   */
FILE *fout;
double pi,xmin,dx,xmax,x,series,func,error;
int kmax,choice;
char wa;
double PSexp(),PScos(),PSsin(),PSarccos(),PSarcsin(),PSln();

pi = 4*atan(1.0);
printf("Power Series Convergence\n");
printf("\nChoose function:");
printf("\n1 exp x      2 cos x      3 sin x");
printf("\n4 acos x     5 asin x     6 ln(1+x): ");
scanf("%i",&choice);
if ( choice < 1 | choice > 6 )
   {
   printf("\n!! Choice=%i is <1 or >6",choice);
   exit(1);
   }
printf("\nWrite over output (w) or Add on (a): ");
scanf("%s",&wa);
switch (choice)
   {
   case 1: fout = fopen("PSEXP",&wa);   break;
   case 2: fout = fopen("PSCOS",&wa);  break;
   case 3: fout = fopen("PSSIN",&wa);   break;
   case 4: fout = fopen("PSACOS",&wa);   break;
   case 5: fout = fopen("PSASIN",&wa);   break;
   case 6: fout = fopen("PSLN",&wa);   break;
   }
xmax = 2;
while ( xmax != 0 )
   {
   printf("\n\nInput xmin,dx,xmax (xmax=0 to end),kmax:\n");
   scanf("%lf%lf%lf%i",&xmin,&dx,&xmax,&kmax);
   if ( xmax == 0 )
     {
```

```
    printf("\nEnd  Power Series Convergence");
    exit(0);
    }
  for ( x = xmin; x <= xmax; x = x+dx )
    {
    switch (choice)
      {
      case 1: series = PSexp(x,kmax);
               func = exp(x);    break;
      case 2: series = PScos(x,kmax);
               func = cos(x);    break;
      case 3: series = PSsin(x,kmax);
               func = sin(x);    break;
      case 4: series = PSarccos(pi,x,kmax,PSarcsin);
               func = acos(x);   break;
      case 5: series = PSarcsin(x,kmax);
               func = asin(x);   break;
      case 6: series = PSln(x,kmax);
               func = log(1+x);   break;
      }
      error = func-series;
    printf("\n%g  %g  %g",x,series,error);
    fprintf(fout,"%g  %g  %g\n",x,series,error);
    }
  }
}

double PSexp(x,kmax)
/*  Power Series for exponential */
double x;
int kmax;
{
double term,sum;
int k;

term = 1;  sum = 1; /* initialize terms & sum */
for ( k =1 ; k <= kmax; k++ )
  {
  term = x*term/k;
  sum = sum+term;
  }
return  sum;
}

double PScos(x,kmax)
```

```
/*  Power Series for cosine */
double x;
int kmax;
{
double xs,term,sum;
int k;

xs = x*x;
term = 1;  sum = 1; /* initialize terms & sum */
for ( k = 1; k <= kmax; k++ )
  {
  term = -xs*term/(2*k*(2*k-1));
  sum = sum+term;
  }
return  sum;
}

double PSsin(x,kmax)
/*  Power Series for sine */
double x;
int kmax;
{
double xs,term,sum;
int k;

xs = x*x;
term = 1;  sum = 1; /* initialize terms & sum */
for ( k = 1; k <= kmax; k++ )
  {
  term = -xs*term/(2*k*(2*k+1));
  sum = sum+term;
  }
return  x*sum;
}

double PSarccos(pi,x,kmax,arcsin)
/*  Arccosine uses PSarcsin series for  arcsin  parameter */
double pi,x;
int kmax;
double (*arcsin)(); /* passing function pointer as parameter */
{

if ( fabs(x) >= 1 )
  {
  printf("\n\n!! In PSarccos x=%g so |x|>=1. Zero returned",x);
```

```
    return 0;
    }
else
    {
    if ( x < 1/sqrt(2) )    return  pi/2-(*arcsin)(x,kmax);
    else    return (*arcsin)(sqrt(1-x*x),kmax);
    }
}

double PSarcsin(x,kmax)
/*  Power Series for arcsine */
double x;
int kmax;
{
double xs,term,sum;
int k;

if ( fabs(x) >= 1 )
    {
    printf("\n\n!! In PSarcsin x=%g so |x|>=1. Set to zero",x);
    return 0;
    }
else
    {
    xs = x*x;
    term = 1;  sum = 1; /* initialize terms & sum */
    for ( k = 1; k <= kmax; k++ )
        {
        term = xs*term*pow((2*k-1),2)/(2*k*(2*k+1));
        sum = sum+term;
        }
    return  x*sum;
    }
}

double PSln(x,kmax)
/*  Power Series for ln(1+x)  */
double x;
int kmax;
{
double power,sum;
int k;

if ( fabs(x) >= 1 )
    {
```

```
printf("\n\n!! In PSln x=%g so |x|>=1. Zero returned",x);
return 0;
}
else
  {
  power = 1;  sum = 1; /* initialize terms & sum */
  for ( k = 1; k <= (kmax-1); k++ )
    {
    power = -x*power;
    sum = sum+power/(k+1);
    }
  return x*sum;
  }
}
```

A few checks should be made on your completed Power Series Convergence program. For example, check that the range of choice is correctly tested as soon as it is input, check that the correct files are then opened, and that the correct function is called for a given value of choice. Also check that the x loop is correctly executed and controlled, including graceful termination if xmax is entered as zero.

With these quick and simple checks out of the way, you are ready to explore the numerical properties of power-series expansions.

Using the program to test series convergence

Having the completed Program 3.4 for testing the convergence of power series of useful functions, we are ready to exercise the computer (and our brains) to test the convergence properties numerically.

Exercise 3.37

(a) For input x values in the range zero to $\pi/2$ for the cosine and sine, and in the range zero to less than unity for the arcsine and logarithm, check the convergence of their power series as a function of varying the upper limit of the summation, kmax. Compare your results with those in Figures 3.9, 3.10, 3.11, and 3.12. What do you conclude about the rate of convergence of each series, and what is the reason for this convergence behavior?

(c) Modify Power Series Convergence so that for a given x the power series for both $\cos x$ and $\sin x$ are computed. The program should then square the series values, add them, then subtract the result from unity. As you change kmax, how rapidly does the result approach zero, which is what the theorem of Pythagoras, (2.43), predicts?

(d) Modify the program so that for a given x the power series expansions of both $\ln(1+x)$ and $\ln(1-x)$ are computed. As you change kmax, how rapidly does each power-series result approach zero, and for $x > 0$ (but less than unity) why is the convergence of the second series much slower? ■

Another numerics question is worthy of exploration with this program. Namely, how sensitive are the power series results to the effects of roundoff errors in the computer arithmetic? If you understand numerical noise in computing, you might investigate this topic now. It is covered in Section 4.2 as a preliminary to discussing numerical derivatives and integrals.

From our extensive discussions in this chapter on power series, from their analysis, numerics, and applications, you will appreciate their importance when computing numerically. Ideas and techniques from power series are used widely in the chapters that follow.

REFERENCES ON POWER SERIES

Cody, W. J., and W. Waite, *Software Manual for the Elementary Functions*, Prentice Hall, Englewood Cliffs, New Jersey, 1980.

Hofstadter, D. R., *Gödel, Escher, Bach: An Eternal Golden Braid*, Basic Books, New York, 1979.

Kellison, S. G., *The Theory of Interest*, Irwin, Homewood, Illinois, 1970.

Kernighan, B. W., and D. M. Ritchie, *The C Programming Language*, Prentice Hall, Englewood Cliffs, New Jersey, second edition, 1988.

Lindstrom, P. A., "Nominal vs Effective Rates of Interest," UMAP Module 474, in *UMAP Modules Tools for Teaching*, COMAP, Arlington, Massachusetts, 1987, pp. 21 – 53.

Miller, W., *The Engineering of Numerical Software*, Prentice Hall, Englewood Cliffs, New Jersey, 1984.

Paeth, A. W., "A Fast Approximation to the Hypotenuse." in A. S. Glassner, Ed., *Graphics Gems*, Academic, Boston, 1990, pp. 427 – 431.

Polya, G., *How to Solve It*, Doubleday Anchor Books, New York, second edition, 1957.

Protter, M. H., and C. B. Morrey, *Intermediate Calculus*, Springer-Verlag, New York, second edition, 1985.

Rucker, R., *The Fourth Dimension*, Houghton Mifflin, Boston, 1984.

Taylor, A. E., and W. R. Mann, *Advanced Calculus*, Wiley, New York, third edition, 1983.

Wolfram, S., *Mathematica: A System for Doing Mathematics by Computer*, Addison-Wesley, Redwood City, California, second edition, 1991.

Chapter 4

NUMERICAL DERIVATIVES AND INTEGRALS

The overall purpose of this chapter is to introduce you to methods that are appropriate for estimating derivatives and integrals numerically. The first, and very important, topic is that of the discreteness of numerical data, discussed in Section 4.2. This is related in Section 4.3 to the finite number of significant figures in numerical computations and to the consequent problem of the increase in errors that may be caused by subtractive cancellation.

With these precautions understood, we are prepared by Section 4.4 to investigate various schemes for approximating derivatives and to develop a C program to compute numerical derivatives, a topic that is explored further in Project 4A in Section 4.5. Numerical integration methods, including derivation of the trapezoid rule and Simpson's formula, are developed and tested in Section 4.6 and applied in Project 4B (Section 4.7) on the electrostatic potential from a line charge. References on numerical derivatives and integrals round out the chapter.

We must emphasize here a profound difference between differentiation and integration of functions. Analytically, for a function specified by a formula its derivatives can almost always be found by using a few direct algorithms, such as those for differentiation of products of functions and for differentiating a function of a function. By contrast, the same function may be very difficult (if not impossible) to integrate analytically. This difference is emphasized in Section 3.5 of Wolfram's book, where he describes the types of integrals that *Mathematica* can and cannot do.

Numerically, the situation with respect to differentiation and integration is usually reversed. As becomes clear in Sections 4.4 and 4.5, the accuracy of numerical differentiation is sensitive to the algorithms used and to the numerical noise in the computer, such as the finite computing precision and subtractive cancellation errors discussed in Section 4.3. On the other hand, almost any function that does not oscillate wildly can be integrated numerically in a way that is not extremely sensitive to numerical noise, either directly or by simple transformations of the variables of integration.

Other methods for differentiating and integrating discretized functions by using splines are presented in Sections 5.4 and 5.5. The numerical methods developed for differentiation in this chapter are useful for differential-equation solution in Sections 7.4 and 7.5 (first-order equations) and in Sections 8.3 – 8.5 (second-order equations). In Section 10.4 we compute the convolution of a Gaussian distribution with a Lorentzian distribution (the Voigt profile) by using a combination of analytical methods and numerical integration.

4.1 THE WORKING FUNCTION AND ITS PROPERTIES

For purposes of comparing various methods, particularly those for numerical derivatives and integrals, it is useful to have a working function and to systematize its analytical properties. A polynomial, which can readily be differentiated and integrated analytically to provide benchmark values, is more appropriate than the functions that we investigated by power series in Section 3.2. Further, most of our approximations are based on the use of low-order polynomials, so by using such a polynomial we can make direct analytical comparison between approximate and exact values. The working function will stress the methods greatest if the polynomial oscillates within the region where the derivatives and integrals are being performed, and such oscillations can be designed into the polynomial by choosing its roots appropriately.

In this section we first derive some analytical and numerical properties of the working function, then we develop a C function for efficient numerical evaluation of polynomials by using Horner's algorithm, then we include this in a program for the working function.

Properties of the working function

We define the working function, $y_w(x)$, by a product expansion in terms of its roots, defining it by the expression

$$y_w(x) \equiv 120\left(x + \frac{1}{2}\right)\left(x + \frac{1}{4}\right)x\left(x - \frac{1}{3}\right)\left(x - \frac{1}{5}\right)(x - 1) \qquad (4.1)$$

which is a sixth-order polynomial having all real roots, and five of these are within the range $-0.5 \le x \le 0.5$ over which we will explore the function properties. This working function, y_w, is shown in Figure 4.1 for a slightly smaller range than this. Outside the range of the graph, the function has another zero at $x = -0.5$ and one at $x = 1.0$.

Although the product expansion (4.1) is convenient for controlling the positions of roots of the working function, and thereby controlling its oscillations, it is not convenient for analytical differentiation or integration. These can be performed by expanding (4.1) into powers of x, then differentiating or integrating term by term. This involves a little algebra and some numerical work, so why don't you try it?

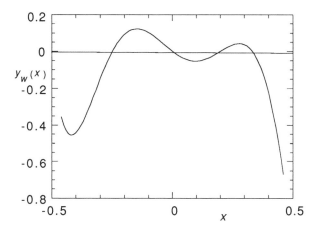

FIGURE 4.1 The working function (4.1), a pathological polynomial having six real roots in the interval [–0.5, 1].

Exercise 4.1

(*a*) Show that the polynomial for the working function, (4.1), can be expanded into

$$y_w(x) = \sum_{k=0}^{6} w_k x^k \qquad (4.2)$$

where the coefficients, w_k, are given in Table 4.1.

(*b*) Starting with the expansion (4.2) with the given w_k, evaluate the successive derivatives of y_w in terms of the expansion coefficients appearing in the formulas

$$y_w^{(i)}(x) = \sum_{k=1}^{6} w_{i,k} x^k \qquad (4.3)$$

in which for the *i*th derivative, $i = 1, 2, ..., 6$, the coefficients are given in Table 4.1.

(*c*) Perform a similar derivation for the integral of the working function $y_w(x)$ in order to show that its indefinite integral is given by

$$\int dx \, y_w(x) = \sum_{k=2}^{7} I_k x^k \qquad (4.4)$$

for which the coefficients I_k are given in the bottom row of Table 4.1. ■

TABLE 4.1 Expansion coefficients of working function (4.1) for the value,w_k, for derivatives, $w_{i,k}$, and for the indefinite integral, I_k. Omitted values are zero.

	Expansion term, k							
	0	1	2	3	4	5	6	7
w_k	0	−1	3	23	−51	−94	120	
$w_{1,k}$	−1	6	69	−204	−470	720		
$w_{2,k}$	6	138	−612	−1880	3600			
$w_{3,k}$	138	−1224	−5640	14400				
$w_{4,k}$	−1224	−11280	43200					
$w_{5,k}$	−11280	86400						
$w_{6,k}$	86400							
I_k	0	0	−1/2	1	23/4	−51/5	−47/3	120/7

This table, transformed into an array, is used in the program at the end of this section. It may also be used for the programs in Sections 4.4, 4.5, and 4.6.

Derivatives of the working function, appropriately scaled, are shown in Figure 4.2. Notice that the derivatives grow steadily in magnitude, but steadily become smoother with respect to x as the order of the derivative increases. Clearly, for our polynomial of sixth order, derivatives past the sixth are zero, as the constancy of $y^{(6)}$ in Figure 4.2 indicates.

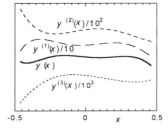
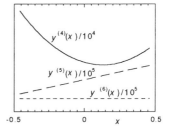

FIGURE 4.2 The working polynomial, (4.1) or (4.2), and its derivatives. Note that the derivatives become larger but smoother as the order of the derivative increases.

It is useful to have available a program for computing the working function (or other polynomial) and its derivatives. Since this function, its derivatives, and its integral are all polynomials, it is practical to have an efficient means of evaluating them. Horner's polynomial algorithm, which we now describe and for which we provide a C function for its implementation, is efficient, accurate, and simple.

A C function for Horner's algorithm

Polynomials occur very frequently in numerical mathematics, for example in any Taylor expansion approximation to a function, as considered in Section 3.2. It is therefore worthwhile to find an accurate and efficient way to evaluate polynomials numerically. Horner's method is one approach. Suppose that the polynomial to be evaluated, $y(x)$, is of order N in the variable x, so that

$$y(x) = \sum_{k=0}^{N} a_k x^k \tag{4.5}$$

We can avoid the inefficient (and sometimes inaccurate) direct evaluation of powers of x by starting with the highest-order term, then adding in the next-lowest coefficient plus the current value of the polynomial multiplied by x. For example, suppose that we have the quadratic polynomial

$$y(x) = a_0 + a_1 x + a_2 x^2 \tag{4.6}$$

We may evaluate this by successive stages as

$$y := a_2 \tag{4.7}$$

$$y := a_1 + y * x \tag{4.8}$$

$$y := a_0 + y * x \tag{4.9}$$

In each of these stages the symbol $:=$ means "assign the right side to the left side." After the last assignment, (4.9), we have completely evaluated the polynomial (4.5) without computing powers of x explicitly. Clearly, this recursive procedure can be generalized to any order of polynomial, N. The Horner method provides an example of the effective use of repetition in computing, as we discuss in the Diversion in Section 3.4.

Program 4.1, Horner Polynomials, implements the Horner algorithm for a polynomial of order N, where N is restricted only by the size of the array, a[MAX], used to hold the values of the a_k, $k = 0, 1, ..., N$. In this program the constant term in the polynomial, a_0, is stored in a[0], and so on for the other coefficients. Thus, we have an exception to our practice, stated in Section 1.2, of starting arrays at element [1]. By breaking this rule, however, we have a direct correspondence between the formula (4.5) and the coding.

The main program consists mainly of bookkeeping to input and test the order of the polynomial N, to test its bounds (a negative value produces a graceful exit from the program), then to input the polynomial coefficients. Once these are chosen by the program user, the polynomial, y, can be evaluated for as many x values as desired, except that input of x = 0 is used to signal the end of this polynomial.

PROGRAM 4.1 A test program and function for evaluating an *N*th-order polynomial by the Horner algorithm.

```c
#include <stdio.h>
#include <math.h>
#define MAX  10

main()
{
/* Horner Polynomials */
double a[MAX];
double x,y;
int N,k;
double Horner_Poly();

printf("Horner Polynomials\n");
N = 1;
while ( N >= 0 )
  {
  printf("\nNew polynomial; Input N (N<0 to end):");
  scanf("%i",&N);
  if ( N < 0 )
    { printf("\nEnd Horner Polynomials");   exit(0); }
  if ( N > MAX-1 )
    {
    printf("\n\n!! N=%i > maximum N=%i",N,MAX-1);
    }
  else
    {
    printf("\nInput a[k], k=0,...,%i (%i values)\n",N,N+1);
    for ( k = 0; k <= N; k++ )        scanf("%lf",&a[k]);
    x = 1;
    while ( x != 0 )
      {
      printf("\n\nInput x (x=0 to end this polynomial):");
      scanf("%lf",&x);
      if ( x != 0 )
        {
        y = Horner_Poly(a,N,x);
        printf("y(%lf) = %lf",x,y);
        }
      } /* end  x  loop */
    }
  } /* end  N  loop */
}
```

```
double Horner_Poly(a,N,x)
/* Horner Polynomial Expansion */
double a[],x;
int N;
{
double poly;
int k;

poly = a[N];
for ( k = N-1; k >= 0; k-- )
  {
  poly = a[k]+poly*x;
  }
return poly;
}
```

The Horner polynomial is in function `Horner_Poly`, which is very compact. Note that the `for` loop is not executed if N = 0, because the value of k is checked before any instructions within the loop is executed. The function is therefore correct even for a zero-order polynomial.

Exercise 4.2

(*a*) Check the correctness of the Horner polynomial algorithm as coded in function `Horner_Poly` in Program 4.1 by using it to check the case $N = 2$ worked in formulas (4.7) through (4.9).

(*b*) Code program `Horner Polynomials`, then test it for a range of N values, x values, and correctness for some low-order polynomials. ■

Note that in the Horner algorithm the order of the polynomial, N, must be known before the polynomial is evaluated. The algorithm is therefore of limited use for computing power-series expansions (as in Chapter 3), unless one has decided beforehand at what order to terminate the expansion.

Program 3.3 in Section 3.2 gives an example of this latter usage for the functions `CosPoly` and `SinPoly`. On the other hand, if a convergence criterion (such as a given fractional change in the series partial sum) is used, then the method of generating successive terms by recurrence (Sections 3.4, 3.5) is appropriate. Program 3.4 in Project 3 (Section 3.5) shows this usage in the functions for the six power series.

The `Horner_poly` function should be of general use to you in future program development. We apply it immediately to program the numerical evaluation of the working function presented above.

Programming the working function

The working function, described in Section 4.1, is used in Sections 4.4 and 4.5 to test numerical derivatives and may be used in Section 4.6 for integrals. It would be possible, but tedious (especially if you use a different polynomial), to evaluate the numerical values by hand or with help from a pocket calculator. Also, if one wants the derivatives of various orders or the integral this is also tedious and error-prone in hand calculations. Here we provide a general-purpose program that, given the order and coefficients of a polynomial, determines the polynomials that represent its non-vanishing derivatives. Then the program allows the user to choose specific x values for evaluating these quantities numerically.

Here we give a preliminary discussion of the program Working Function: Value and Derivatives (henceforth abbreviated to Working Function). We defer discussion of the functions for numerical derivatives (function names starting with Der) to Sections 4.4 and 4.5. Working Function is structured as follows. The main program outermost loop is over the order of the polynomial, Npoly. Within this loop are input and testing sections for Npoly, a section for the input of the polynomial coefficients, a section where derivative coefficients are calculated, then a loop for choosing x values at which to evaluate the function and its first and second derivatives.

The polynomials are evaluated using the Horner algorithm and a modified version of the C function from the preceding subsection. Function Horner_Poly_2 has a two-dimensional array (rather than one-dimensional, as in Horner_Poly), and an additional argument, i, to control the first index of the array. For the function and its derivatives the array elements are the w_k in (4.2) as the zeroth elements of the array, and the $w_{i,k}$ in (4.3) as the elements for the ith derivative. At the expense of a slight awkwardness, this array is also two-dimensional, which allows the Horner-polynomial program function to be used also to evaluate coefficients for the integral.

Note that the labeling of arrays from zero, rather than unity, in the C language is a great advantage in Working Function. Therefore, to maintain a closeness between the analysis and the numerics, we break with our general rule (Section 1.2) of starting arrays at unity (as is common practice among Fortran programmers).

Exercise 4.3

Code and test Working Function. For testing it may be useful to add an output section for the coefficient elements w[i][k] and Integ[0][k]. Test coefficients for all these are given in Table 4.1 for the working function (4.2), which is equivalent to (4.1). The latter provides x values at which the function, but not its derivatives, should evaluate to zero. ∎

We now have a working function and a program for its numerical properties to use in examples and tests of the numerical methods in this chapter. After listing the program, we consider some limitations of numerical computation.

PROGRAM 4.2 The working function (4.1) or (4.2); its values and derivatives.

```
#include <stdio.h>
#include <math.h>
#define MAXDER 7


main()
{
/* Working Function: Value and Derivatives */

double w[MAXDER][MAXDER];
double x,ywDeriv_i,h;
double ym2,ym1,y0,yp1,yp2,yF,yC,yCCD,y3CD,y5CD;
int Npoly,k,i;
char polyOK,choose_x;
double Horner_Poly_2(),Der_1F(),Der_1C(),
       Der_2CCD(),Der_23CD(),Der_25CD();

printf("Working Function: Value and Derivatives\n");
Npoly = 1;
while ( Npoly >= 0 )
  {
  printf("\nNew polynomial;Input order,Npoly (<0 to end): ");
  scanf("%i",&Npoly);
  if ( Npoly < 0 )
    {
    printf("\nEnd  Working Function Value and Derivatives");
    exit(0);
    }
  polyOK = 'y';
  if ( Npoly >= MAXDER )
    {
    printf("\n!! Order=%i>max_order=%i\n",Npoly,MAXDER-1);
    polyOK = 'n';
    }
  else
    {
    printf("\nInput %i coefficients:\n",Npoly+1);
    for ( k = 0; k <= Npoly; k++ )
      {
      scanf("%lf",&w[0][k]);
      }
```

```
    /* Generate all derivative coefficients */
    if ( Npoly > 0 )
      {
      for ( i = 1; i <= Npoly; i++ ) /* derivative order */
        {
        for ( k = 0; k <= Npoly-i; k++ )/* polynomial terms */
          {
          w[i][k] = (k+1)*w[i-1][k+1];
          }
        }
      }
    }
  if ( polyOK == 'y' )
    {
    /* Evaluate working function at input  x  values */
    choose_x = 'y';
    while ( choose_x == 'y' )
      {
      printf("\n\nWant an  x  ( y or n)? ");
      scanf("%s",&choose_x);
      if ( choose_x == 'y' )
        {
        printf("Input x and stepsize h: ");
        scanf("%lf%lf",&x,&h);
        for ( i = 0; i <= Npoly; i++ )
          {
          ywDeriv_i = Horner_Poly_2(w,Npoly-i,i,x);
          if ( i == 0 )
          printf("\nPolynomial value is %lf",ywDeriv_i);
          else
          printf("\ni=%i, derivative is %lf",i,ywDeriv_i);
          }
        /* Numerical derivatives;  function values first */
        ym2 = Horner_Poly_2(w,Npoly,0,x-2*h);
        ym1 = Horner_Poly_2(w,Npoly,0,x-h);
        y0 = Horner_Poly_2(w,Npoly,0,x);
        yp1 = Horner_Poly_2(w,Npoly,0,x+h);
        yp2 = Horner_Poly_2(w,Npoly,0,x+2*h);
        /* First derivatives:  */
        yF = Der_1F(y0,yp1,h);  /* Forward */
        yC = Der_1C(ym1,yp1,h);    /* Central */
        /* Second derivatives:  */
        yCCD = Der_2CCD(ym2,y0,yp2,h);/* Central */
        y3CD = Der_23CD(ym1,y0,yp1,h);/* 3-point */
        y5CD = Der_25CD(ym2,ym1,y0,yp1,yp2,h);/* 5-point */
```

```
            printf("\nNumerical first derivatives:");
            printf("yF = %lf, yC = %lf",yF,yC);
            printf("\nNumerical second derivatives:");
            printf("\nyCCD=%lf y3CD=%lf y5CD=%lf",yCCD,y3CD,y5CD);
            }
        } /* end while loop for x */
     }
   } /* end Npoly loop */
}

double Horner_Poly_2(a,N,i,x)
/* Horner Polynomial Expansion */
/* Coefficients have two subscripts;first subscript is i */
double a[MAXDER][MAXDER],x;
int N,i;
{
double poly;
int k;
poly = a[i][N];
for ( k = N-1; k >= 0; k-- )
  {
  poly = a[i][k]+poly*x;
  }
return poly;
}


double Der_1F(y0,yp1,h)
/* Forward-difference first derivative */
double y0,yp1,h;
{
double y1F0;
y1F0 = (yp1-y0)/h;
return y1F0;
}


double Der_1C(ym1,yp1,h)
/* Central-difference first derivative */
double ym1,yp1,h;
{
double y1C0;
y1C0 = (yp1-ym1)/(2*h);
return y1C0;
}


double Der_2CCD(ym2,y0,yp2,h)
```

```
/* Central-difference first derivative */
double ym2,y0,yp2,h;
{
double y2CCD;
y2CCD = ( (yp2-y0)-(y0-ym2) )/(4*h*h);
return y2CCD;
}


double Der_23CD(ym1,y0,yp1,h)
/* Three-point central-difference derivative */
double ym1,y0,yp1,h;
{
double y23CD;
y23CD = ( (yp1-y0)+(ym1-y0) )/(h*h);
return y23CD;
}


double Der_25CD(ym2,ym1,y0,yp1,yp2,h)
/* Five-point central-difference derivative */
double ym2,ym1,y0,yp1,yp2,h;
{
double diff1,diff2,y25CD;
diff1 = (yp1-y0) + (ym1-y0);
diff2 = (yp2-y0) + (ym2-y0);
y25CD = ( 4*diff1 - diff2/4 )/(3*h*h);
return y25CD;
}
```


4.2   DISCRETE DATA AND NUMERICAL MATHEMATICS

We wish to emphasize briefly the importance of the discreteness of data and of numerical mathematics. Both topics place fundamental limits on the completeness of our interpretation of quantitative information.

The discreteness of data

In experimental work and also in numerical calculations, data are always obtained at discrete values of the independent variables, even though we may believe that the quantity being measured is a smooth function of the parameters (independent variables) of the problem. Similarly, in numerical applications of mathematical analysis the values are always obtained at finite parameter spacing. Such values are often, quite appropriately, called "numerical data." Subsequent operations on data, such as approximations to derivatives and integrals of the presumed smooth function underlying them, will always be limited in accuracy by the discreteness of the data.

Numerical mathematics

In this chapter, and in much of this book, we have a different emphasis than in pure mathematics. The latter emphasizes the existence of mathematical entities, the logical consistency of results, and the analytic solution and properties of quantities such as derivatives and integrals. Numerical analysis, such as we are developing, must always be guided by pure mathematics, especially on questions of the existence and convergence of quantities that we attempt to evaluate numerically. This viewpoint explains why in Figure 1.1 we showed mathematical analysis as the foundation upon which are built numerical methods and computer programming for various scientific applications.

When evaluating a formula numerically, discrete values of the variables are always required, rather than the continuous range of values often allowed by the formula definition. For practical reasons, such as minimizing computer storage and execution time, one often characterizes a function by calculating accurately as few values as possible with large stepsizes for the independent variables. Related properties of the function, such as values at intermediate values of the variables (interpolation), slopes (differentiation), and area under the function between two limits (integration), are then estimated by numerical procedures.

One often refers to mathematics that emphasizes numerical properties as *numerical analysis*, and the process of replacing continuous functions by discrete values is called *discretization*. They are to be distinguished from two other distinct branches of mathematics, number theory and discrete mathematics.

4.3 NUMERICAL NOISE IN COMPUTING

In a digital computer, most numbers can be represented to only a finite number of significant figures, unless they can be expressed exactly in binary (base-2) arithmetic requiring fewer than the number of bits used for each number. For example, in a 32-bit computer word the largest integer is $2^{32} - 1$ (allowing for a sign bit), and there can be at most a precision of 1 part in 2^{32}, that is, less than 1 part in 10^{10}. The decimal fraction 0.5 is exactly represented, since $0.5 = 2^{-1}$, so that in base-2 arithmetic $0.5_{10} = 0.1_2$. On the other hand, however, the precise decimal fraction $0.4_{10} = 0.0110011001..._2$, which is an imprecise binary fraction.

The term "computer arithmetic," in which the number of digits is finite, is often used to make a distinction from formal or exact arithmetic. Note that errors in numerical calculations introduced by computer arithmetic must be clearly distinguished from random errors introduced in computations by computer hardware (or software) malfunctions. The type of error that we discuss should be completely reproducible over time, provided that you execute exactly the same program as previously. (In the worst possible case of program bugs, this may be difficult to do unless the computer is performing *exactly* the same from run to run.) Note that the random error rate in the arithmetic hardware of a computer is typically less than 1 in 10^{12}, which is less than one error per year of continuous running.

Two main methods of handling the least-significant digits after an arithmetic operation are roundoff and truncation. We now discuss briefly these types of numerical noise.

Roundoff and truncation errors

When a computer uses *roundoff* of numbers the least-significant digit is rounded to the nearest least-but-one significant digit. For example, 0.12345676 rounded to 7 significant figures is 0.1234568, while 0.12345674 rounded to 7 significant figures is 0.1234567; the maximum error from such roundoff is therefore about 10^{-7}. As another example, the numbers 1.99999 and 2.00001 are both 2.0000 when rounded to 5 significant figures.

In *truncation*, sometimes referred to as "chopping," the least-significant digit is dropped. For example, both 0.12345676 and 0.12345674 truncate to 0.1234567. When converting floating-point variable values to integer-variable values, truncation is always used by scientific programming languages. Thus, both 3.9999999 and 3.0000001 truncate to integer 3. For this reason, counting should *never* be programmed in terms of floating-point variables.

A simple program to determine both the number of significant figures carried in a computer's floating-point calculations, and also whether roundoff or truncation is used, is given as Program 4.3. The program Significant Digits in Floating Point will estimate the number of significant digits to a given base, and will signal 'y' or 'n' depending on whether the final value was truncated or not. A nonzero test number close to unity, value, is input, a given floating-point base (input as base) is chosen, then a while loop is executed until value is unchanged by adding in successively smaller powers. The number of digits is (approximately, for base ≠ 2) the number of significant digits carried.

Exercise 4.4

(*a*) Explain how the algorithm coded in Significant Digits in Floating Point works.

(*b*) Code and run the program on your computer. Does the number of significant binary digits and decimal digits agree with what is stated in your computer manual? Is the floating-point arithmetic done by truncating or by rounding? Note that you should run several nearly equal values through the test. If any of them returns truncate = 'n', then your computer does rounding arithmetic. Explain this remark. ∎

For the double-precision variables declared in Program 4.3, my workstation gave 63 binary bits and did truncating arithmetic. The 63 bits are presumably stored in two 32-bit words, with one bit allowed for the sign. When using base-10 the program declared that 19 decimal digits were carried: indeed 2^{63} is just less than 10^{19}, and 2^{64} is just greater than 10^{19}.

PROGRAM 4.3 A program for estimating, for the computer on which it is run, the number of significant digits to a given base.

```
#include <stdio.h>
#include <math.h>

main()
{
/* Significant Digits in Floating Point */
double value,power,newvalue;
int base,digits;
char truncate;

printf("Significant Digits in Floating Point\n");
value = 1;
while ( value != 0 )
   {
   printf("\n\nInput test value near 1 (0 to end):");
   scanf("%lf",&value);
   if ( value == 0 )
      {
      printf("\nEnd Significant Digits in Floating-Point");
      exit(0);
      }
   printf("\nInput floating-point base:");
   scanf("%i",&base);
   power = 1;  digits = 0;  newvalue = 0;  truncate = 'y';
   while ( value != newvalue )
      {
      power = power/base;
      digits = digits+1;
      newvalue = value+power;
      }
   if ( newvalue > value+power*base )  truncate = 'n';
   printf("Number of digits is %i",digits-1);
   printf("\nValue of truncate is '%c'",truncate);
   }
}
```

The same algorithm can be used with a pocket calculator to determine its number of significant digits and whether it uses roundoff or truncation. The most common scheme in both computers and calculators is truncation, because this allows simpler hardware to be used.

Now that you know the number of significant digits in your computer or calculator, it is interesting to see some consequences of the finite number of digits. This exercise is probably easiest done on a pocket calculator rather than on a computer.

Exercise 4.5

(*a*) Demonstrate by examples that finite-precision arithmetic does not satisfy the associative law of addition: $(a + b) + c = a + (b + c)$. Exhibit cases in which the signs of a, b, and c differ.

(*b*) Prove that for addition the ordering that produces the smallest error from loss of significant figures is to arrange the numbers by increasing order of magnitude, that is, smallest numbers first.

(*c*) Demonstrate that the identity $b(a/b) = a$ for $b \neq 0$ is not always satisfied in numerical arithmetic. For what ranges of a, b, and c are the relative errors largest in your arithmetic unit? ∎

Part (*b*) of this exercise suggests that convergence of power series, as considered in Section 3.5, is not performed in the most accurate way. The Horner polynomial expansion, considered later in this section, which considers the highest powers of a variable first, may often be preferred for numerical accuracy whenever feasible.

Unstable problems and unstable methods

The effects of the finite accuracy of computer arithmetic that we have just considered arise partly from the use of *unstable methods*, which are sensitive to the accuracy of the calculation steps. Examples are subtractive cancellation in calculating variances or the solutions of quadratic equations, as we investigate in the next subsection. Such unstable methods should be clearly distinguished from *unstable problems*, in which the exact solution is very sensitive to slight variations in the data, independently of the accuracy of the arithmetic.

An example of an unstable problem is the solution of the pair of linear equations

$$x + y = 2 \qquad x + 1.02 \, y = 2.02 \tag{4.10}$$

which are shown graphically in Figure 4.3 as the heavy solid and solid lines, respectively. These have an exact and unique solution $x = 1$, $y = 1$, which is the intersection point (solid circle) of the two lines in the figure.

The solution of a pair of equations that are very similar to (4.10), namely,

$$x + y = 2 \qquad x + 1.01 \, y = 2.03 \tag{4.11}$$

shown dotted in Figure 4.3, is at $x = -1$, $y = 3$, the intersection point (dotted circle) in the figure. If one attempts to solve such an unstable problem numerically, then the results will be very sensitive to numerical noise. Geometrically, we see from Figure 4.3 that all three lines have almost the same slope, so that their intersection points are very sensitive to these slopes.

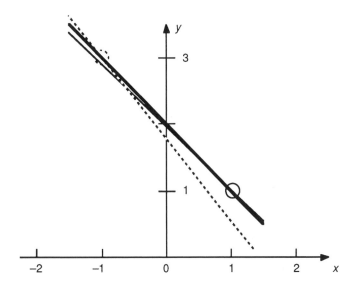

FIGURE 4.3 Lines that represent (4.10) (heavy solid and solid lines) and (4.11) (heavy solid and dotted lines). Determination of their intersections is an unstable problem, since the intersection is at (1, 1) for (4.10) but at (1, 3) for (4.11).

From the viewpoint of linear algebra and the numerical solution of matrix inverses, the determinant of the two-by-two matrices made from the left-hand sides in (4.10) and (4.11) is very much less than the quantities on the right-hand sides of these equations, so that the matrices are ill-conditioned.

Exercise 4.6
(*a*) Generalize the above example from the equation pair (4.10) or (4.11) to the pair

$$x + y = a \qquad x + by = c \qquad (4.12)$$

Eliminate x between this pair of equations to show that

$$y = \frac{c - a}{b - 1} \qquad (4.13)$$

in which the denominator is the determinant from the left-hand side of the pair of equations. Explain why values of b close to 1 produce y values that are very sensitive to the value of b.

(*b*) Check out (4.13) by first choosing $a = 2$, $b = 1.000001$, $c = 2.001$, to show that $y = 1000$ and therefore $x = -998$. Second, choose $a = 2$, but $b = 0.999999$, $c = 2.000001$, to show that now $y = -1$, $x = 3$, which is a thousand-fold change in the y solution for a change of only two parts per million in the b coefficient.

(*c*) Show that the problem cannot be alleviated by first solving for x, then using the first equation in (4.12) to calculate y. First give an algebraic demonstration, then rotate the book through a right angle and look at Figure 4.3 to devise a geometrical explanation. ∎

In many numerical problems differences in coefficients such as those in Exercise 4.6 (*b*) may arise from roundoff or truncation errors in steps of the computation that generate the coefficients. Because there are so many numbers flying around in the computer, we seldom inspect the intermediate results to check on this numerical malfunction, unlike the situation when most of the calculations are done using a pocket calculator with manual input and output for many of the steps.

Therefore, numerical methods developed in the olden days of hand-cranked calculators, when numerical errors could easily be monitored and sometimes controlled, are often quite unsuitable for computer applications. Nowadays, the user of a numerical recipe might not even know how the algorithm that is being used was coded, and whether the method used has been tested for numerical stability. This distinction between the suitability of numerical methods for hand calculation versus computer calculation is discussed further in Section 3.6 in the diversion on computers and spline methods.

The difficulties with unstable problems and with unstable methods are related to instability of equilibrium and to chaos in mechanical systems, as well as to noise amplification and feedback in electrical systems. These topics are lucidly discussed in Chapter 9 of Pippard's book on the physics of vibration, and are developed in more detail in his book on response and stability. Unstable problems and unstable methods also occur when solving differential equations numerically, often because of extreme sensitivity to boundary conditions. This is discussed in Sections 8.4 and 8.6 for the second-order Euler method and for stiff differential equations, respectively. Such mechanical, electrical, and numerical instabilities are intimately related.

Errors from subtractive cancellation

Among other problems facing those who would compute numerically are errors arising from subtractive cancellation, which is the reduction in the number of significant figures that may occur when two numbers (assuming they have the same signs) are subtracted. For example, if $x_1 = 1000000$ and $x_2 = 999999$, then their difference $x_1 - x_2 = 1$, has only one significant figure, whereas x_1 and x_2 had about six significant figures. We now examine two examples in which such perils of subtractive-cancellation errors may arise, variance calculations and the solution of quadratic equations.

Variance calculations (and thence standard deviations) among large numbers provide good examples of difficulties from subtractive cancellation. Suppose that the average of a set of x values is calculated as

$$\bar{x} = \frac{\sum_{j=1}^{N} x_j}{N} \tag{4.14}$$

The variance among this set, V, is defined by

$$V \equiv \sum_{j=1}^{N} (x_j - \bar{x})^2 \tag{4.15}$$

Analytically, this can be expanded to produce

$$V = \sum_{j=1}^{N} x_j^2 - N\bar{x}^2 \tag{4.16}$$

Exercise 4.7
Verify the algebraic equivalence of these two expressions for the variance, V. ∎

Notice that in (4.15) all N of the x_j values must be input and their average must be calculated before their variance can be calculated, whereas in (4.16) the sum of squares and the sum of values (needed to compute the average) may be accumulated as soon as each x value is entered. If (4.15) is used for large N, more storage than such devices as pocket calculators have available for data will often be required. So, for calculators the second method is almost always used, because it requires storage only for the running sums. Numerically, the outcome of using (4.16) can be complete loss of significance, as you will discover if you work the following exercise.

Exercise 4.8
(a) Use a pocket calculator to calculate the average and variance for the three data values $x_1 = 999999$, $x_2 = 1000000$, $x_3 = 1000001$. Show that using (4.16) you should get $V = 2$. What result does your calculator get, and why?
(b) Reduce the order of magnitude of the three data by successive powers of 10 until the two formulas agree. From this numerical experiment estimate the number of significant figures carried by your calculator. Does this agree with its instruction manual? ∎

By using my pocket calculator and the data in Exercise 4.8 (a), I obtained a variance of zero according to (4.16). With values of 99999, 100000, and 100001, however, I obtained the correct variance by using either (4.15) or (4.16). My calculator manual advertises 10 digits for entry or display and 12 digits for each step of a calculation: I now believe it, but I don't like the subtractive cancellation effects. An extensive discussion of roundoff errors in computers is given in Chapter 1 of the textbook by Maron.

Quadratic equations provide another example where subtractive cancellation can be a severe problem with a numerical calculation as simple as finding their roots accurately. Write the quadratic equation with two real roots as

$$(x - 1)(x - \varepsilon) = 0 \qquad (4.17)$$

The choice of unity for the first root, $x = x_1 = 1$, can always be achieved by appropriate scaling of x, and is convenient when comparing the second root, $x = x_2 = \varepsilon$, with the first, especially when ε is small in magnitude. By expanding (4.17) into standard quadratic-equation form, we have

$$x^2 - (1 + \varepsilon)x + \varepsilon = 0 \qquad (4.18)$$

The usual way that this is solved is to write the solution as

$$x = \frac{(1 + \varepsilon) \pm \sqrt{(1 + \varepsilon)^2 - 4\varepsilon}}{2} \qquad (4.19)$$

Since we know the roots, we could simplify this analytically and obtain the two solutions given above. Proceeding numerically, however, is quite another story. I used my pocket calculator to input ε, then calculate the two numerical solutions in (4.19), x_{n1} and x_{n2}, as written. This calculator enters and displays 10 decimal digits but it uses 12 digits at each step of a calculation. For values of ε between 10^{-11} and 10^{-5} (and presumably for larger ε values) the first root, x_{n1}, was exact, but the relative error in the second root, x_{n2}, behaved quite erratically, as Table 4.2 shows.

TABLE 4.2 Relative errors in root of (4.18) calculated using (4.19) and a 10-digit calculator.

	ε						
	10^{-11}	10^{-10}	10^{-9}	10^{-8}	10^{-7}	10^{-6}	10^{-5}
$(x_{n2} - \varepsilon)/\varepsilon$	1.0	1.0	0.5	1.0	0.0	0.5×10^{-6}	0.0

This example illustrates that when ε is less than the square root of the accuracy of the calculator ($\varepsilon < 10^{-5}$ for a 10-digit calculator), then the smaller root cannot be reliably computed, because the squared factor under the root in (4.19) is not calculated accurately enough. Significant digits are then lost when the root is subtracted from the first term in the denominator of (4.19).

When solving quadratic equations, how can we reduce this numerical noise from subtractive cancellation? The easiest way is to avoid subtracting. To show how this can be accomplished, and to cast the quadratic-equation problem in a more conventional form, let us write

$$ax^2 + bx + c = 0 \qquad (4.20)$$

The general analytical solution of this quadratic when $a \neq 0$ is

$$x = \frac{-b \pm \sqrt{b^2 - 4ac}}{2a} \qquad (4.21)$$

Exercise 4.9

(a) Multiply numerator and denominator of (4.21) by factor $b \pm \sqrt{b^2 - 4ac}$ in order to show that the roots of a quadratic are equivalently given by

$$x = \frac{-2c}{b \pm \sqrt{b^2 - 4ac}} \qquad (4.22)$$

in which the denominator is assumed to be nonzero.

(b) In order to make the derivation of (4.22) less *ad hoc*, consider the following transformation of the quadratic equation (4.20). Assume that the roots are non-zero $(c \neq 0)$, so that (4.20) can be divided throughout by x. You now have a quadratic equation in the variable $1/x$ with the roles of a and c interchanged. Write down the solution for this quadratic by using (4.21), then take its reciprocal. *Voilà!* the solution (4.22) is obtained. ∎

Between the use of the two equations (4.21) and (4.22) for the roots, we can avoid subtractive cancellation completely. For, if $b > 0$ (4.22) may be used with the $+$ sign and (4.21) may be used with the $-$ sign. If $b < 0$ the opposite choice of signs is appropriate.

Program for roots of quadratic equations

The program that we now provide uses the algorithm devised in Exercise 4.9. Our Program 4.4, `Quadratic Equation Roots`, computes the roots of quadratic equations (4.20) for which the coefficients a, b, and c are each real numbers.

Program `Quadratic Equation Roots` has a small driver program to control program execution, and function `quadroots` to solve for the roots. The main program reads a, b, and c; then if any one of them is nonzero it attempts a solution. Most of the function `quadroots` is occupied with taking care of the case $a = 0$, including the subcase that gives rise to inconsistent solutions, namely $b = 0$. In this subcase if the coefficients of the powers of x in (4.20) are both zero, then the roots are indeterminate if $c = 0$ and the equation is inconsistent if $c \neq 0$.

If a is nonzero, then subtractive cancellation can be a problem only if the square root is real, since there is no subtraction when the real and imaginary parts of a complex number have opposite signs, as discussed in Section 2.1. Therefore, a test for a positive discriminant is made before deciding whether to use formulas (4.21) and (4.22) for the case of real roots. For complex roots the usual form of the solution, (4.21), is used for the two roots, which are related through complex conjugation.

PROGRAM 4.4 Quadratic equation roots avoiding subtractive cancellation.

```
#include <stdio.h>
#include <math.h>

main()
{
/* Quadratic Equation Roots */
double a,b,c,ReXp,ImXp,ReXm,ImXm;
int error;
void quadroots();

printf("Quadratic Equation Roots\n");
a = 1;
while ( a != 0  |  b != 0  |  c!= 0 )
  {
  printf("\nInput a,b,c (all zero to end):\n");
  scanf("%lf%lf%lf",&a,&b,&c);
  if ( a == 0  &&  b == 0  &&  c == 0 )
    {
    printf("\nEnd  Quadratic Equation Roots");  exit(0);
    }
  quadroots(a,b,c,&ReXp,&ImXp,&ReXm,&ImXm,&error);
  printf("\nRoots are:");
  printf("(%lf)+i(%lf) & (%lf)+i(%lf)",ReXp,ImXp,ReXm,ImXm);
  printf("\nError flag = %i",error);
  }
}

void  quadroots(a,b,c,ReXp,ImXp,ReXm,ImXm,error)
/* Quadratic equation roots; Real and Imaginary */
double a,b,c,*ReXp,*ImXp,*ReXm,*ImXm;
int *error;
{
double Disc;

{
*ReXp = *ImXp = *ReXm = *ImXm = 0;
*error = 0; /* no problems so far */
if ( a == 0 )
  {
  if ( b == 0 )
    {
    if ( c != 0 ) *error = 2; /* inconsistent equation */
      else    *error = 1; /* roots indeterminate */
```

```
    }
  else  /* a is zero, b is not zero */
    {
    *ReXp = -c/b; *ReXm = *ReXp; /* degenerate */
    }
  }
else  /* a is not zero */
  {
  Disc = b*b-4*a*c; /* discriminant */
  if ( Disc >= 0 )  /* real roots */
    {
    if ( b >= 0 )
      {
      *ReXp = -2*c/(b+sqrt(Disc));
      *ReXm = (-b-sqrt(Disc))/(2*a);
      }
    else  /* b is negative */
      {
      *ReXp = (-b+sqrt(Disc))/(2*a);
      *ReXm = -2*c/(b-sqrt(Disc));
      }
    }
  else  /* complex roots */
    {
    *ReXp = -b/(2*a);           *ReXm = *ReXp;
    *ImXp = sqrt(-Disc)/(2*a);  *ImXm = -*ImXp;
    }
  }
}
}
```

After function `quadroots` has been executed, the two complex numbers ReXp + i ImXp and ReXm + i ImXm are returned, together with the integer variable `error`, which is 0 for no problem, 1 for indeterminate roots, and 2 for an inconsistent equation. This value is output with the roots before the program requests another set of coefficients. If all three coefficients are input as zero, then execution of Quadratic Equation Roots terminates.

Exercise 4.10

(*a*) Write down the solution of (4.20) when $a = 0$, and also when both a and b are zero. Check that these solutions agree with what is coded in the function quadroots in the program Quadratic Equation Roots.

(*b*) Test the program with some simple (say integer) values of a, b, and c, including the special cases of $a = 0$, as well as when both a and b are zero.

(*c*) Verify that subtractive cancellation is not a problem with the present algorithm. To do this, choose in `Quadratic Equation Roots` values of $a = 1$, $b = -(1 + \varepsilon)$, $c = \varepsilon$, where $\varepsilon \ll 1$, as in the examples in Table 4.2. ∎

Now that we have explored subtractive cancellation with the practical examples of variance calculations and of solving quadratic equations, we should be prepared to recognize the possibilities for subtractive-cancellation effects as a source of noise in many numerical computations. These effects can be especially troublesome in the numerical approximation of derivatives (Sections 4.4, 4.5) and in the related topic of numerical solution of differential equations (Sections 7.4, 7.5, and 8.3 – 8.6).

4.4   HOW TO APPROXIMATE DERIVATIVES

For computing derivatives numerically it is important to find stable methods that are insensitive to errors from subtractive cancellation. The possibilities of such errors are implicit in the definition of derivatives. For the first derivative of y at the point x, $y^{(1)}$, one has the definition

$$y^{(1)}(x) \equiv \lim_{h \to 0} \left[\frac{y(x+h) - y(x)}{h} \right] \qquad (4.23)$$

From this definition we see that numerical evaluation of derivatives is inherently an unstable problem, as defined in Section 4.3. Our challenge is to find a method that is insensitive to such instability.

We introduce an abbreviated notation in which the ith derivative evaluated n steps (each of equal size h) away from the point x is written as

$$y_n^{(i)} \equiv y^{(i)}(x + nh) \qquad (4.24)$$

If $i = 0$, then we are dealing with y itself, so the superscript is usually omitted. In the notation in (4.24) it is understood that both x and h are fixed during a given calculation. The stepsize h is assumed to be positive, whereas the integer n may be of either sign. The function y is assumed to be well-behaved, in the sense that for whatever order of derivative we wish to estimate numerically there exists a corresponding analytical derivative at that x value.

The general formulas we derive are polynomial approximations, and they are often derived from the viewpoint of fitting a polynomial through a given number of points ($N + 1$, say), then finding an analytical expression for the polynomial derivatives. There is no direct way of estimating the error incurred in the various derivatives in such a viewpoint. On the other hand, if we consider that we are making a Taylor expansion of a function in the neighborhood of some point and that we are truncating the expansion after $N + 1$ terms, then we have an N th-order polynomial approximation. Additionally, an estimate of the remainder term (3.7) serves to estimate the error incurred. Here we use the first neglected Taylor term as an estimate of the remainder, and therefore as an approximation to the error.

For our working function, $y_w(x)$, discussed in Section 4.1, the analytical derivatives are obtained from the expansion coefficients in Table 4.1. If these derivatives are evaluated at $x = 0$, they are just the coefficients in the $k = 0$ column. It is also useful to record the derivatives divided by the factorial of the same order, since we need these for our improved polynomial approximation below. These modified derivatives are just the expansion coefficients w_k with $k = i$ from the first row of Table 4.1, but we repeat them for convenience. Table 4.3 shows the coefficients that are needed.

TABLE 4.3 Expansion coefficients for derivatives of the working function, evaluated at $x = 0$.

	Order of the derivative, i						
	0	1	2	3	4	5	6
$y_0^{(i)}$	0	−1	6	138	−1224	−11280	86400
$y_0^{(i)}/i! = w_i$	0	−1	3	23	−51	−94	120

Now that we have notation established, and a working function with its check values well understood, we investigate two simple schemes for approximating derivatives numerically. We also discuss some programming aspects of the derivative functions (those starting with Der) in Program 4.1.

Forward-difference derivatives

The most obvious estimate of the first derivative of y is just to consider the quantity in the brackets on the right-hand side of (4.23), with h small but finite. Graphically, we use the slope of the line segment FF in Figure 4.4. The function shown is actually our working function (4.1) and we have chosen $x = 0$ and $h = 0.1$. Thus we have the *forward-difference derivative* estimate, given by

$$y_0^{(1)} \approx y_{F0} \equiv \frac{y_1 - y_0}{h} \tag{4.25}$$

Note that the estimate clearly depends upon the value of h, but we hope that this dependence is not too strong. We can check this out for the working function.

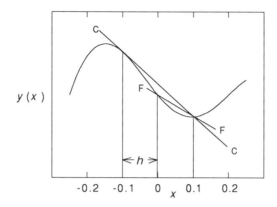

FIGURE 4.4 Schemes for estimating first derivatives are shown for the working function, (4.1) and Figures 4.1 and 4.2, near $x = 0$. The forward-difference derivative is taken along FF. The central-difference derivative is taken along CC.

Exercise 4.11

(a) Use the working function (4.2), a more convenient form than (4.1) for this purpose, to calculate its forward-difference derivative defined in (4.25) at $x = 0$ for the values of h in Table 4.4. If you have coded up and tested Program 4.2, `Working Function`, you can get the forward-difference derivative estimate as variable yF from the function `Der_1F`.

(b) Write down the first few terms of the Taylor expansion of any function $y(x + nh)$ in terms of the function $y(x)$ and its derivatives at x in order to show that the first additional term beyond the first derivative itself in the forward-difference derivative estimate (4.25) is $y^{(2)}(x) h/2$.

(c) Show that this neglected term is a fair estimate of the error in the numerical derivatives in Table 4.4, especially for the smallest h value. From Table 4.3 or the program `Working Function`, the analytical derivative at $x = 0$ has the value -1. ■

TABLE 4.4 Forward-difference first derivatives for the working function, evaluated at $x = 0$. The exact first derivative has value -1.

	h		
	0.1	0.05	0.025
y_{F0}	−0.529	−0.799	−0.911

From Exercise 4.11 and from Figure 4.4 we can see why the forward-difference derivative estimate is inaccurate for the h values that we used. The second derivative is of the opposite sign to the first derivative, so its inclusion in the forward-difference estimate makes the magnitude of the slope too small. It is therefore appropriate to make an estimate for the first derivative that is less sensitive to the presence of second derivatives.

Derivatives by central differences

In choosing a numerical estimator of the first derivative in the preceding subsection, we could as well have used a backward difference. We would have run into similar problems as for the forward difference. So why not make a more balanced approach, using *central differences,* as follows:

$$y_0^{(1)} \approx y_{C0} \equiv \frac{y_1 - y_{-1}}{2h} \tag{4.26}$$

This scheme is shown graphically in Figure 4.4 by the line segment CC. It is straightforward to extend the method of analysis made for the forward-derivative estimate.

Exercise 4.12

(*a*) Use the working function (4.2) to calculate its central-difference derivative estimate defined by (4.26) at $x = 0$ for the values of h in Table 4.5. If you have Program 4.2 (Working Function) running, you can get the central-difference derivative estimate as variable yC from the function Der_1C.

(*b*) Write down the first few terms of the Taylor expansion of a function $y(x + nh)$ in terms of the function $y(x)$ and its derivatives at x. Thus show that the first term beyond the first derivative itself in the central-difference derivative estimate (4.26) is $y^{(3)}(x) h^2/6$.

(*c*) Show that this neglected term is a very good estimate of the error in the numerical derivatives in Table 4.5, especially for the smallest h value.

(*d*) Use the program Working Function to show that if one wants accuracy of about 6 significant figures for the first derivative at $x = 0$, then the stepsize for the forward-difference derivative must be about $h = 10^{-6}$, whereas the same accuracy in the central-difference derivative requires a stepsize of only about $h = 10^{-4}$. ∎

TABLE 4.5 Central-difference first derivatives for the working function, evaluated at $x = 0$. The exact first derivative is -1.

	h		
	0.1	0.05	0.025
y_{C0}	-0.779	-0.943	-0.997

From these examples, especially from a comparison of Tables 4.4 and 4.5, we see that the central-difference derivative estimate should be used whenever feasible. It is less likely to run into problems from subtractive cancellation effects than is the forward-difference method because its stepsize can be significantly larger for the same accuracy. Occasionally, only the forward-difference (or backward-difference) method can be used, as when the function is not defined for x values less than (or greater than) the value at which the derivative is required. This may occur in starting the numerical solution of differential equations or at the endpoints of runs of data whose slopes we are trying to estimate.

In the remainder of this treatment of numerical derivatives we use only central-difference estimates because of their superior accuracy.

Numerical second derivatives

The numerical estimation of second derivatives can be based on the analytical definition

$$y^{(2)}(x) \equiv \lim_{h \to 0} \left[\frac{y^{(1)}(x + h) - y^{(1)}(x)}{h} \right] \qquad (4.27)$$

Immediately we see, from the preceding discussion of first derivatives, that the numerical estimation problems may be much more severe for second derivatives than for first derivatives, especially if we take differences of numerical first derivatives rather than doing as much of the calculation as possible by analytical methods.

One possibility is suggested by looking at Figure 4.5. Here we could estimate a first central derivative at $x + h$ (in the figure at $x = 0.1$ by the slope of C_+C_+), then a similar derivative at $x - h$ (at $x = -0.1$ by the slope of line C_-C_-).

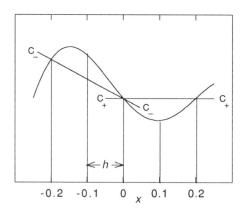

FIGURE 4.5 Schemes for numerical second derivatives, illustrated for the working function (4.1) and Figure 4.1 near $x = 0$. Consecutive central derivatives (CCD) are obtained from central-difference first derivatives along C_+C_+ and C_-C_-. The 3CD estimate (4.35) and the 5CD estimate (4.37) use the central 3 points and all 5 points, respectively.

This method, *consecutive central derivatives*, or CCD, can be analyzed algebraically, as follows. First, write the formula (4.26) for central derivatives of the function which is already the first derivative, thus

$$y_0^{(2)} \approx y_{CCD}^{(2)} = \frac{y_1^{(1)} - y_{-1}^{(1)}}{2h} \tag{4.28}$$

Next, use the central-difference estimates for the first derivative, (4.26), in this expression, to obtain

$$y_{CCD}^{(2)} \approx \frac{(y_2 - y_0) - (y_0 - y_{-2})}{4h^2} \tag{4.29}$$

For optimum accuracy, minimizing subtractive cancellation, it is better to compute the second-derivative estimate this way rather than from further algebraic simplification of the expression.

Exercise 4.13

(*a*) For the working function (4.2) calculate its central-difference derivative estimate (4.29) at $x = 0$ for the values of h in Table 4.6. In Program 4.2, Working Function, this second-derivative estimate is variable yCCD from the function Der_2CCD.

(*b*) Write down the first few terms of the Taylor expansion of a function $y(x + nh)$ with $n = -2, 0$, and 2, in terms of the function $y(x)$ and its derivatives at x. Thus show that the first additional term beyond the second derivative itself in the central-difference derivative estimate (4.29) is $y^{(4)}(x) h^2/2$.

(*c*) Show that this neglected term is only a fair estimate of the error in the numerical derivatives in Table 4.6, except for the smallest h value. (You can obtain the value of the fourth derivative at $x = 0$ from Table 4.3.) For this example and values of h, why is the estimate of the error inaccurate?

(*d*) Use the program Working Function (or a program of your own devising) to show that if one wants an accuracy of about 6 significant figures in the second derivative at $x = 0$, then the stepsize required is about $h = 10^{-4}$. Show that this agrees with the error estimate derived in (*b*). ∎

TABLE 4.6 Central-difference second derivatives for the working function, evaluated at $x = 0$. The exact second derivative is 6.

	h		
	0.1	0.05	0.025
y_{CCD}	2.304	5.004	5.746

Compare Table 4.6 for the second derivative with Table 4.5 for the central-difference first derivative, to note that about the same value of h is required for the same accuracy in first and second derivatives. This is not a usual behavior. By inspecting the derivative values in Table 4.3 you can see that, although the increasing powers of h (which is < 1) bring down the error estimates, the steadily increasing derivatives increase them. Our choice of the wiggly sixth-order polynomial, defined by (4.1) and shown in Figures 4.1 and 4.2, is responsible for this pathological behavior. By contrast, the derivatives at $x = 0$ for the exponential function are all equal to unity, and the convergence is therefore much more rapid, as we discover in Section 4.5.

Better algorithms for second derivatives

The algorithm for numerical second derivatives, (4.29), can be significantly improved without increased complexity, as we now show. The method that we use also shows how optimized formulas can be produced and how their errors can be estimated from Taylor series. We derive two formulas, one using three points and the other using five points, all having the x value of interest as the central value.

Suppose that we go n steps, each of size h, to the left or right of the point x at which we want a numerical estimate of the second derivative. Let us use Taylor's theorem (3.6) for up to five points surrounding x, as in Figure 4.5. We then have

$$y_{\pm 1} - y_0 = \pm v_1 + v_2 \pm v_3 + v_4 \pm v_5 + v_6 \tag{4.30}$$

$$y_{\pm 2} - y_0 = \pm 2v_1 + 4v_2 \pm 8v_3 + 16v_4 \pm 32\, v_5 + 64v_6 \tag{4.31}$$

where the coefficients

$$v_i = \frac{y_0^{(i)} h^i}{i!} \tag{4.32}$$

contain the desired (and undesired) derivatives. For second derivatives, we need to eliminate the term v_1 and as many higher derivatives as possible. All odd-order derivatives (all v_i with i odd) can be eliminated by summing values that are symmetric about the midpoint. Thus

$$y_1 - y_0 + y_{-1} - y_0 = 2(v_2 + v_4 + v_6) \tag{4.33}$$

$$y_2 - y_0 + y_{-2} - y_0 = 8(v_2 + 4v_4 + 16v_6) \tag{4.34}$$

From these two formulas, depending on how many points are available, we obtain various estimates for the second derivative that is hiding in v_2.

Exercise 4.14
By using (4.33) show that an estimate of the second derivative that uses three points is y_{3CD}, given by

$$y_{3CD} \approx \frac{(y_1 - y_0) + (y_{-1} - y_0)}{h^2} \tag{4.35}$$

with an error, ε_{3CD}, estimated as

$$\varepsilon_{3CD} \approx -\frac{y_0^{(4)} h^2}{12} \tag{4.36}$$

which is the first omitted term of the Taylor expansion. ∎

Notice that this formula is predicted to be a factor of 6 more accurate than our CCD formula (4.29).

If we may use all five points centered on x, as in Figure 4.5, then the estimate of the second derivative can be significantly improved.

Exercise 4.15
Use both (4.33) and (4.34) to eliminate the fourth-derivative term between them, and thence to show that an estimate of the second derivative that uses five points is y_{5CD}, given by

$$y_{5CD} \approx \frac{4\left[(y_1 - y_0) + (y_{-1} - y_0)\right] - \left[(y_2 - y_0) + (y_{-2} - y_0)\right]/4}{3h^2} \tag{4.37}$$

with an error, ε_{5CD}, estimated as

$$\varepsilon_{5CD} \approx -\frac{y_0^{(6)} h^4}{90} \tag{4.38}$$

which is the first omitted term of the Taylor expansion. ∎

Formula (4.37) predicts many improvements over the two other estimates of the second derivative, in that the error depends upon the sixth derivative, which is usually smaller than the fourth derivative (except for our pathological working function!), and scales as h^4 rather than h^2 so it will become relatively much more accurate as h is decreased.

Now that we have done much intricate algebra, let us try out these second-derivative formulas on our working function. The example is again for the derivative at $x = 0$.

Exercise 4.16

(a) For the working function (4.1) calculate its central-difference derivative estimate (4.29) at $x = 0$ for the values of h in Table 4.7. In Program 4.2, Working Function, this second-derivative estimate is variable yCCD from the function Der_2CCD.

(b) Show that if you want the second derivative of the working function (4.1) at $x = 0$ accurate to about 6 significant figures, then you have to choose $h \approx 2 \times 10^{-4}$ for the three-point second derivative, but only $h \approx 1 \times 10^{-2}$ for the five-point derivative. ∎

TABLE 4.7 Central-difference second derivatives and error estimates for the working function evaluated at $x = 0$. The exact second derivative is 6.

			h	
	0.1	0.05	0.025	10^{-3}
y_{3CD}	5.0040	5.7465	5.93634	5.999898
ε_{3CD}	1.0200	0.2550	0.06375	0.000102
y_{5CD}	5.9040	5.9940	5.99625	6.000000
ε_{5CD}	0.0960	0.0060	0.00375	10^{-9}

We now have considerable experience with estimating first and second derivatives numerically for the very badly behaved working function (Figures 4.1 and 4.2). In the next section we explore the derivatives of some functions that are better behaved.

4.5   PROJECT 4A: COMPUTING DERIVATIVES NUMERICALLY

Although in Section 4.4 we acquired considerable experience with numerical first and second derivatives of the badly behaved working function (Section 4.1), it is now interesting to explore the numerics with functions that are more regular. In this project we explore two common functions that are paradigms for many other functions and data, the exponential function and the cosine function. These characterize steadily increasing (or decreasing, depending on your point of view) and oscillatory functions, respectively.

Derivatives of the exponential function

The exponential function $y = e^x$ is straightforward to work with, especially at the origin $x = 0$, because there its derivatives have value unity.

Exercise 4.17

(*a*) Modify the program `Working Function` by replacing the sections for polynomial evaluation (including the Horner-polynomial function) by the exponential function, $y = e^x$. Keep the five functions for estimating derivatives. (Since all the derivatives of the exponential are just the function itself, a single use of the function `exp(x)` will produce all the derivatives you need.) Add program statements to compute the difference between these derivatives and those obtained from the five formulas for first and second derivatives that are already programmed. Test this new program, called `Exponential Function Numerical Derivatives`.

(*b*) Run this program with $x = 0$ and successively smaller values of stepsize h, say $h = 0.2$, 0.1, 0.01, 0.001, calculating all five numerical derivatives and their errors for each h. Make a log-log plot of the absolute value of the error versus h, as in Figure 4.6.

(*c*) To the extent that the graph for the error in each method versus stepsize h is a straight line on a log-log plot, argue that the error has a power-law dependence on stepsize h, and that the overall scale of the error estimates the factor preceding the power of h. Do the powers estimated for the five derivative methods from your graph agree with those predicted from our Taylor series analyses? ■

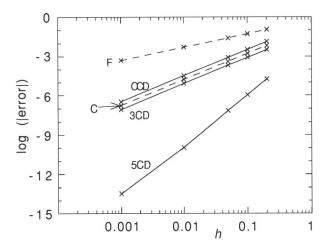

FIGURE 4.6 Errors in derivatives of the exponential function at $x = 0$ as a function of stepsize h for various derivative estimators. First-derivative estimators are forward (F) from (4.25) and central (C) from (4.26). Second-derivative estimators are consecutive central derivatives (CCD) from (4.29), three-point derivatives (3CD) from (4.35), and five-point derivatives (5CD) from (4.37).

Now that you have a working program for estimating numerical derivatives, by replacing the exponential function in Exponential Function Numerical Derivatives by the function of your choice, you can explore its application to various other functions, such as the cosine function in the next subsection. Also the program may be used directly with data (if they are equally spaced) to estimate slopes and higher-order derivatives.

Differentiating the cosine function

The cosine function, $y = \cos x$, is interesting to investigate because it characterizes oscillatory behavior, such as we obtain from the numerical solution of differential equations in Chapters 7 and 8, as well as from the Fourier expansions in Chapters 9 and 10. Indeed, an understanding of numerical derivatives provides an essential foundation for developing numerical solutions of differential equations.

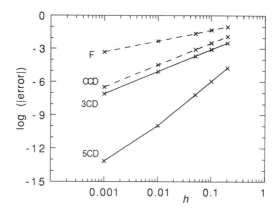

FIGURE 4.7 Errors in derivatives of the cosine function at $x = 0$ as a function of stepsize h for the first-derivative forward (F) from (4.25), and for second derivatives – consecutive central derivatives (CCD) from (4.29), three-point derivatives (3CD) from (4.35), and five-point derivatives (5CD) from (4.37). The errors in the central-difference first-derivative estimate (4.26) are zero.

The procedure for the second part of Project 4A is similar to that in the first part, and it can be done almost effortlessly.

Exercise 4.18

(*a*) Modify either the program Working Function or (more readily) the program Exponential Function by writing sections for evaluating the cosine function, $y = \cos x$, for its value, for its first derivative $-\sin x$, and for its second derivative $-\cos x$. Keep the five functions for estimating derivatives. Include program statements to compute the difference between these derivatives and those obtained from the five formulas for first and second derivatives.that are already programmed. Test this new program, called Cosine Function Numerical Derivatives, for example against hand calculations.

(b) Run the program with $x = 0$ and successively smaller values of stepsize h, say $h = 0.2, 0.1, 0.01, 0.001$, calculating all five numerical derivatives and their errors for each h. Except for the central-difference first derivative (which has zero error at $x = 0$), make a table and a log-log plot of the absolute value of the error against h, as in Table 4.8 and Figure 4.7.

(c) Assuming that the graph for each method on a log-log plot is a straight line, show that the error has a power-law dependence on stepsize h, and that the overall scale of the error estimates the factor preceding the power of h. How well do the powers estimated for the five derivative methods from your graph agree with those predicted from our Taylor-series analyses? ∎

TABLE 4.8 Central-difference first and second derivatives errors for the cosine function evaluated at $x = 0$. The exact second derivative is 6.

	h				
	0.2	0.1	0.05	10^{-2}	10^{-3}
ε_F	1.0×10^{-1}	5.0×10^{-2}	2.5×10^{-2}	5.0×10^{-3}	5.0×10^{-4}
ε_{CCD}	-1.3×10^{-2}	-3.3×10^{-3}	-8.3×10^{-4}	-3.3×10^{-5}	-3.3×10^{-7}
ε_{3CD}	-3.3×10^{-3}	-8.3×10^{-4}	-2.1×10^{-4}	-8.3×10^{-6}	-8.3×10^{-8}
ε_{5CD}	-1.8×10^{-5}	-1.1×10^{-6}	-6.9×10^{-8}	-1.1×10^{-10}	-7.5×10^{-14}

Notice in Figure 4.7 that the most accurate numerical method, the five-point central-difference second derivative, does not continue to improve as rapidly as predicted when h becomes of order 10^{-3}. For this small an h value the error in the numerical approximation approaches my computer's roundoff error of about 10^{-14}.

Now that you have experience in numerical differentiating with the badly behaved working function, with a monotonic exponential, and with the oscillatory cosine, you should be confident to tackle many functions and the differentiation of numerical data.

There are several other practical methods for numerical estimating derivatives. For example, cubic-spline fitting through a set of points will also produce, as part of the process, first through third derivatives, as described in Section 5.4. Another very effective method is to consider the derivatives as functions of the stepsize h, then to extrapolate the derivative values at finite h to the limit $h \rightarrow 0$. This method, called Richardson extrapolation, is described in Section 7.2 of Maron's book.

4.6 NUMERICAL INTEGRATION METHODS

We remarked at the beginning of this chapter that integrals of functions are much less likely to be able to be calculated analytically than are derivatives, but that integration is usually a more stable problem (in the sense discussed in Section 4.3) than is differentiation. We therefore seek efficient numerical-integration methods because we will probably have to do many such integrals (absent analytical results), but a fairly simple numerical algorithm may perform surprisingly well.

The differentiation-integration question can be approached from another viewpoint, namely that the integral of a function is formally the "antiderivative" of that function. Since we saw in Section 4.4 that successively higher-order derivatives become numerically more difficult, therefore (at least formally) successive integrals should become numerically less difficult. As we discussed for derivatives at the beginning of Section 4.4, we also base our integration formulas on Taylor expansions of the integrand. If we truncate the expansion to a given order, N, we are using an Nth-order polynomial approximation. The first neglected term in the Taylor expansion serves to estimate the error in the polynomial approximation.

The working function, introduced at the beginning of this chapter as (4.1), is used for tests and for comparisons of the methods we develop. Recall that this function was designed as a highly oscillatory but bounded function near $x = 0$, as shown in Figures 4.1 and 4.2. The expansion of its integral is given by (4.4), using the coefficients I_k in Table 4.1. The coefficients of the derivatives of various orders, also in this table, are useful for estimating errors in integral approximations.

We derive the two most common integration formulas, the trapezoid and Simpson approximations. The emphasis in our discussion is on understanding how to derive these formulas and how to estimate their errors, with a view to indicating how to generalize the procedure for such approximations. Our approach is also consistent with that used in Section 4.4 for obtaining the derivative estimates.

Integration algorithms using constant stepsize h are readily derived by using a Taylor expansion of the integrand within the region to be integrated. Suppose that the integration is centered on x_0, so that the appropriate Taylor expansion of an integrand, $y(x)$, through the nth power of $(x - x_0)$ can be obtained by using Taylor's theorem, (3.6), as

$$y(x) = y(x_0) + \sum_{k=1}^{n} \frac{y^{(k)}}{k!} (x - x_0)^k + R_n \tag{4.39}$$

in which the derivative is evaluated at x_0 and the remainder, R_n, is formally given by (3.7), but it may be approximated by the next term in the power series, which is what we shall use. The indefinite integral over x of $y(x)$ is obtained by term-by-term integration of the series (4.39), a procedure which is usually valid for integrals of interest.

Exercise 4.19

Show that integration of the series terms in (4.39) leads to

$$\int dx\, y(x) = y(x_0)x + \sum_{k=1}^{n} \frac{y^{(k)}}{(k+1)!} (x - x_0)^{k+1} + \int dx R_n \tag{4.40}$$

for the indefinite integral. ∎

The presence of n derivatives in this formula, ignoring the remainder integral, requires that y be known at $(n + 1)$ values for each basic integral. We now show the construction explicitly for $n = 1$ (trapezoid rule) and for $n = 2$ (Simpson rule).

Trapezoid formula and program for integration

Trapezoid rule formula. The lowest-order formula obtained using (4.40) is for $n = 1$, but let us carry the expressions to $n = 2$ then neglect the integral over the remainder. The $n = 2$ term will then be an estimate of the error in the $n = 1$ approximation. Suppose that we wish to integrate from $x = a$ to $x = a + h$. The most rapid convergence of the series (4.40) will then probably be achieved by choosing $x_0 = a + h/2$.

Exercise 4.20

(a) Make the indicated substitutions of variables in (4.40) in order to show that

$$\int_a^{a+h} dx \; y(x) \approx y(a+h/2)h + \frac{y^{(2)}(a+h/2)h^3}{24} \tag{4.41}$$

in which derivatives of fourth and higher order have been neglected.

(b) Write out the Taylor expansions (4.39) of $y(a)$ and $y(a + h)$ about $x_0 = a + h/2$, then solve for $y(a + h/2)$ in terms of these y values and the second derivative at $a + h/2$. Thus show that (4.41) becomes

$$\int_a^{a+h} dx \; y(x) \approx [y(a) + y(a+h)] \frac{h}{2} - \frac{y^{(2)}(a+h/2)h^3}{12} \tag{4.42}$$

which, with neglect of the derivative term, is the *basic trapezoid formula.*

(c) Show that if y is linear in x, then the trapezoid rule is exact. ∎

This derivation might seem a bit long-winded, because all we have ended up doing is to average the y values at the beginning and end of the integration interval, then multiplying by the length of the interval, which is h. However, if we did not use the more complete analysis that you have just made (haven't you?), then we would not be able to estimate the error in the approximation.

Notice that this error is better than you might have guessed, since the effects of the first derivative at the midpoint of the interval has vanished, leaving the second derivative as the leading term in the error estimate. By using the midpoint of the integration interval as the expansion point, all odd-order derivatives vanish from the integration formula. Therefore, the next neglected term in the integral, and therefore in our error estimate, involves $y^{(4)}h^5$.

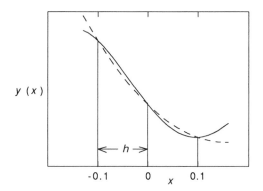

FIGURE 4.8 Integration formulas, illustrated for the working function (4.1) with stepsize $h = 0.1$ in $[-0.1, 0.1]$. The trapezoid rule uses a straight-line segment in each panel. The Simpson rule uses the parabolic section (dashed) over two panels.

Graphical interpretation of the trapezoid rule clarifies the analysis. Figure 4.8 shows our working function, (4.1), near $x = 0$, the same region where we investigate derivatives in Sections 4.4 and 4.5. The geometric interpretation of Exercise 4.20 (c) is that we approximate the function between the tabulated values at $x = a$ and $x = a + h$ by a straight-line segment. For example, in Figure 4.8 we have a trapezoid between $x = -0.1$ and $x = 0$, with another between $x = 0$ and $x = 0.1$.

Starting from the basic trapezoid formula (4.42), we may use additivity of integration to find an approximation to the integral between $x = a$ and $x = a + nh$. Thereby we produce the *composite trapezoid formula*

$$\int_a^{a+nh} dx\, y(x) \approx h \left[\frac{y(a) + y(a+nh)}{2} + \sum_{j=1}^{n-1} y(a+jh) \right] + \varepsilon_n^{(T)} \qquad (4.43)$$

in which the error estimate is

$$\varepsilon_n^{(T)} \approx -\frac{n\, y^{(2)} h^3}{12} \qquad (4.44)$$

Here the second derivative is not larger in magnitude than the maximum second derivative in the entire interval. The error scales as the quantity $n\, y^{(2)} h^3/y$, so halving the stepsize h (while doubling n to cover the same x range) should produce about a factor of 4 greater accuracy in the integral estimated by the trapezoid rule.

Trapezoid rule program. Programming the trapezoid function is quite straightforward. As usual, the code for the driver program to make the function testable is longer for the code than for the function itself. Also, we anticipate including the Simpson rule in the next subsection, so some code is included for this. Here we emphasize parts relevant to the trapezoid rule.

PROGRAM 4.5 Trapezoid and Simpson integrals, composite formulas, applied to the working function (4.1).

```c
#include <stdio.h>
#include <math.h>


main()
{
/*  Trapezoid and Simpson Integrals */
double a,h,IntAnalyt,IntByTrap,error,IntBySimp;
int n,NTrap,NSimp;
double SimpInt(),TrapInt(),y(),YWInt();

printf("Trapezoid and Simpson Integrals\n");
n = 2;
while ( n != 0 )
  {
  printf("\n\nInput n, a, h (n=0 to end): ");
  scanf("%i%lf%lf",&n,&a,&h);
  if ( n == 0 )
    {
    printf("\nEnd  Trapezoid and Simpson Integrals");
    exit(0);
    }
  NTrap = n;
  if ( n == 2*(n/2) )   NSimp = n;
  else  NSimp = 0;/* to bypass Simpson for n odd */
  IntAnalyt = YWInt(a+n*h)-YWInt(a);
  printf("Analytical integral=%lg",IntAnalyt);
  IntByTrap = TrapInt(a,NTrap,h);/* trapezoid rule */
  error = IntAnalyt-IntByTrap;
  printf("\nTrapezoid integral=%lg",IntByTrap);
  printf(" Error=%lg",error);
  if ( NSimp != 0 )
    {
    IntBySimp = SimpInt(a,NSimp,h);/* Simpson rule */
    error = IntAnalyt-IntBySimp;
    printf("\nSimpson integral=%lg",IntBySimp);
    printf(" Error=%lg",error);
    }
  } /* end  n  loop */
}
```

```
double SimpInt(a,n,h)
/* Simpson integral */
double a,h;
int n;
{
double sum;
double y();
int n2,j;
n2 = n/2;
if ( (2*n2 != n) || (n2 == 1) )
  {
  printf("\n!! SimpInt has n=%i; not allowed",n);
  return 0;
  }
else
  {
  sum = 0;
  for ( j = 1; j < n2; j++ )
    {
    sum = sum+y(a+2*j*h)+2*y(a+(2*j-1)*h);
    }
  return h*(y(a)+4*y(a+(n-1)*h)+y(a+n*h)+2*sum)/3;
  }
}


double TrapInt(a,n,h)
/* Trapezoid integral */
double a,h;
int n;
{
double sum;
double y();
int j;
sum = ( y(a) + y(a+n*h) )/2;
for ( j = 1; j < n; j++ )
  {
  sum = sum+y(a+j*h);
  }
return h*sum;
}
```

```
double y(x)
/* Working function value*/
double x;
{
double prod;
prod=120*(x+0.5)*(x+0.25)*x*(x-(1.0/3.0))*(x-0.2)*(x-1);
return prod;
}

double YWInt(x)
/* Working function integral */
double x;
{
double sum;
sum=(-0.5+x+(23.0/4.0)*x*x-10.2*pow(x,3)
    -(47.0/3.0)*pow(x,4)+(120.0/7.0)*pow(x,5))*x*x;
return sum;
}
```

The structure of Trapezoid and Simpson Integrals can be summarized as follows. The program uses a while loop controlled by n, the number of steps in the integration interval, with a zero for n being used to terminate program execution. The analytical value for the working-function integral, IntAnalyt, is obtained from (4.4) and Table 4.1, which are coded in function YWInt. The trapezoid rule is coded in function TrapInt and its value in the main program is IntByTrap.

The function TrapInt integrates the function y, which here is coded as the working-function expression (4.1). In the C function for y, the roots are given either as exact decimals or as ratios of (floating-point) integers. The roots are thereby computed to machine accuracy. Similarly, maximum accuracy is gained in the C function YWInt for the working-function integral.

Exercise 4.21

(*a*) Code and test the trapezoid parts of program Trapezoid and Simpson Integrals. You can test the program by temporarily substituting for the working function any linear function of *x*, and substituting in the analytic-integral function the corresponding integral. The safest way to code these temporary changes is to add them just before the return statements in the two C functions. By doing this you do not have to disturb the code for the working function. Integration of a linear function should produce results accurate to machine roundoff error.

(*b*) Run the program for the integral from a = -0.2 up to 0.2, which Figure 4.1 suggests has a very small value because of the zero of the function at *x* = 0. Thus, for h = 0.1 you need to input n = 4, and so on for three successively halved values of h. Check your results against mine in Table 4.9.

(*c*) Use (4.44) to estimate the error in the trapezoidal rule for each choice of n and h, using the second derivative given in Table 4.3, namely the value at $x = 0$, as an estimator of the appropriate derivative. Show that this overpredicts the actual error by a factor of about 6. Suggest why. ∎

TABLE 4.9 Trapezoid integrals for the working function (4.1) from –0.2 to 0.2 with various stepsizes *h*. The analytical value of the integral is 9.9109×10^{-3}.

		h		
	0.1	0.05	0.025	0.0125
Trapezoid integral ($\times 10^3$)	9.6120	9.8291	9.8900	9.9056
Error ($\times 10^3$)	0.2989	0.0818	0.0209	0.0053

Notice, from comparing Table 4.9 with Table 4.4, relatively how much more accurate the simplest formula for numerical integration (trapezoid rule) is compared with the simplest formula for numerical differentiation (central-difference formula). This justifies our remarks at the start of this section about integrals being considered as antiderivatives.

We return to the trapezoid rule and consider its application to functions that are less pathological than our working function at the end of the next subsection, where it is also compared with the Simpson rule.

Simpson formula and program for integrals

In the trapezoid integration formula we assumed straight-line segments between pairs of "knots" (points where the function is to be evaluated), and this collapsed the general integration formula to the basic trapezoid formula. If we instead assume parabolic segments between triples of knots, as shown in Figure 4.8, we can derive the Simpson formula and an estimate of its error.

Simpson rule formula. The method of deriving this formula is the same as for the trapezoid formula (Exercise 4.20), except that having three values of the function allows more of the (usually) unknown derivatives to be eliminated. Try it and see.

Exercise 4.22

(*a*) To derive the Simpson integration rule, start from the general integral expression (4.40), expand about $x_0 = a + h$, then solve for the second derivative in terms of the function values at $x = a$, $x = a + h$, and $x = a + 2h$. Keep the term containing the fourth derivative. Thus derive the *basic Simpson formula* for integration

$$\int_a^{a+2h} dx\, y(x) \approx \frac{h}{3}[y(a) + 4y(a+h) + y(a+2h)] + \frac{y^{(4)} h^5}{90} \qquad (4.45)$$

(*b*) Show that if y is a cubic or lower-order polynomial, then the Simpson formula is exact.

(*c*) Use additivity of integration and the basic integration formula (4.45) to derive the *composite Simpson rule* for n intervals, where n must be even,

$$\int_a^{a+nh} dx\, y(x) \approx \frac{h}{3}\left\{ \begin{array}{l} [y(a) + 4y(a+\overline{n-1}h) + y(a+nh)] \\[2ex] + 2\sum_{j=1}^{n_2-1} [y(a+2jh) + 2y(a+\overline{2j-1}h)] \end{array} \right\} + \varepsilon_n^{(S)} \qquad (4.46)$$

where $n_2 = n/2$, and the error estimate for the composite Simpson rule is

$$\varepsilon_n^{(S)} \approx -\frac{n\, y^{(4)} h^5}{180} \qquad (4.47)$$

in which the second derivative is bounded by its maximum value in the interval $x = a$ to $x = a + nh$. ∎

The parabolic shape that the Simpson formula assumes for the function in the basic interval of width $2h$ is shown as the dashed curve in Figure 4.8 for our working function (4.1). A different parabola is assumed for each triple of points, so that the rather poor fit of the dashed to the solid curve outside the range $[-0.1, 0.1]$ is not of immediate concern. The discontinuities of the parabolic curves between successive panels of width $2h$ that occur in the Simpson formula are removed if one uses the cubic-spline technique developed in Chapter 5.

Simpson rule program. The overall program structure for `Trapezoid and Simpson Integrals` was described in the preceding subsection on the trapezoid formula and program, so here we describe only parts specific to the Simpson formula. On input the value of n is checked; only if n is even is the Simpson algorithm used to produce the value `IntBySimp`, which is compared with `IntAnalyt` to obtain the value `error` for output.

The function `SimpInt` uses the composite Simpson rule (4.46) to estimate numerically the integral of the function y. In case this function is to be used in another program, there is a trap to ensure that the number of points, n, is even. For the composite rule (4.46) n > 2 is also required by the program. (For n = 2 the basic Simpson formula (4.45) may be used.) The check for even n in `SimpInt` can be tested by temporarily disabling the test made in the main program. Note that the function is coded to resemble closely the formula (4.46), rather than to be very efficient. If you need a speedy version, you should recode most of SimpInt.

Now that you know how the Simpson integration formula is supposed to work, it's time to plug and play.

Exercise 4.23

(a) Test the Simpson rule parts of the program Trapezoid and Simpson Integrals. Temporarily substitute for the working function any function of x as high as third order, and substitute in the analytic-integral function the corresponding integral. Code these temporary changes just before the return statements in the two C functions, so as not to disturb the code for the working function. Integration of a cubic or lower-order polynomial should produce results accurate to machine roundoff error, according to (4.47).

(b) Run the program for the Simpson integral from a = -0.2 up to 0.2, which has a very small value because of the zero of the working function at $x = 0$, as seen in (4.1) and Figure 4.1. For h = 0.1 input n = 4, and so on for three successively halved values of h. Check your results against those in Table 4.10.

(c) Use (4.47) to estimate the error in the Simpson rule for each choice of n and h, using the fourth derivative in Table 4.3 at $x = 0$ as an estimator of the appropriate derivative. Show that this overpredicts the actual error by a factor of about 2. Suggest why. ∎

TABLE 4.10 Simpson-rule integrals for the working function (4.1) from –0.2 to 0.2 with various stepsizes h. The analytical integral has value 9.9109×10^{-3}.

	h			
	0.1	0.05	0.025	0.0125
Simpson integral ($\times 10^3$)	9.7440	9.9015	9.9103	9.9109
Error ($\times 10^3$)	0.1669	0.0094	0.0006	0.0000

By comparing Tables 4.9 and 4.10 for the trapezoid and Simpson formulas, respectively, notice how the latter improves much more rapidly than the former as h decreases, since there is an h^5 dependence of the error in the Simpson rule, compared with an h^3 dependence in the trapezoid-rule error. This occurs in spite of the much larger fourth derivative than second derivative (Table 4.3) for the very wiggly polynomial that is our working function.

Integrals with cosines

Now that we have a working program that has been torture tested against a badly behaved function, it is interesting to try a more-usual function such as the cosine. Since the cosine has derivatives that are bounded in magnitude by unity, the error estimates for the trapezoid and Simpson rules, (4.44) and (4.47), will be dominated by their dependence on h. How about trying your integration program on the cosine function?

Exercise 4.24

(a) Modify the program Trapezoid and Simpson Integrals by replacing in the C function $y(x)$ the working function (4.1) by the cosine function, $y = \cos x$. Run the program for the integration range –2 to 2 and successively smaller stepsizes h, say $h = 1, 0.5, 0.25, 0.125$, and calculate the errors for each h. Make a table and a log-log plot of the absolute value of the error against h, as in Table 4.11 and Figure 4.9.

(b) Assume that the graph for each method (trapezoid and Simpson) on the log-log plot is a straight line. Thus show that the error has a power-law dependence on stepsize h, and that the overall scale of the error estimates the factor preceding the power of h. Show that the trapezoid formula error is consistent with an h^3 dependence, as in the estimate (4.44), and with a value of the second derivative of about –0.5, just half the maximum-magnitude value of –1 (at $x = 0$).

(c) Similarly, for the Simpson formula error show that there is an h^5 dependence, consistent with estimate (4.47), and show that an average fourth derivative of about 0.5 (again half the maximum value) would make a good match between prediction and calculation. ■

TABLE 4.11 Trapezoid and Simpson integrals for the cosine function from –2 to 2 with various stepsizes h. The analytical integral has value $2 \sin (2) = 1.81859$.

			h		
	1	0.5	0.25	0.125
Trapezoid error	1.5×10^{-1}	3.8×10^{-2}	9.5×10^{-3}	2.4×10^{-3}
Simpson error	-1.1×10^{-2}	-6.5×10^{-4}	-4.0×10^{-5}	$< 10^{-6}$

In Section 4.5 we investigated the numerical first derivative of the cosine function, and we showed (Table 4.8 and Figure 4.7) that an accuracy of a few parts in 10^5 is attained for a method comparable to the Simpson integrator (central derivatives, CCD) only for h of order 10^{-2}, whereas for integration $h = 0.25$ gives about the same accuracy (Table 4.11). This contrast emphasizes the much greater difficulty of computing accurate numerical derivatives than numerical integrals.

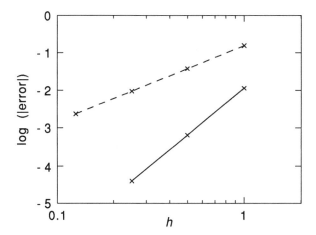

FIGURE 4.9 Errors in the integral of the cosine from –2 to 2 for various stepsizes h. The trapezoid-rule errors are indicated by the dashed line, and the Simpson-rule errors are shown by the solid line.

Higher-order polynomial integration

The numerical-integration technique of approximating a function by a polynomial of a given order, then evaluating the integral of this polynomial can be made quite general. If one does not require an estimate of the error involved, then a Lagrange interpolating polynomial, as described in Sections 6.1 and 7.1 of Maron's book, may be useful. Various recipes for polynomial integration algorithms, with error estimates of the kind that we derived, are derived in Chapter 4 of the text by Nakamura and listed in the handbook of Abramowitz and Stegun. The spline method of integration we describe in Section 5.5 allows a low-order polynomial to be used for each stepsize, but accuracy is achieved by requiring continuity of values and derivatives across adjacent steps.

The technique of evaluating the integrals as a function of h, then taking the limit as $h \to 0$ analytically, is called Romberg integration, and it can be very effective, especially if the integrands are difficult to calculate, so that the number of integrand values that one wishes to evaluate is to be minimized. Romberg integration is discussed in detail in Vandergraft's text on numerical computation.

If unequal steps of the integrand are allowed and the integrand can be evaluated to arbitrary accuracy at any points, then the technique of Gaussian quadrature is capable of providing great accuracy efficiently. Chapter 4 of Nakamura's text on applied numerical methods has a description of such integration algorithms, and Chapter 4 of the numerical recipes book of Press et al. has a function that performs Gaussian quadrature. A compendium of numerical-integration methods by Davis and Rabinowitz includes modern techniques (such as the use of splines) and very complete coverage of the whole spectrum of methods.

4.7   PROJECT 4B: ELECTROSTATIC POTENTIAL FROM A CHARGED WIRE

It is interesting to compute the electrostatic potential from a charged wire because it involves analysis, numerics, and application to a scientific problem. It is also simple enough that you can understand the science without extensive physics background. Suppose that we have electric charge distributed along a finite segment of the y axis, as shown in the insert to Figure 4.10.

In the figure insert the charge is indicated as being uniformly distributed along the wire, but we assume only that its distribution, $\lambda(y)$, goes from $y = -1$ to $y = 1$ and is zero for y beyond this and for any nonzero x. Our task is to compute the potential at any point in space, (x, y), not including the wire itself. Such points are called field points.

From Coulomb's law of electrostatic interactions, a small element of charge on a segment of length $\Delta y'$ of wire located near y' on the wire produces at (x, y) a potential $\Delta V = \lambda(y') \Delta y'/r$, where r is the distance between the charge element and the field point. (In various systems of electrical units there will be an overall factor between charge and potential, which we ignore, but electricians should not.) The total potential at this point is obtained by superimposing the potentials from all points along the wire in the limit that the lengths of the segments shrink to zero. Thus, the potential is the Riemann integral of $\lambda(y')/r$. Explicitly, the potential is given by

$$V(x,y) = \int_{-1}^{1} dy' \frac{\lambda(y')}{\sqrt{(y'-y)^2 + x^2}} \qquad (4.48)$$

in which we used the theorem of Pythagoras to express r in terms of the coordinates.

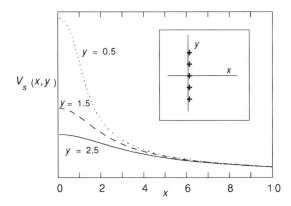

FIGURE 4.10 Potential from a wire having a uniform distribution of charge along the wire, as indicated in the insert. The potential is shown as a function of x distance from the wire for three different heights, y, above the center of the wire.

Now that the physics problem has been set up, we should concern ourselves with evaluating this integral. Suppose that the charge density has a simple power-law dependence on position along the wire as

$$\lambda(y') = \lambda_0 y'^p \tag{4.49}$$

where p is any integer that is not negative. The potential formula can readily be transformed so that it is expressed in terms of some simpler integrals.

Exercise 4.25

(a) Make a change of variables in (4.48) and (4.49) in order to simplify the denominator in (4.48), namely, set

$$z = \frac{y' - y}{x} \tag{4.50}$$

($x = 0$ doesn't occur in this problem), then show that the scaled potential, V_s, is given by

$$V_s(x,y) \equiv \frac{V(x,y)}{\lambda_0} = \sum_{k=0}^{p} \frac{p!\, x^k\, y^{p-k}}{k!\,(p-k)!} I_k \tag{4.51}$$

where the binomial expansion is used to expand $(y + xz)^p$. The units of λ_0, and therefore of V_s, depend upon the exponent p. The integrals I_k are defined by

$$I_k \equiv \int_{-(1+y)/x}^{(1-y)/x} dz\, \frac{z^k}{\sqrt{z^2 + 1}} \tag{4.52}$$

in which k is an integer. The potential problem has been replaced by the actual problem of evaluating these integrals.

(b) Show (by your own analytical skills, by consulting a table of integrals, or by using a computer-algebra system) that the indefinite integrals in (4.52) for $k = 0$ and 1 are given by

$$I_0 = -\ln\left(\sqrt{z^2 + 1} - z\right) \tag{4.53}$$

$$I_1 = \sqrt{z^2 + 1} \tag{4.54}$$

(c) To derive the analytical form of the integral for higher values of k, write the factor z^k in (4.52) as $z \times z^{k-1}$. Then use integration by parts, the result for I_1 in (4.54), and some algebraic manipulations, to derive the recurrence relation

$$I_k = \left[\sqrt{z^2 + 1}\; z^{k-1} - (k-1)I_{k-2}\right]/k \qquad k > 1 \qquad (4.55)$$

Check this result by differentiating both sides with respect to z in order to verify that you recover the integrand in (4.52). ∎

From the results of this exercise we have the component integrals that can be combined by using (4.51), after inserting the limits in (4.52), to produce the total scaled potential V_s at any field point (x, y).

We now have a choice of methods for calculating the electrostatic potential, analytical or numerical. The former method, although it has the elegant results from Exercise 4.25 at its command, is restricted to integer powers p. On the other hand any positive p may be used with numerical integration of (4.48), provided that $|y'|$, rather than y', is used in (4.49). However analytical integration is useful for testing the accuracy of the numerical methods, so we start with it.

Potentials by analytical integration

The simplest analytical potentials are obtained if the electric charge is distributed uniformly along the wire. Then in the above integrals we have $k = p = 0$ only. The scaled potential is then given by

$$V_s(x,y) = \ln\left[\sqrt{\left(\frac{y-1}{x}\right)^2 + 1}\; - \frac{y-1}{x}\right] - \ln\left[\sqrt{\left(\frac{y+1}{x}\right)^2 + 1}\; - \frac{y+1}{x}\right] \qquad (4.56)$$

This potential is plotted in Figure 4.10 as a function of x for three values of y, halfway up the wire ($y = 0.5$), above the level of the wire ($y = 1.5$), and far above the wire ($y = 2.5$). As you would guess from Coulomb's law of the inverse-distance dependence of the potential, V_s dies away rapidly as the field point is moved farther away from the source of the charge on the wire.

If the electric charge is distributed linearly along the wire (which would have to be made of a nonconducting material, so that the charges didn't flow and neutralize each other), with positive charges along the top half and negative charges along the bottom half, then the scaled potential can be calculated from the integral expansion (4.51).

Exercise 4.26

(a) Write a small C function to evaluate the analytical potential, $V_s(x,y)$, for the linear charge distribution ($p = 1$) in terms of the integrals I_k obtained in Exercise 4.25 and the expression (4.51) for V_s.

(b) Make a plot similar to Figure 4.10 for the potential due to this linear distribution of charge. ∎

With two check cases in hand, we are ready to develop the numerical applications.

Potentials by numerical-integration methods

In the preceding subsection we showed that analytical expressions for the potentials from a line charge with various (positive-integer) power-law distributions for the charge, given by (4.49), can be derived. In the numerical-integration methods we use these as test cases, then explore the situation with noninteger power laws.

First we need a way to handle the numerics. It is straightforward to modify Program 4.5, Trapezoid and Simpson Integrals, to do this. The resulting program, Electrostatic Potentials by Numerical Integration, is given in the following program.

PROGRAM 4.6 Electrostatic potentials by numerical integration, trapezoidal and Simpson formulas, for a uniform charge distribution.

```
#include <stdio.h>
#include <math.h>

main()
{
/*  Electrostatic Potentials
    by Numerical Integration */
double X,Y,zmin,zmax,h,VAnl,VByTrap,error,VBySimp;
int n;
double Vs(),SimpInt(),TrapInt(),y();

printf("Potentials by Numerical Integration\n");
X = 1;
while ( X != 0 )
  {
  printf("\n\nInput X, Y, n (X=0 to end): ");
  scanf("%lf%lf%i",&X,&Y,&n);
  if ( X == 0 )
    {
    printf("\nEnd Potentials by Numerical Integration");
    exit(0);
    }
  n = 2*(n/2); /* truncate n to nearest even value */
  zmin = (X-1)/Y;   zmax = (X+1)/Y;
  h = (zmax-zmin)/n;
  VAnl = Vs(X,Y);
  printf("Analytical potential=%lg",VAnl);
  VByTrap = TrapInt(zmin,n,h);/* trapezoid rule */
  error = VAnl-VByTrap;
  printf("\nTrapezoid integral=%lg",VByTrap);
  printf(" Error=%lg",error);
  VBySimp = SimpInt(zmin,n,h);/* Simpson rule */
```

```
  error = VAnl-VBySimp;
  printf("\nSimpson integral=%lg",VBySimp);
  printf(" Error=%lg",error);
  } /* end  X  loop */
}


double SimpInt(a,n,h)
/* Simpson integral */
double a,h;
int n;
{
double sum;
double y();
int n2,j;
n2 = n/2;
if ( ( 2*n2 != n ) || ( n2 == 1) )
  {
  printf("\n!! SimpInt has n=%i; not allowed",n);
  return 0;
  }
else
  {
  .sum = 0;
  for ( j = 1; j < n2; j++ )
    {
    sum = sum+y(a+2*j*h)+2*y(a+(2*j-1)*h);
    }
  return h*(y(a)+4*y(a+(n-1)*h)+y(a+n*h)+2*sum)/3;
  }
}


double TrapInt(a,n,h)
/* Trapezoid integral */
double a,h;
int n;
{
double sum;
double y();
int j;
sum = ( y(a) + y(a+n*h) )/2;
for ( j = 1; j < n; j++ )
  {
  sum = sum+y(a+j*h);
  }
return h*sum;
}
```

```
double y(x)
/* Potential integrand value*/
double x;
{
double prod;
prod = 1/sqrt(x*x+1);
return prod;
}

double Vs(x,y)
/* Analytical potential integral */
double x,y;
{
double sum;
sum = -log(sqrt(pow((x+1)/y,2)+1)-(x+1)/y)
      -log(sqrt(pow((x-1)/y,2)+1)+(x-1)/y);
return sum;
}
```

In Program 4.6 there has been some renaming of variables in the main program, slight modifications of input, but no changes to integrator, functions SimpInt and TrapInt. These two functions may therefore be copied from their parent, Program 4.5. The price paid for this is that the variable names do not agree with those used in formulating our electrostatic-potential problem, so care is needed if the programs are modified. Note that the method of signalling termination of program execution, input X = 0, is convenient rather than necessary. Provided that $|y| > 1$, so the field point is not touching the wire (how shocking!), the potential is well-defined on the x axis. If you don't like this way of stopping execution, devise your own.

The functions for the potential integrand, y(x), and for the analytical potential, Vs(x,y), have been coded to agree with (4.48) and (4.56), respectively, for the case $p = 0$. Now you should be ready to code and test the potential program.

Exercise 4.27

(a) Code and test Electrostatic Potentials by Numerical Integration. The program can be tested by coding below the formulas in y(x) and Vs(x,y) an integrand and an analytic integral, respectively, that can easily be checked by hand, such as a low-order polynomial.

(b) Run the program for the coded example of the uniform charge density and a reasonable stepsize, say $h = 0.1$, to check your results against those displayed in Figure 4.10 at $y = 0.5$, 1.5, and 2.5 for $x > 0$.

(c) If you have a program that displays contours, modify Electrostatic Potentials to compute a mesh of (x,y) points fine enough that smooth contours (equipotentials) are produced by the Simpson method. Thus generate the contours for the potential from a uniformly charged wire. Do the contours make physical sense, in that electrostatic energy would be constant along a contour?

(*d*) Modify the program to allow positive integer powers, *p*, for the charge density (4.49). This requires only minor changes to the potential-integrand function y(x), but substantial changes are needed to the potential integral function, Vs(x,y), to include the recurrence formula (4.56) and the summation (4.51). (You may have already done this in Exercise 4.26.) Then this function has to be used for both limits of integration. The numerical integration functions should be scarcely modified, and thereby they serve to identify any errors in your analytical results. As *p* is increased, the charges should pile up (positive and negative) at the two ends of the wire, and the potential should resemble that from two point charges of opposite sign at the two ends of the wire. Is this what you find?

(*e*) As a final topic of exploration, turn off the analytical solution and try the numerical integrations with noninteger values of *p*, such as *p* = 0.5, 0.75, and a charge distributed as $|y|^p$ rather than y^p, in order to avoid complex results. Your results should roughly interpolate between those for adjacent integer values of *p*, such as *p* = 0, 1. To make such a comparison you will need to modify the solutions to accommodate the change in the charge-distribution formula (4.49) from y^p to $|y|^p$. For example, just split the integral into contributions from above and below the origin. ■

With the experience in numerical integration that you have gained from this applied-physics project and from the exercises in Sections 4.4 – 4.6, you should be confident to compute many kinds of numerical derivatives and integrals.

REFERENCES ON NUMERICAL DERIVATIVES AND INTEGRALS

Abramowitz, M., and I. A. Stegun, *Handbook of Mathematical Functions*, Dover, New York, 1964.

Davis, P. J., and P. Rabinowitz, *Methods of Numerical Integration*, Academic Press, Orlando, Florida, second edition, 1984.

Maron, M. J., *Numerical Analysis*, Macmillan, New York, second edition, 1987.

Nakamura, S., *Applied Numerical Methods with Software*, Prentice Hall, Englewood Cliffs, New Jersey, 1991.

Pippard, A. B., *Response and Stability*, Cambridge University Press, Cambridge, England, 1985.

Pippard, A. B., *The Physics of Vibration*, Cambridge University Press, Cambridge, England, 1989.

Press, W. H., B. P. Flannery, S. A. Teukolsky, and W. T. Vetterling, *Numerical Recipes in C,* Cambridge University Press, New York, 1988.

Vandergraft, J. S., *Introduction to Numerical Computations*, Academic Press, New York, 1978.

Wolfram, S., *Mathematica: A System for Doing Mathematics by Computer*, Addison-Wesley, Redwood City, California, second edition, 1991.

Chapter 5

FITTING CURVES THROUGH DATA

In this chapter we emphasize making essentially exact fits of curves through data values, as if the data were mathematically precise. Therefore, you will notice as you skim the section headings that the applications (interpolation, derivatives, integrals) take the data as found and estimate other quantities from them. This emphasis is different from data-fitting methods developed in Chapter 6, where the best-fit curves do not usually pass through each data value, and where our attitude is that the data are imprecise and we make our best effort to find a more exact representation of them.

Spline fitting and least-squares fitting are thus complementary. For example, a spline fit may be used to preprocess data, for example to interpolate them so that data from different sources have their independent variables sampled at the same points, then a least-squares fit will be made on the interpolated data. (It is feasible, but complicated, to combine the two methods and to make a spline-least-squares fit.)

For the exact fits to data considered in this chapter, one method would be to use a truncated Taylor series to approximate $y(x)$ and to fit a single polynomial of order $n - 1$ through the n points, then to estimate the desired properties from the polynomial. Although this method is hallowed by tradition (see, for example, Abramowitz and Stegun), it has the severe disadvantage that if n becomes large the polynomial fit will generally have more and more wiggles in it, especially if the data are experimental or arise from an approximate numerical procedure. We encountered this problem with polynomials in Sections 4.1 and 4.4 – 4.6.

Exercise 5.1
Explain why, both from analytical and graphical considerations, a polynomial of order n may have as many as $n - 1$ changes of direction over its whole range of variation. ∎

For example, the working function (4.1), a sixth-order polynomial all of whose roots are real and lie in the interval $[-0.5, 1]$, has four changes of direction in the interval $[-0.5, 0.5)$, as seen in Figure 4.1. Such wiggly behavior, sometimes called the "polynomial wiggle problem," is usually inconsistent with the problem at hand. Can we develop a curve-fitting method that guarantees a smooth fit through a sequence of points, and whose behavior does not change much as we add data?

In this chapter we develop the method of curve fitting by splines. For simplicity, while still being practicable, we consider the *cubic spline,* in which the behavior of the fitting polynomial is a cubic in the interval between consecutive points, but there is a different cubic polynomial in successive intervals. The major reason why the local fitting polynomial is of order three (a cubic) is that this order guarantees continuity of the slopes of the spline tangents at the knots.

Exercise 5.2
Show that a necessary and sufficient condition for a piecewise polynomial to have a continuous second derivative (slope of its tangent) at the matching knots is that it be at least of third order. ∎

If we used a polynomial of order higher than third the spline equations would be more complex to solve and to use, usually without much gain in accuracy of the approximation except for the higher-order derivatives.

This chapter is organized as follows. We derive the fitting algorithm in Section 5.1, and we investigate in Section 5.2 how the results depend on the spline behavior near the endpoints. In Project 5 (Section 5.3) you are shown how to program the algorithm derived in Sections 5.1 and 5.2, and there are exercises so that you can check the program correctness. We show in Section 5.4 how to interpolate with cubic splines and how to use them to estimate derivatives, then in Section 5.5 we show their usefulness for integration and provide several examples.

As a diversion, we discuss in Section 5.6 the close relation between the development of computers and the extensive growth in the use of spline methods in applications from data analysis through the design of type fonts for printing and the description of surfaces in computer graphics. References on spline fitting round out the chapter.

5.1 HOW TO FIT CURVES USING SPLINES

In this section we derive the properties that a spline fit is required to satisfy, then we derive and solve the spline equations and develop an algorithm for spline fitting that can readily be coded in the C language. The code itself is presented and first used in Section 5.3.

What is a spline?

The various methods of spline fitting are the numerical analogs of using a thin, uniform, flexible strip (a drafting spline) to draw smooth curves through and between points on a graph. The cubic spline method that we develop is locally smooth and is insensitive to adding data. We begin by introducing some terminology used in spline fitting. When needed, we will be mathematically more precise in the formula derivations.

- A spline of Nth *order* is one in which $N + 1$ adjacent points define a polynomial of order N to pass through the middle point. For example, one most often uses a cubic spline ($N = 3$), in which the spline behaves locally as a cubic polynomial. However, there will be different cubics going through successive points. Therefore, the spline is a composite curve, as shown in Figure 5.1.

- The *knots* of the spline are the points through which the spline curve is required to pass. A sample of n data therefore has n knots. Thus, spline algorithms assume exact fitting at the knots, rather than the compromise made in least-squares methods.

- Spline *endpoint conditions* are the constraints imposed on the behavior of the spline near the first and last data points. You can see that some special constraints are needed, because if you try to use a drafting spline you can bend it at the ends without changing the fit near its middle very much. Usually, a drafting spline just sticks out straight at the ends, so it has second derivatives zero. This is called the *natural-spline* endpoint condition.

Examples of spline fits with two different endpoint conditions are shown in Figures 5.1 and 5.2. We investigate spline boundary conditions in detail in Section 5.2.

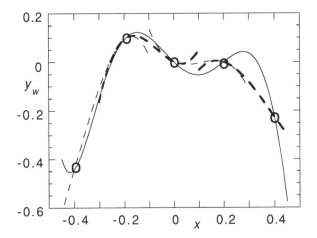

FIGURE 5.1 The pathological working function (4.1) and its fit by a cubic spline through five knots (indicated by the circles) using natural endpoint conditions. The solid curve is the working function and the dashed curves are the cubics used between each pair of knots. Each cubic is shown beyond the region in which it is used in order to show the continuity of the spline curvature.

If the curvature of the spline fit (which is indicated by the second derivative) is to be smooth, but we are willing to compromise on higher derivatives implied by the $y(x)$, then local approximation by a cubic spline, $s(x)$, is appropriate. Our development assumes that the spline is of third order and (for simplicity of explanation and ease of use) that the knots are equally spaced.

With the definitions taken care of, we are ready to analyze the spline properties, then to set up and solve the spline equations.

Properties for spline fits

We require for a cubic-spline fit that the fitting function, $s(x)$, satisfy the following properties:

1. Within each subinterval $x_j \leq x \leq x_{j+1}, j = 1,2,...,n - 1$, $s(x)$ is a cubic polynomial.
2. Each of the derivatives with respect to x, namely $s, s^{(1)}, s^{(2)}$, is continuous over the whole range $x_1 \leq x \leq x_n$.
3. At each knot, $x = x_j$, the spline fit goes exactly through the data, so that $s(x_j) = y_j, \; j = 1,2,...,n$.

You can see in Figure 5.1 the cubics that are used to go through each pair of points, and how the slopes of successive splines match at the knots. Note, however, that each cubic is used for only two points. In the figure the extensions beyond the knots are drawn to indicate how use of a given cubic segment outside its range would often give a poor representation of the curve.

With the above three conditions on the spline fit, we are ready to derive an appropriate algorithm. For simplicity, and as often encountered in practice, we assume equal spacing of the independent variable values x_j by an amount h.

Deriving the spline equations

Since the fitting conditions relate to derivatives, we make Taylor (polynomial) expansions (Sections 3.1, 3.2) in the variable x about the point x_j for the jth interval. For a cubic polynomial, derivatives past the third are zero, so we can immediately calculate the third derivative at x_j in terms of the second derivatives at the knots:

$$s_j^{(3)} = \left(s_{j+1}^{(2)} - s_j^{(2)}\right)/h \tag{5.1}$$

Thus, the cubic spline can be written in terms of its first two derivatives, using also the requirement that it pass through the knot value y_j,

$$s(x) = y_j + s_j^{(1)}(x - x_j) + s_j^{(2)}(x - x_j)^2/2 + \left(s_{j+1}^{(2)} - s_j^{(2)}\right)(x - x_j)^3/6h \tag{5.2}$$

Our goal is to solve for the derivatives in terms of the y_j by using the spline continuity conditions, item (2) in the above list. By differentiating (5.2) we obtain

$$s^{(1)}(x) = s_j^{(1)} + s_j^{(2)}(x - x_j) + \left(s_{j+1}^{(2)} - s_j^{(2)}\right)(x - x_j)^2/2h \tag{5.3}$$

and, on differentiating this slope equation,

$$s^{(1)}(x) = s_j^{(2)} + \left(s_{j+1}^{(2)} - s_j^{(2)}\right)/h \tag{5.4}$$

We can relate the first derivatives at the knots by using (5.3) with j stepped down to $j - 1$, then $x \to x_j$, to find that

$$s_j^{(1)} - s_{j-1}^{(1)} = \left(s_{j-1}^{(2)} + s_j^{(2)}\right)h/2 \tag{5.5}$$

Using property (3) enables us to equate the datum y_{j+1} with the spline fit at x_{j+1}, so that in (5.2) we have

$$y_{j+1} = y_j + s_j^{(1)}h + s_j^{(2)}h^2/2 + \left(s_{j+1}^{(2)} - s_j^{(2)}\right)h^2/6 \tag{5.6}$$

Stepping back one point gives similarly

$$y_j = y_{j-1} + s_{j-1}^{(1)}h + s_{j-1}^{(2)}h^2/2 + \left(s_j^{(2)} - s_{j-1}^{(2)}\right)h^2/6 \tag{5.7}$$

We can now obtain a relation between adjacent spline second derivatives by subtracting the second equation from the first, substituting (5.5), and rearranging, to obtain for $j = 2, 3, \ldots, n - 1$

$$s_{j-1}^{(2)} + 4s_j^{(2)} + s_{j+1}^{(2)} = d_j \tag{5.8}$$

where the second differences of the data are

$$d_j = 6\left[y_{j+1} - y_j - (y_j - y_{j-1})\right]/h^2 \tag{5.9}$$

at the jth point. This formula is arranged so that roundoff errors are minimized without increasing the number of arithmetic operations, as discussed in Section 4.4.

The equations in (5.8) form a set of $n - 2$ linear equations which can be solved once the derivatives at the endpoints have been chosen by the user. As we discuss in more detail in Section 5.2, the most natural spline is one in which these derivatives are zero, that is, the fit sticks out straight at each end. Whatever our choice of endpoints, the same method of solving the equations (5.8) can be used. We may think of the relations (5.8) as defining the elements of matrices, with those on the left forming a banded matrix, with elements only on the diagonal, j, and one place above and below the diagonal, $j \pm 1$. This property has arisen because we are using a spline of cubic order, whereas, one of higher order would have more off-diagonal elements. Because, also for simplicity, we are assuming equally spaced data, it is straightforward to solve the equations directly, rather than using the full power of methods useful for matrix inversion.

We solve equations (5.8) iteratively by first eliminating $s_{j-1}^{(2)}$ between pairs of these equations. Then we set

$$a_2 = 4 \qquad b_2 = d_2 - s_1^{(2)} \tag{5.10}$$

$$a_j = 4 - 1/a_{j-1} \qquad b_j = d_j - b_{j-1}/a_{j-1} \qquad j = 3,4,...,n-1 \tag{5.11}$$

The spline equations (5.8) then become

$$a_j s_j^{(2)} + s_{j+1}^{(2)} = b_j \qquad j = 2,3,...,n-1 \tag{5.12}$$

From this, starting at $j = n - 1$ and working downward to smaller j, we find the spline second derivatives at the knots

$$s_{n-1}^{(2)} = b_{n-1}/a_{n-1} \qquad s_j^{(2)} = (b_j - s_{j+1})/a_j \qquad j = n - 2,...,2 \tag{5.13}$$

The reader who is experienced in matrix operations may notice that we have just performed Gaussian elimination to find the inverse of a tridiagonal matrix. This connection is used in more general treatments of spline fitting, such as those in the monographs by De Boor and by Schumaker.

Exercise 5.3 Verify all of the above steps in the spline-fitting derivation. (Spare the spline and spoil the child.) ∎

We have now completed the analysis required for deriving cubic-spline fits, and it is time to cast the formulas into an algorithm suitable for programming.

The spline algorithm

The algorithm for cubic spline fitting of n data values $y_1, y_2, ..., y_n$ can now be summarized as follows:

(1) Compute the coefficients a_j according to the iteration scheme

$$a_2 = 4 \qquad a_j = 4 - 1/a_{j-1} \qquad j = 3,4,...,n-1 \tag{5.14}$$

These spline coefficients depend only on the number of data points, n, and not on the data values. They could be computed and stored for reuse, provided that n did not change.

(2) With j increasing compute the first differences among the data values

$$e_j = y_j - y_{j-1} \qquad j = 2,3,...,n \tag{5.15}$$

and thence the second-difference quantities

$$d_2 = 6(e_3 - e_2)/h^2 - s_1^{(2)} \tag{5.16}$$

$$d_j = 6\,(e_{j+1} - e_j)/h^2 \qquad j = 3,4,...,n-2 \tag{5.17}$$

$$d_{n-1} = 6\,(e_n - e_{n-1})/h^2 - s_n^{(2)} \tag{5.18}$$

and the coefficients

$$b_2 = d_2, \quad b_j = d_j - b_{j-1}/a_{j-1} \quad j = 3,4,...,n-1 \tag{5.19}$$

Note that the second derivatives at the boundaries must be supplied. They are zero for the natural cubic spline.

(3) With j decreasing, compute the second derivatives at the spline knots

$$s_{n-1}^{(2)} = b_{n-1}/a_{n-1} \tag{5.20}$$

$$s_j^{(2)} = \left(b_j - s_{j+1}^{(2)}\right)/a_j \quad j = n-2,...,2 \tag{5.21}$$

(4) With j increasing, compute the first derivatives at the knots

$$s_1^{(1)} = e_2/h - s_1^{(2)}h/3 - s_2^{(2)}h/6 \tag{5.22}$$

$$s_j^{(1)} = s_{j-1}^{(1)} + \left(s_{j-1}^{(2)} + s_j^{(2)}\right)h/2 \quad j = 2,...,n \tag{5.23}$$

and the third derivatives at the same points

$$s_1^{(3)} = 0 \quad s_n^{(3)} = 0 \tag{5.24}$$

$$s_j^{(3)} = \left(s_{j+1}^{(2)} - s_j^{(2)}\right)/h \quad j = 2,...,n-1 \tag{5.25}$$

Steps (1) through (4) summarize the algorithm for generating the spline-fitting coefficients. In the next two sections we discuss the boundary conditions for cubic-spline fits, then the coding and testing of a program in the C language.

5.2   BOUNDARY CONDITIONS FOR SPLINE FITTING

Because a spline is a piecewise-continuous function, but with different coefficients for the cubic between each pair of knots, the derivatives at the boundaries x_1 and x_n (which are usually called the "endpoints") have significant effects on the behavior of the fit even at the interior points. Therefore, we have to discuss the boundary conditions for spline fitting, for example (for a cubic spline) the second derivatives at each endpoint.

Natural splines

If you use a flexible strip of wood and force it to pass through several points on a curve, it bends between the points but it sticks out straight at both ends, so that its second derivatives are zero. This condition, when applied at both ends, is called the *natural-spline endpoint condition*. Note that this naturalness comes from the drafting analogy, rather than being related to the properties of the underlying function that we are trying to approximate.

An interesting property of cubic splines with the natural endpoint conditions imposed is that they have a minimal value of the curvature, C, where C is defined by

$$C \equiv \int_{x_1}^{x_n} dx \left[s^{(2)}(x) \right]^2 \tag{5.26}$$

For the natural-spline fit, $s_N(x)$, let us call the curvature C_N, so that

$$C_N \equiv \int_{x_1}^{x_n} dx \left[s_N^{(2)}(x) \right]^2 \tag{5.27}$$

Our claim is then that $C \geq C_N$ for any cubic-spline fit. This condition is also called "minimum tension" because it corresponds to a the behavior of a flexible strip (the mechanical drafting spline) when constrained to touch the knots but otherwise to be unconstrained.

Exercise 5.4
Derive the following result for the second derivatives (curvature) of spline fits:

$$C = C_N + \int_{x_1}^{x_n} dx \left[s^{(2)}(x) - s_N^{(2)}(x) \right]^2 \tag{5.28}$$

To do this, compute the second integral on the right-hand side by expanding the square, then split the integral into a sum of integrals over segments of length h, integrate by parts, and resum. Complete the proof by using the natural-spline endpoint conditions. ∎

Thus, since the integral of the squared difference of spline curvatures in (5.28) is necessarily nonnegative, the least curvature is obtained when the spline fit, s, coincides with the natural-spline fit, s_N, at all x in the range of integration.

The effects of the endpoint conditions on the overall closeness of the spline fit can be seen for the pathological case of our working function (4.1) with its closely spaced roots and consequent very wiggly behavior over a small range of x values. With natural endpoint conditions the spline fit is as shown in Figure 5.1. By comparison, Figure 5.2 shows the working function with the same knots fitted by the cubic spline with the exact endpoint second derivatives, obtained by using in (4.3) the coefficients $w_{2,k}$ from Table 4.1.

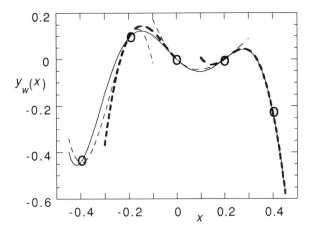

FIGURE 5.2 The working function (4.1) and its fit by a cubic spline through five knots (indicated by the circles) using exact endpoint conditions. The solid curve is the working function and the dashed curves are the cubics used between each pair of knots, with the cubic shown beyond the region in which it is used, thus showing the continuity of the spline curvature.

We make further comparisons of natural-spline and exact endpoint conditions, for both the badly behaved working function and the well-behaved cosine function, in Sections 5.4 and 5.5.

5.3   PROJECT 5: PROGRAM FOR SPLINE FITTING

We now have the spline algorithm summarized at the end of Section 5.1, together with some understanding of endpoint conditions (Section 5.2). Therefore, we are ready to program, test, and use the program for spline fitting presented as Program 5.1.

PROGRAM 5.1 Cubic-spline interpolation, differentiation, and integration with equally spaced knots and user-controlled endpoint conditions.

```
#include <stdio.h>
#include <math.h>
#define MAX 101

main()
{
/* Cubic Splines */
double ylist[MAX],sd1[MAX],sd2[MAX],sd3[MAX];
double xmin,xmax,h,x,sx,sd1x,sd2x,sd3x;
```

```c
double yw1,yw2,yw3,yerr,y1err,y2err,y3err;
double xL,xU,Intx,yInterr;
int n,j;
char Interp,Integ;
void SplineFit(),SplineInterp();
double yw(),SplineInt(),ywInt();

printf("Cubic Splines");
n = 3;
while ( n > 0 )
  {
  printf("\nInput n <=%i (n=0 to end): ",MAX-1);
  scanf("%i", &n);
  if ( n < 1 )
    {
    printf("\nEnd  Cubic Splines");  exit(0);
    }
  if ( n > MAX-1 )
    {
    printf("\n!! More than MAX = %i data points\n", MAX-1);
    }
  else
    {
    if (n < 4)
      {
      printf("\n!! Fewer than 4 data points\n");
      }
    else
      {
      printf("\nInput %i data: ",n);
      for ( j = 1; j <= n; j++ )
        {
        scanf("%lf",&ylist[j]);
        }
      printf("\nInput second derivatives at j=1 & j=%i: ",n);
      scanf("%lf%lf",&sd2[1], &sd2[n]);
      printf("\nInput minimum & maximum x: ");
      scanf("%lf%lf",&xmin,&xmax);
      h = (xmax-xmin)/(n-1); /* uniform step size */
      SplineFit(n,h,ylist,sd1,sd2,sd3);
      x = (xmin+xmax)/2;
      while ( x >= xmin && x <= xmax )
        {
        printf("\nInterpolate? (y or n):");  scanf("%s",&Interp);
        if ( Interp == 'y' )
          {
```

```
      printf("Input x (x<%g or x>%g to end): ",xmin,xmax);
      scanf("%lf",&x);
      if ( x >= xmin && x <= xmax )
        {
        SplineInterp(x,xmin,h,ylist,sd1,sd2,sd3,
                      &sx,&sd1x,&sd2x,&sd3x);
        printf("Interpolated s=%g, s'=%g, s''=%g, s'''=%g\n",
                sx,sd1x,sd2x,sd3x);
        yerr = yw(x,&yw1,&yw2,&yw3)-sx;   y1err = yw1-sd1x;
        y2err = yw2-sd2x;                 y3err = yw3-sd3x;
        printf("Error in y=%g, y'=%g, y''=%g, in y'''=%g\n",
                yerr,y1err,y2err,y3err);
        }
      }
    printf("\nIntegrate? (y or n):");  scanf("%s",&Integ);
    if ( Integ == 'y' )
      {
      printf("Input xL (>=%g)  xU (<=%g): ",xmin,xmax);
      scanf("%lf%lf",&xL,&xU);
      if ( xL >= xmin && xU <= xmax )
        {
        if ( xU >= xL )
          {
          Intx = SplineInt(xL,xU,h,xmin,ylist,sd1,sd2,sd3);
          yInterr = (ywInt(xU)-ywInt(xL))-Intx;
          printf("Integral = %g & error = %g\n",Intx,yInterr);
          }
        else
          printf("\n!! Negative range (%g to %g)\n",xL,xU);
        }
      else printf("\n!! A limit is outside spline range\n");
      }
    if ( (Interp != 'y') && (Integ != 'y') )
      {
      x = 2*(fabs(xmin)+fabs(xmax)); /* forces x loop exit */
      }
    }/* end  while x  interpolation loop */
    }
  }
} /* end  while n  loop */
}

void SplineFit(n,h,ylist,sd1,sd2,sd3)
/* Cubic spline fit; uniform step size (h) in x */
double ylist[],sd1[],sd2[],sd3[];
double h;
```

```
int n;
{
  double as[MAX],es[MAX],ds[MAX],bs[MAX];
  double hfact,hd2;
  int j;

  hfact = 6/(h*h);   hd2 = h/2;
  /* Spline coefficients */
  as[2] = 4;
  for ( j = 3; j < n; j++ )
    {
    as[j] = 4-1/as[j-1];
    }
  for ( j = 2; j <= n; j++ )
    { /* First differences */
    es[j] = ylist[j]-ylist[j-1];
    }
  for ( j = 2; j < n ; j++ )
    { /* Second differences */
    ds[j] = hfact*(es[j+1]-es[j]);
    }
  ds[n-1] = ds[n-1]-sd2[n];
  /* Make b coefficients */
  bs[2] = ds[2]-sd2[1];
  for ( j = 3; j < n; j++ )
    {
    bs[j] = ds[j]-bs[j-1]/as[j-1];
    }
  /* Second derivatives of spline */
  sd2[n-1] = bs[n-1]/as[n-1];
  for ( j = n-2; j >= 2; j-- ) /* downward recurrence */
    {
    sd2[j] = (bs[j]-sd2[j+1])/as[j];
    }
  /* First derivatives of spline */
  sd1[1] = es[2]/h-(sd2[1]/3+sd2[2]/6)*h;
  for ( j = 2; j <= n; j++ )
    {
    sd1[j] = sd1[j-1]+(sd2[j-1]+sd2[j])*hd2;
    }
  /* Third derivatives of spline */
  for ( j = 1; j < n; j++ )
    {
    sd3[j] = (sd2[j+1]-sd2[j])/h;
    }
  sd3[n] = sd3[n-1];
```

```
}

void SplineInterp(x,xmin,h,ylist,sd1,sd2,sd3,
                  sx,sd1x,sd2x,sd3x)
/* Cubic spline interpolation */
double ylist[],sd1[],sd2[],sd3[];
double x,xmin,h;
double *sx,*sd1x,*sd2x,*sd3x; /* interpolated values */
{
double e; /* distance above i-th point */
int i; /* interpolation index */

i = (x-xmin)/h+1.1;  e = x-xmin-(i-1)*h;
/* Interpolated value, sx, */
*sx = ylist[i]+(sd1[i]+(sd2[i]/2+sd3[i]*e/6)*e)*e;
*sd1x = sd1[i]+(sd2[i]+sd3[i]*e/2)*e; /* first derivative */
*sd2x = sd2[i]+sd3[i]*e; /* second derivative */
*sd3x = sd3[i]; /* third derivative */
}

double SplineInt(xL,xU,h,xmin,ylist,sd1,sd2,sd3)
/* Cubic spline integral */
double ylist[],sd1[],sd2[],sd3[];
double xL,xU,h,xmin;
{
double eL,eU,ysum,sum1,sum2,sum3;
double yU,s1U,s2U,s3U,yL,s1L,s2L,s3L,splint;
int jL,jU,j;
double Horner4Poly();

jL = (xL-xmin)/h+1.01;   eL = xL-xmin-(jL-1)*h;
jU = (xU-xmin)/h+1.01;    eU = xU-xmin-(jU-1)*h;
ysum = sum1 = sum2 = sum3 = 0; /* sums over whole strips */
for ( j = jL; j < jU; j++ )
  {
  ysum = ysum + ylist[j];
  sum1 = sum1 + sd1[j];
  sum2 = sum2 + sd2[j];
  sum3 = sum3 + sd3[j];
  }
splint = Horner4Poly(ysum,sum1,sum2,sum3,h);
/* Contributions from partial strips at ends */
yU = ylist[jU]; s1U = sd1[jU]; s2U = sd2[jU]; s3U = sd3[jU];
yL = ylist[jL]; s1L = sd1[jL]; s2L = sd2[jL]; s3L = sd3[jL];
splint = splint + Horner4Poly(yU,s1U,s2U,s3U,eU)
                - Horner4Poly(yL,s1L,s2L,s3L,eL);
```

```
return splint;
}

double Horner4Poly(y,s1,s2,s3,e)
/* Horner 4th-order polynomial algorithm */
double y,s1,s2,s3,e;
{
double poly;
poly = (y+(s1/2+(s2/6+s3*e/24)*e)*e)*e;
return poly;
}

double yw(x,yw1,yw2,yw3)
/* Working function and its first 3 derivatives */
double x,*yw1,*yw2,*yw3;
{
double y;
y = 120*(x+0.5)*(x+0.25)*x*(x-(1.0)/(3.0))*(x-0.2)*(x-1);
*yw1 = -1+(6+(69+(-204+(-470+720*x)*x)*x)*x)*x;
*yw2 = 6+(138+(-612+(-1880+3600*x)*x)*x)*x;
*yw3 = 138+(-1224+(-5640+14400*x)*x)*x;
return y;
}

double ywInt(x)
/* Working function integral */
double x;
{
double yint;
yint = (-0.5+(1+(5.75+(-10.2+((-47.0/3.0)
        +(120.0/7.0)*x)*x)*x)*x)*x)*x*x;
return yint;
}
```

We now describe the structure of the major parts of the spline program, which are the main program and the fitting function `SplineFit`. Detailed discussions of the interpolation function, `SplineInterp`, and of the integration function, `Spline-Int`, are deferred to the sections in which they are used, namely Sections 5.4 and 5.5, respectively.

The main program, Cubic Splines

The structure of `Cubic Splines` is as follows. The control loop in the main program is controlled by n, the number of points in the fit. Since at least 4 points are needed to describe a cubic, we input n < 1 to signal program termination.

If n is greater than the number of points assigned for the arrays (starting at 1 to follow Fortran usage) or if $0 < n < 4$, then a new value of n is requested. Otherwise, it's time to input the n data values to be splined. Then the second derivatives at the endpoints, sd2[1] and sd2[n], are input. If you want natural splines, just give both of these derivatives as zero. The final input quantities are the minimum and maximum x values. Since the program already knows the number of steps and they are uniform, the stepsize (h) is readily computed from xmin and xmax.

Exercise 5.5

(a) Code and compile program Cubic Splines. At this stage you may wish to bypass coding or use of the integration function SplineInt. If so, replace it by a "stub," which will either just print a message that the function was entered or will return a value of zero, or both.

(b) Run the program in order to test a variety of combinations of the control loops, such as the allowed and terminating values of n, the allowed range of x for interpolation (which is, by definition, in the interval [xmin, xmax]), and (for the same reason) the range allowed for integration. In addition, in order to simplify programming, the upper limit on an integral, xU, is not allowed to be less than the lower limit, xL, even though this is mathematically permissible and could be programmed, at some increase in coding complexity. ∎

The function SplineFit

The function SplineFit, which does the main work, is organized very much as the spline algorithm subsection in Section 5.1 describes. The only aspect of coding the algorithm that may be unusual to Fortran programmers is in computing the second derivatives of the spline. As (5.20) and (5.21) show, at this stage in the algorithm one makes a *downward* recurrence on j. Since the C language allows for loops to decrement, it is most natural to use the decrementing operation (coded as j − −). The required results can also be achieved, with great loss of clarity and naturalness, by using an incrementing loop-control variable (a DO loop in Fortran), then computing the decrementing array index from this variable. It's messy and error-prone, believe me.

We now describe the control structures for using the function SplineInterp. In Section 5.4 we discuss how to program this function. Because splines provide a means of interpolating, it is reasonable to control the interpolation loop by terminating its execution if the x value at which the spline is to be interpolated is beyond the input-data range [xmin, xmax], which is how the while loop over x in the main program is controlled. At any chosen x within this range the function SplineInterp returns the interpolated value, and also the interpolated first, second, and third derivatives. They are then printed by the main program before it requests input of another value of x.

Exercise 5.6
Modify the `Cubic Splines` program, Program 5.1, to be convenient for your computing environment. In particular, the simple but general-purpose input given there is not convenient for many interactive systems. A practical alternative is to input the data from files or from a spreadsheet. Similarly, modify the output so that it is convenient for your system and graphics software. ■

Now that we have an algorithm and an understanding of the main parts of `Cubic Splines`, we are ready to learn how to use it to interpolate, differentiate, and integrate — techniques that we develop and practice in the next three sections.

5.4   INTERPOLATING BY SPLINES

Spline fitting is a very useful technique, because it allows us to make a fit to the data which is locally smooth and that has decreasing influence from points further away. It is therefore particularly suitable for interpolation between data before using other techniques for data massaging and analysis.

Interpolating values and derivatives

To estimate values at a point x in the range x_1 to x_n, locate the appropriate interpolation index i by

$$i = \text{trunc}[(x - x_1)/h] + 1 \tag{5.29}$$

Next, find the remainder,

$$e = x - x_1 - (i - 1)h \tag{5.30}$$

The *interpolated value* is then simply obtained from

$$s(x) = y_i + \left[s_i^{(1)} + \left(s_i^{(2)}/2 + s_i^{(3)}e/6\right)e\right]e \tag{5.31}$$

The *first derivative* from the spline interpolation is obtained from (5.31) (noting that the derivative with respect to x is the same as the derivative with respect to e)

$$s^{(1)}(x) = s_i^{(1)} + \left(s_i^{(2)} + s_i^{(3)}e/2\right)e \tag{5.32}$$

The *second derivative* estimate is thence obtained from (5.32) as

$$s^{(3)}(x) = s_i^{(2)} + s_i^{(3)}e \tag{5.33}$$

Finally (for a cubic), the *third derivative* estimate is merely a constant in each interval, so that

$$s^{(3)}(x) = s_i^{(3)} \qquad (5.34)$$

and it is usually discontinuous across intervals.

Exercise 5.7

(*a*) Verify the correctness of each of the derivative formulas (5.32) – (5.34) starting with the spline equation (5.31).

(*b*) Suppose that $y(x)$ is exactly a cubic in the range of spline fitting. Verify that the spline derivatives and the derivatives of y are the same, at least analytically. ■

The C function SplineInterp

The function $\texttt{SplineInterp}$ is coded to compute formulas (5.29) through (5.34). Both the formulas and the code are written in the Horner-polynomial form (Section 4.1) for efficiency. This function also illustrates the use in C of dereferencing, so that the values of the interpolated derivative estimates, $\texttt{*sd1}$, $\texttt{*sd2}$, $\texttt{*sd3}$, are passed back through the function to the program using them. Here are some suggestions for testing the interpolation function.

Exercise 5.8

(*a*) Patch in some code in function $\texttt{yw}$ just above the $\texttt{return}$ statement. This code should look like that immediately above, but you should use formulas for a function that is a simple cubic or lower-order polynomial whose values and derivatives you can check easily and accurately by hand calculations. Then compile and run this version of $\texttt{Cubic Spline}$, including the interpolation option. The interpolated and input values should agree within computer roundoff errors. Explain why.

(*b*) A nice way to see the input values at desired x values is to give a list of zeros as input to be fitted. Then the error in the interpolated fit at the knots will just be the values of your cubic (or lower-order) polynomial. Clever, eh?

(*c*) If your test runs fail to give agreement with a cubic, drop off the cubic term contributions from the value and derivative formulas, then try again. If discrepancies beyond computer roundoff error persist, keep decreasing the order of the test polynomial. The coding bug will be in the derivative one order higher than the lowest successful order. Note carefully that you must use exact endpoint conditions for each order polynomial. ■

Once the interpolation sections of $\texttt{Cubic Spline}$ are running correctly, it is time to test them on our working function presented in Section 4.1.

Interpolating working-function values and derivatives

As our first test of spline interpolation, consider the working function (4.1), which was also considered extensively in Sections 4.1 and 4.4. Program 5.1, Cubic Splines, contains the function yw that evaluates the working function and its first three derivatives and passes them to the main program.

For testing interpolation by cubic splines, it is clear from Figures 5.1 and 5.2, which used only five knots, that reasonable accuracy of values and derivatives for such a wiggly (sixth-order) polynomial requires considerably more points, even if exact endpoint second derivatives (which are usually not known) are used, as Figure 5.2 shows. The choice of n = 9 knots is a compomise between ease of input and accuracy of interpolated values and some derivatives of the working function. The knots are chosen at x = -0.4 by steps of h = 0.1 up to x = 0.4. We interpolate values to points midway between the knots. Either exact (subscript E on the error) or natural-spline (Section 5.2, subscript N on the error) boundary conditions are used.

Table 5.1 shows the results numerically, and they are displayed graphically in Figure 5.3.

TABLE 5.1 Nine-point cubic spline input y_w, errors in interpolated values using exact endpoint second derivatives (e_E) and natural-spline endpoints (e_N).

x	$y_w(x)$	$e_E \times 10^2$	$e_N \times 10$	x	$y_w(x)$	$e_E \times 10^2$	$e_N \times 10$
−0.4	−0.4435			0.00	0.0000		
−0.35		0.59	−0.24	0.05		−0.05	−0.01
−0.30	−0.1482			0.10	−0.0529		
−0.25		0.00	0.08	0.15		−0.05	0.02
−0.20	0.0922			0.20	0.0000		
−0.15		0.06	−0.02	0.25		−0.05	−0.08
−0.10	0.1030			0.30	0.0370		
−0.05		−0.03	0.05	0.35		0.04	0.30
				0.40	−0.2246		

Notice from Table 5.1 and Figure 5.3 that the errors in fitting the function values are largest near the endpoints, where there is least information about the function. Thus, whenever reliable second derivatives are available near the endpoints they should be used. For our pathological working function and small number of spline knots, using exact endpoint conditions (e_E in Figure 5.3) results in almost one order of magnitude better fit to the function than obtained using natural endpoints (e_N).

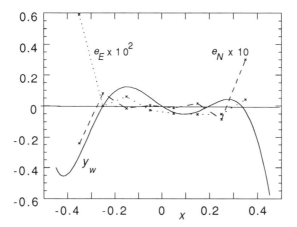

FIGURE 5.3 Spline interpolation using nine points of the working function, y_w. Scaled errors are shown for the natural-spline fit (dashed curve, e_N) and for the exact-endpoint spline fit (dotted curve, e_E).

Exercise 5.9

Run the program Cubic Splines using the values of the working function at the nine knots, as given in Table 5.1. First make a run with exact second derivatives at the endpoints, then make a second run with zero endpoint derivatives for the natural splines. Second derivatives can be obtained as suggested in Exercise 5.8 (*b*). Verify the two sets of numerical results for the errors, as given in Table 5.1. ∎

Now that we have some idea how spline fitting works for function values, let us learn how to estimate function derivatives by splines. Spline fitting provides a particularly simple, convenient, and accurate way of estimating derivatives of functions or of data, because the method is derived from considerations of matching derivatives at the knots, as discussed in Section 5.1. Notice that the interpolated third derivative will jump between adjacent knots, often quite quickly, as for our working function, since it has nonzero derivatives up through sixth order.

As our first, and very demanding, test of spline derivatives we use the working function,(4.1) or (4.2), a sixth-order polynomial whose properties are discussed extensively in Section 4.1. The coefficients for calculating its analytical derivatives are given in Table 4.1. In Section 4.4 we explored polynomial approximations for first and second derivatives of functions, including the working function. The results appear in Tables 4.4 – 4.7. We can now use the spline methods to estimate the same derivatives, so that we can judge the relative merits of the two techniques.

With the same nine-point spline as used for interpolating values, one obtains the first derivatives at the knots summarized in Table 5.2 and Figure 5.4.

TABLE 5.2 First derivatives of the working function estimated by a nine-point cubic-spline fit using either exact endpoints (errors e_E) or natural-spline endpoints (errors e_N).

x	$y_W(x)$	$y_W^{(1)}(x)$	$e_E^{(1)}$	$e_N^{(1)}$
−0.4	−0.4435	1.291	0.219	−1.668
−0.3	−0.1482	3.361	−0.086	0.420
−0.2	0.0922	1.210	0.002	−0.134
−0.1	0.1030	−0.760	−0.016	0.023
0.0	0.0000	−1.000	−0.005	−0.024
0.1	−0.0529	0.046	−0.001	0.037
0.2	0.0000	0.806	0.001	−0.134
0.3	0.0370	−0.555	0.016	0.518
0.4	−0.2246	−5.275	−0.015	−1.888

Notice in Table 5.2 and Figure 5.4 that by using exact endpoint second derivatives one obtains nearly a one order of magnitude improvement in the first derivatives at the knots, and that with both boundary conditions the accuracy of the first derivatives is much better near the middle of the spline range than near the ends.

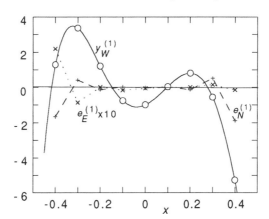

FIGURE 5.4 Spline derivatives using nine points of the working function, y_W. The exact first derivative is shown by the solid curve. Scaled errors are shown for the natural-spline derivatives (dashed curve) and for the exact-endpoint spline derivatives (dotted curve).

Exercise 5.10

Use the program Cubic Splines to make a nine-point cubic-spline fit to the data in the second column of Table 5.2, which are the working-function values at the knots. Run two sets of data, one with exact endpoint derivatives and the other with natural-spline endpoints. Verify that the errors in the first derivatives are as given in Table 5.2 and compare these derivatives graphically with those shown in Figure 5.4. ∎

You may be pessimistic about the accuracy of spline fitting, on the basis of the low accuracy of the spline descriptions of the working function (4.1). This is, however, a pathological function, with derivatives that keep on increasing through the sixth derivative. Its main use is to stress-test the methods and to make inaccuracies of approximations glaringly obvious. We now switch to a more harmonious function, the cosine, and explore how well it and its derivatives can be described by spline fitting.

Interpolating cosine values and derivatives

In Project 4A (Section 4.5) we explored the numerical derivatives of the cosine function at $x = 0$, using methods based on power series around the point where the derivatives are required. We now repeat the analysis for an extended range of x, namely 0 to $\pi/2$, and we compare the accuracy of the two methods.

Exercise 5.11

(a) Consider the cosine function in the range 0 to $\pi/2$. Use exact endpoint conditions of second derivatives equal to -1 and zero, respectively. Give the exact values of the cosine at $x = j\pi/16$, where $j = 0, 1, ..., 8$, so that there are 9 knots. Compare the spline estimates of the first and second derivatives with the analytical derivatives at the same points, $-\sin x$ and $-\cos x$, respectively. The results are listed in Table 5.3 and are displayed in Figure 5.5.

(b) Compare the spline estimates of derivatives with the results for numerical differentiation of the cosine at $x = 0$ obtained by the methods in Section 4.5. Those estimates of the first and second derivatives are given in Table 4.8. ■

Table 5.3 Cubic-spline 9-knot fit to the cosine function between $x = 0$ and $x = \pi/2$ with exact endpoint conditions. The errors in first derivatives have superscript (1) and in second derivatives the superscript is (2).

$16x/\pi$	$y(x)$	$e_E^{(1)} \times 10^4$	$e_E^{(2)} \times 10^2$
0	1.0	0.00	0.00
1	0.980785	-0.52	0.40
2	0.923880	0.08	0.28
3	0.831470	-0.06	0.27
4	0.707107	-0.03	0.23
5	0.555570	-0.05	0.18
6	0.382683	-0.08	0.12
7	0.195090	-0.09	0.06
8	0	-0.01	0.00

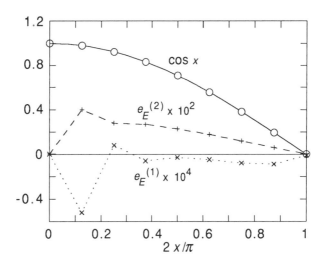

FIGURE 5.5 Spline interpolation using nine points of the cosine function, as shown. Scaled errors in first derivatives (dotted curve) and in second derivatives (dashed curves) are shown for the exact-endpoint spline fit.

The standard deviation of the errors in the first derivatives in Table 5.3 is 2×10^{-5}, which is about the same as the error in the cosine first derivative from the three-point central-derivative method at $x = 0$, given in Table 4.8 for a stepsize $h = 0.2$, which is about the spacing between the data ($\pi/16 = 0.196350$) given in Table 5.3. Similarly for the errors in the second derivatives, the standard deviation of these in Table 5.3 is 2×10^{-3}, comparable to that for the three-point method at $x = 0$. Overall, a cubic-spline fit of a well-behaved function such as a cosine (a rapidly convergent series, as explored in Section 3.2) is three orders of magnitude more accurate than a fit to our badly behaved working function considered in the preceding subsection.

We have also discovered that derivatives computed using cubic splines can easily be of comparable accuracy to those obtained from the power-series methods in Sections 4.4 and 4.5. Although the spline fitting initially requires more number crunching than do the power-series methods, once the spline coefficients have been found the computation of derivatives is very speedy.

The unstable problem (Section 4.3) of differentiation solved by an unstable method involving subtractive cancellation has apparently been bypassed. This merit of splines is more apparent than real because the spline-algorithm formulas (5.15) – (5.18) show that we must take first and second differences of the y_j; therefore, subtractive-cancellation errors are possible. However, such errors are less likely with splines because the spline procedure does not involve (either explicitly or implicitly) finding a limiting value of the differences as the spacing between the knots decreases to zero.

5.5   INTEGRATION METHODS USING SPLINES

Even though spline fitting usually emphasizes interpolation properties (as in Section 5.4), splines are also very useful for integration, which is especially straightforward for the cubic spline. In this section we derive the algorithm for integration by cubic splines with equal spacing of the knots. We then test the formulas and apply them to the working function and to the cosine. This allows us to compare the relative accuracy of splines with the trapezoid and Simpson formulas in Section 4.6.

Deriving the integration algorithm

Equation (5.31) shows that the spline fit in the jth interval, $s(x)$, can be expressed as

$$s_j(x) = y_j + s_i^{(1)}e + s_j^{(2)}e^2/2 + s_j^{(3)}e^3/6 \qquad (5.35)$$

where e is given by (5.30). Therefore, in order to integrate the spline fit from x_L to x_U we need only integrate $s(x)$ over each subinterval, then sum the results. The indefinite integral of s_j for any x in the range x_j to x_{j+1} is

$$I_j(x) = \int dx'\ s(x') = (y_j + (s_j^{(1)}/2 + (s_j^{(2)}/6 + s_j^{(3)}e/24)e)e)e \qquad (5.36)$$

In this formula we have expressed it efficiently in the Horner-polynomial algorithm form (Section 4.1), at the expense of loss of clarity in the typography. An estimate of the total integral from x_L to x_U that involves no further approximation can therefore be obtained by summing the partial integrals given by (5.36).

Note several improvements of cubic spline integration over the trapezoidal or Simpson rules in Section 4.6, namely:

- Within each interval the spline uses a higher-order polynomial — a cubic rather than linear over the interval (trapezoidal rule) or quadratic over double intervals (Simpson rule).

- Spline fitting provides smooth connections between intervals: there are no such constraints for the other methods.

- Integration with splines uses approximations that are consistent between interpolation and estimation of derivatives. So, if these operations are also required, splines may be the method of choice.

The cost of spline methods compared with the other integration methods is that the initial fitting algorithm is more time-consuming. This may not be a penalty overall, since for a specified accuracy of the integral the stepsize h can usually be chosen much larger in this method than in the others, resulting in fewer y_i values being needed. Therefore, if the y_i are costly to obtain, either as experimental data or as computed values, spline methods will often be preferred. Such methods have been under extensive development recently, as described in Chapter 2 of the compendium on numerical-integration methods by Davis and Rabinowitz, which also gives references to appropriate computer programs.

The C function for spline integration

Now that we have an algorithm for integration by using cubic splines, we describe the functions SplineInt, Horner4Poly, and ywInt in program Cubic Splines (Program 5.1). Equation (5.36) gives the integral from the start of the *j*th segment to a distance *e* into the segment. A direct way to combine the integrals from several segments is indicated in Figure 5.6.

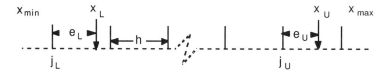

FIGURE 5.6 Segments and partial segments for the spline integration.

The integration segments are combined as follows. In the notation used in Figure 5.6 and in function SplineInt, the integration is to be from x_L to x_U, and the main program has already checked that these lie within the input range [xmin,xmax] and that x_U is not less than x_L. The function therefore first computes the integer knot positions j_L and j_U and the extra pieces e_L and e_U, both of which are positive, because conversion to integers is done by truncation. The next step is to make the sums over the whole strips, each of which is of the same stepsize h. It is therefore most efficient (for both programmers and computers) first to combine coefficients of the powers of h, then to use the Horner polynomial algorithm to compute the integral contribution, called splint in SplineInt. Then the function *adds* the contribution above j_U (which is of length e_U) and *subtracts* the contribution from the lower segment of length e_L to obtain the complete integral.

The function Horner4Poly is a customized version of the algorithm described and programmed in Section 4.1. It is adapted to express the integral of a polynomial in terms of its value at the start of the interval of integration (y), the first through third spline derivatives (s1, s2, s3), and the interval e, according to (5.36). The last function needed for tests of integration is ywInt, the comparison analytical formula for the working-function indefinite integral, obtained from (4.4) and Table 4.1.

Exercise 5.12

Code the integration routines, following Program 5.1. Then test the routines as follows. First, use a constant input function of value equal to unity, and as many knots as you wish. The integral should be equal to the length of the interval (even if this length is not a integer number of steps h) and it should be exact to within roundoff errors. If this test is successful, use any cubic polynomial and give the exact endpoint second derivatives. Again results should agree with analytical integrals to within roundoff. ■

With coding and testing completed, it is time to use the spline-integration function.

Integrating the working function and cosine

We investigate two examples of integrating by cubic splines, the pathological working function and the well-behaved cosine. Both of these functions are integrated by the trapezoid and Simpson methods in Section 4.6, where the results are given in Tables 4.9 – 4.11. Here we repeat some of these integrals so that we can compare the accuracy and convenience of the algorithms. On the basis of our discussion and experience in Chapter 4 that numerical integration should be more accurate than numerical differentiation, we may expect agreeable results.

In Table 5.4 we show the results of integrating the working and cosine functions using the nine-knot cubic-spline fits that are shown in Figures 5.3 and 5.5, respectively.

TABLE 5.4 Nine-knot cubic-spline integrals for working function and cosine with indicated endpoint conditions. The analytical value of the working-function integral is 9.9109×10^{-3}.

Function	Endpoint conditions	x_L	x_U	$\int dx \ s(x)$	Error $\times 10^3$
$y_w(x)$	Exact	−0.2	0.2	9.94779×10^{-3}	−0.0369
	Natural	−0.2	0.2	9.94862×10^{-3}	−0.0378
$\cos(x)$	Exact	0	$\pi/2$	$1 - 3 \times 10^{-6}$	0.003
	Natural	0	$\pi/2$	$1 - 1.8 \times 10^{-4}$	0.18

Exercise 5.13

Run Cubic Splines (Program 5.1) as a nine-point fit from –0.4 to 0.4 to check out the numerical results in Table 5.4 for the working function, y_w, and for the cosine function. Use both exact and natural-spline endpoint conditions. Choose the integration ranges as shown to allow comparison with results for the trapezoidal and Simpson rules in Section 4.6. ■

By comparing the spline-integration errors for the working function in Table 5.4 with those in Tables 4.9 and 4.10 for the trapezoid and Simpson methods (errors of 0.2989×10^{-3} and 0.1669×10^{-3}, respectively, for stepsize $h = 0.1$), we see that for either endpoint condition the spline method is a factor of 5 more accurate than Simpson's rule and a factor of nearly 10 better than the trapezoid rule. Admittedly there is much more arithmetic involved in the spline fitting, but it will also allow interpolating, estimating derivatives, as well as integrating.

We have now learned how to fit curves through data using a method that produces fits that are locally very smooth, and that allows interpolation, estimation of derivatives, and integration to be carried out reliably. With all this power of spline fits, it is interesting to enquire why such methods have not been emphasized as part of the repertoire of numerical methods.

5.6   DIVERSION: COMPUTERS, SPLINES, AND GRAPHICS

The practical applications of curve fitting by splines have been closely related to the development of computers. To understand why, notice that the five steps involved in spline fitting require a computing device with significant random-access memory, because for an n-point spline about $9n$ quantities are computed, and more than half of these must be available after the fourth step. Further, unlike polynomial interpolation formulas, which have coefficients depending only upon the order of the polynomial assumed, in spline fitting the last four steps must be completely recomputed whenever even a single function value, y_j, is changed. In hand calculations, using pencil and paper, memory is at a premium and frequent re-recording of intermediate steps is to be avoided.

Before electronic computers, the large number of arithmetic operations required in spline fitting was a severe hindrance, since any error propagates through the calculation in a complicated way that is difficult to correct. Spline methods were therefore seldom used for practical work until computers were widely available and convenient to use.

Historically, splines were given most attention in pure mathematics, until about 1960 when using digital computers became practical and common. Thus, the formal mathematics of splines is very well developed, as exemplified in the monographs by Schultz and by Schumaker. The practical applications of splines are still under intensive development, as described in the influential book by De Boor, which contains extensive Fortran programs. The near disjointness of the two sets of literature on splines, in pure mathematics and in computing, makes their juxtaposition on library bookshelves confusing if not amusing. A history of the development of spline methods is given in Schumaker's book.

Computer graphics and computer-aided design make extensive use of a variety of splines in one or more variables. Splines for interpolating and smoothing curves and surfaces are described in Chapter 11 of the treatise on computer graphics by Foley et al., in Farin's practical guide to curves and surfaces for computer-aided geometric design, and in the extensive treatment by Bartels et al. Fast interpolation formulas for cubic splines with Bézier control points is described in the article by Rasala. A treatment of these topics that is more mathematically oriented is provided in the book by Lancaster and Salkauskas. For representing surfaces in three dimensions one requires so-called bicubic splines with two independent variables, or splines with other properties than those we have developed. The design and description of type fonts for printing, a specialized application of two-dimensional graphics, also makes use of spline curves and their variants, as you can read in Chapter 4 of Rubinstein's book on digital typography.

By contrast with splines, Fourier expansion techniques (such as we present in Chapters 9 and 10) started in the early 1800s in applied mathematics and were only later developed in mathematical rigor and generality. From the two examples of splines and Fourier expansions we see that the three disciplines of mathematical analysis, computing, and applications form a research triangle, each contributing to the development of the others.

A lesson to be learned from the history of splines is that, generally speaking, numerical methods developed in the days of laborious hand calculation will often not be optimal for computer calculation. Older numerical methods usually emphasize minimum storage, easy error detection and correction, and minimal logical complexity. On the other hand, digital-computer methods may be relatively profligate of storage, they should be numerical robust because the intermediate results are seldom inspected, and they may have quite involved logic – provided that it can be rigorously checked to ensure correctness.

REFERENCES ON SPLINE FITTING

Abramowitz, M., and I. A. Stegun, *Handbook of Mathematical Functions*, Dover, New York, 1964.

Bartels, R. H., J. C. Beatty, and B. A. Barsky, *An Introduction to Splines for Use in Computer Graphics and Geometric Modeling*, Morgan Kaufmann, Los Altos, California, 1987.

Davis, P. J., and P. Rabinowitz, *Methods of Numerical Integration*, Academic Press, Orlando, Florida, second edition, 1984.

De Boor, C., *A Practical Guide to Splines*, Springer-Verlag, New York, 1978.

Farin, G., *Curves and Surfaces for Computer Aided Geometric Design,* Academic Press, San Diego, California, 1988.

Foley, J. D., A. van Dam, S. K. Feiner, and J. F. Hughes, *Computer Graphics*, Addison-Wesley, Reading, Massachusetts, third edition, 1990.

Lancaster, P., and K. Salkauskas, *Curve and Surface Fitting*, Academic Press, New York, 1986.

Rasala, R., "Explicit Cubic Spline Interpolation Formulas," in A. S. Glasner, Ed., *Graphics Gems*, Academic, Boston, 1990, pp. 579 – 584.

Rubinstein, R., *Digital Typography*, Addison-Wesley, Reading, Massachusetts, 1988.

Schultz, M. H., *Spline Analysis*, Prentice Hall, Englewood Cliffs, New Jersey, 1973.

Schumaker, L. L., *Spline Functions: Basic Theory*, Wiley, New York, 1981.

Chapter 6

LEAST-SQUARES ANALYSIS OF DATA

In Chapters 4 and 5 we considered derivatives, integrals, and curve fitting of functions such that at each $x = x_j$ the fitted values are required to agree exactly with the given $y(x_j) = y_j$. By contrast, in applications the y_j are usually data from measurements with associated uncertainties that should influence the fit made to them. Therefore, the most realistic fitted curve to the whole set of data may not necessarily pass through each y_j. For example, if we expect that the data would lie on a straight line in the absence of any measurement uncertainty, then a fit that passes near the data but not through them may be acceptable.

The emphasis in this chapter is therefore on analyzing data in which we take into account some sources of data uncertainty, namely random errors in measurement. We therefore discuss criteria for a best fit (Section 6.1), the use of orthogonal functions in least-squares fitting (Section 6.2), and we derive formulas for straight-line least squares when both variables have errors (Section 6.3). In Section 6.4 we pose and solve the important problem of normalization factors determined from a least-squares fit. Logarithmic transformations of data are very common in science and engineering, and Section 6.5 shows how such a transformation biases the fitted parameters and how such bias may often be removed. At last, in Project 6 (Section 6.6) there is a program for straight-line least squares, based on the formulas in Section 6.3, as well as suggestions for using the program. We round out the chapter with references on least-squares analysis.

Background to this chapter, with emphasis on the underlying statistics and probability ideas is found in the books by Barlow, by Siegel, and by Snell. Required reading for all those who would use (or misuse) statistics is the book by Jaffe and Spirer. The monograph by Draper and Smith on applied regression analysis has several interesting examples and interpretations. Many basic techniques for error analysis in the physical sciences are described in the books by Taylor and by Lichten, as well as in the practical guide by Lyons. For topics in advanced physics, especially high-energy physics, the book by Lyons is suitable.

6.1 INTRODUCTION TO THE LEAST-SQUARES CRITERION

If we give up the condition that the fitted function must pass through each data point, then we must decide on objective criteria for a best fit. Let the fitting function be $Y(x)$. Suppose that our best-fit criterion were to minimize the (signed) difference between the y data, y_j, and the fitting values $Y_j = Y(x_j)$.

Exercise 6.1
Prove that if the signed difference between data and fitting function is to be minimized, then the best-fit function is just $Y_j = \bar{y}$, the average of the y data, the same value at all points. ■

The average of a set of data values is not, however, usually a good representative of the behavior of the data. For example, the annual *average* temperature in the town of Yakutsk on the Lena River in Siberia is a comfortable 10°C, but with a range from –40°C to +20°C.

Thus, given that signed differences are usually inappropriate, how about absolute values of differences? Unfortunately, this fitting criterion is fraught with difficult mathematical analysis, since the fitting methods are based on derivatives and the absolute value function is singular at the origin, which is the point that a perfect fit would reach. We therefore turn to a reasonable alternative, namely minimizing sums of differences squared.

Minimization of a sum of squares of differences between data and fitting function is called a *least-squares fit*. The method was developed by Gauss and Legendre in the early nineteenth century in their analysis of astronomical orbits from imprecise observations. Gauss also related the least-squares principle to the theory of probability, as we now outline in the context of *maximum-likelihood* analysis.

Maximum likelihood and least squares

We first introduce the Gaussian probability distribution, which is often called the *normal distribution* in statistics. Under quite general and plausible conditions (see, for example, Snell's book) the probability that the value of variable v is obtained if it has true average value $\bar{v}$ is $P(v)$, given by

$$P(v) = \frac{1}{\sigma\sqrt{2\pi}} \exp\left[-\frac{(v-\bar{v})^2}{2\sigma^2}\right] \tag{6.1}$$

Here σ is the standard deviation of v, with the mean-square deviation of the v values from $\bar{v}$ is σ^2. Also, $P(v)$ is normalized so that the total probability over all v values is unity. Strictly speaking, P is a probability density. In accord with common practice, we will use the term probability. One way to characterize a probability distribution, $P(v)$, is to specify its moments, which are the values of the positive integer powers of v, weighted by $P(v)$, then averaged over all values of v. Thus, for a continuous distribution (such as the Gaussian) the nth moment about the mean, M_n, is given by

$$M_n = \int_{-\infty}^{\infty} dv \; v^n P(v) \tag{6.2}$$

in which the origin of v has been moved to coincide with its average value. Notice that if the probability distribution is symmetric, which is quite common, then all the moments for odd values of n must be zero; you don't need to attempt the integration to see this. For the even moments, there are often clever methods of calculating them, as suggested in Exercise 6.2 (c). There are several interesting and relevant properties of the Gaussian distribution, which you may like to derive in the following exercise.

Exercise 6.2

(a) In order to calculate the mean-square deviation (the second moment) of the Gaussian, consider the value of $v^2 P(v)$ integrated over v from $-\infty$ to ∞. By using the method of parametric integration (described in many calculus books), show that this integral is σ^2, as claimed below (6.1).

(b) Show that the Full Width at Half Maximum (FWHM) of the Gaussian distribution, the range of v over which it goes from half maximum through maximum to half maximum again, is $v_{FW} = \sigma \sqrt{8 \ln 2} = 2.355\sigma$.

(c) Use the technique of parametric integration to derive a formula for the even moments of the Gaussian distribution, namely the integrals of $v^{2n} P(v)$ with n a positive integer. ■

The Gaussian distribution, with its FWHM, is illustrated in Figure 6.1.

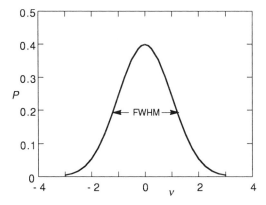

FIGURE 6.1   The Gaussian probability distribution with unity total probability and unity standard deviation.

Now that we have some understanding of a probability distribution that will be associated with errors in data, let us apply it to relate random errors in x and y data to the derivation of a fitting criterion. Suppose that at the jth data point a measurement gives values x_j and y_j with standard deviations σ_{xj} and σ_{yj}, respectively. We also suppose that their average values over many measurements would be X_j and Y_j, respectively, just the fitting values of x and y that we would like to know. Thus, we assume that there are no systematic errors and that the data, if free of their random errors, could be exactly describable by the fitting function. In the real world, it is seldom true that either assumption is likely to be strictly justifiable. By assuming further that uncertainties in x and y measurements are independent, the probability of obtaining such values is obtained as P_j, given by

$$P_j = P(x_j)\,P(y_j) \tag{6.3}$$

What is the likelihood, L, that n independent measurements, $j = 1, 2, ..., n$, would produce the data actually obtained? This likelihood is just

$$L = \prod_{j=1}^{n} P_j \tag{6.4}$$

Exercise 6.3
Show that *maximizing* the likelihood function L in (6.3) for Gaussian distributions of x_j and y_j values is equivalent to *minimizing* the χ^2 function, given by

$$\chi^2 = \sum_{j=1}^{n} \left[(x_j - X_j)^2 / \sigma_{xj}^2 + (y_j - Y_j)^2 / \sigma_{yj}^2 \right] \tag{6.5}$$

if it assumed that the standard deviations are not fitting parameters. ∎

Thus, we have derived the least-squares criterion, minimization of the sum of squares in (6.4), from the assumptions of independent measurements from point to point and independent Gaussian-distributed random errors at each point. Note that both x and y variables may have errors, so that we do not make any distinction between independent and dependent variables. This is a common situation in the life sciences and in astronomy, where pairs of observations are associated. In physics and engineering one can often control x relatively well ($x_j \equiv X_j$), so that it becomes the independent variable, and χ^2 in (6.4) collapses to the y-dependent terms. For generality, we retain the full form (6.4) in the following. The relation between probability distributions and random errors ("stochastic processes") is developed in Chapter 10 of Solomon's book.

One disadvantage of a least-squares best-fit criterion arises from data with relatively small errors that, if not included in the fit, would lie far from the best-fit function value. When included in the fitting, such "outliers" may have a very strong influence on the fitting function because they contribute as the *square* of their distance from it. Although this is as it should be for independent random Gaussian errors that are correctly estimated, real data and their errors may not satisfy all the conditions discussed above that are necessary to relate χ^2 to the maximum-likelihood criterion. Therefore, the presence of outliers are usually considered quite problematic in least-squares fitting.

Least squares and the objective function

Although the maximum-likelihood model for normally-distributed errors leads to a least-squares condition when the weights are chosen as the inverses of the data variances, this is not the only justification for a least-squares criterion. For example, the Gaussian error distribution is not appropriate when the errors arise from counting statistics. Then one has Poisson error distributions (Chapter 3 in Barlow, Chapter 6 in Solomon), for which an estimate of the standard deviation for a mean value of y counts is $\sqrt{y}$. I have discussed elsewhere (Thompson, 1992) Poisson statistics in the context of finding least-squares normalization factors.

In many experiments one adjusts the measurement procedure so that the statistical uncertainties are proportional to the measured values, that is, one has constant percentage errors. Under this condition the analysis often becomes tractable, as we find for the analysis of parameter bias in logarithmic transformations in Section 6.5.

The function to be minimized is therefore usually generalized from χ^2 in (6.4) to the *objective function*, O, defined by

$$O = \sum_{j=1}^{n} \left[w_{xj} (x_j - X_j)^2 + w_{yj} (y_j - Y_j)^2 \right] \tag{6.6}$$

in which w_{xj} and w_{yj} are weights assigned to the jth x and y points, respectively. The maximum-likelihood formula (6.4) is recovered if the weights are equal to the inverses of the squares of the standard deviations. Only then is $O = \chi^2$ the function occurring in statistics to which confidence limits (C.L.) can be assigned. If weights other than inverses of error variances are used, then any overall scaling factor applied to them or to O does not affect the values of X_j and Y_j at which the minimum will occur. The statistical significance of the value of the objective function depends strongly on how relative weights are assigned, as well as on their overall magnitude. This fact is often ignored by scientists and engineers, especially by physicists.

Because we have gone beyond the probability origin for least-squares fitting, alternative motivations for using a least-squares criterion are worth discussing, as we do in the next section.

6.2 ORTHOGONAL FUNCTIONS AND LINEAR LEAST SQUARES

The general least-squares condition, minimization of the objective function O defined by (6.5), serves as the starting point for least-squares fitting using common functions. Of greatest practical importance are *orthogonal functions*. They also connect least-squares fitting with the Fourier expansions discussed in Chapters 9 and 10. In this section we introduce orthogonal functions, then show their connection to least-squares fitting.

What are orthogonal functions?

Orthogonal functions are sets of functions such that with weight factor w_j and summation over the n observation points labeled by j, any two functions in the set are constrained by

$$\sum_{j=1}^{n} \phi_K(x_j) \, \phi_L(x_j) w_j = 0 \qquad K \neq L \qquad (6.7)$$

For example, in Fourier expansions the $\phi_K(x)$ are the complex-exponential functions e^{iKx}, or cos (Kx) and sin (Kx), which can all be made orthogonal over the x_j range if the x_j are equally spaced and if the weights are all equal. You are probably already familiar with orthogonality in the context of vector geometry. To understand the analogy, work the following exercise.

Exercise 6.4
Suppose that **k** and **l** are two three-dimensional vectors with components expressed in Cartesian coordinates.

(*a*) Write down, in a form analogous to (6.6), the orthogonality condition for these two vectors.

(*b*) If **k** is a fixed vector, what is the set of all vectors orthogonal to **k**? ■

In the following subsection we relate this orthogonality property to the key concepts of least-squares fitting.

For linear least-squares fits with polynomials it is advantageous that the polynomials be orthogonal, but this property does not hold if just simple powers of the independent variable x are used. You can investigate this problem and its cure in the following exercise.

Exercise 6.5
(*a*) Make a sketch of the powers of x, $\phi_L(x) = x^L$, for x between -1 and $+1$, then use this to indicate why these polynomials cannot be orthogonal over this range if positive weights are assumed.

(*b*) Generate polynomials of orders $L = 0,1,2,...$ by the following recurrence relations, called the *Schmidt orthogonalization procedure*:

$$\phi_0(x) = 1 \qquad (6.8)$$

$$\phi_{L+1}(x) = (x - \alpha_{L+1})\phi_L(x) - \beta_{L+1}\phi_{L-1}(x) \quad L = 1,2,... \tag{6.9}$$

where the coefficients are

$$\alpha_{L+1} = \sum_{j=1}^{n} x_j \, [\phi_L(x_j)]^2 w_j / S_L \tag{6.10}$$

with the denominator sum given by

$$S_L = \sum_{j=1}^{n} [\phi_L(x_j)]^2 w_j \tag{6.11}$$

and the second set of coefficients is given by

$$\beta_1 = 0 \quad \beta_{L+1} = S_L/S_{L+1} \tag{6.12}$$

The normalization scale of these polynomials is such that the power of x^L has unit coefficient.

(c) Prove that the polynomials generated in (b) are orthogonal by applying, for a given L, the method of induction over I to the sequence of sums over the data

$$\sum_{j=1}^{n} \phi_{I+1}(x_j)\phi_L(x_j)w_j = 0 \tag{6.13}$$

Thus we have a general procedure for generating orthogonal polynomials, with their coefficients depending on the weights associated with each datum. ∎

This result may seem rather abstract, but it may surprise you that it provides a straightforward route to the Legendre polynomials, a favorite set of fitting functions, especially in problems which involve three-dimensional polar coordinates, where usually $x = \cos\theta$ and θ is the polar angle. Another reason for using Legendre polynomials is that, with a suitable choice of weight factors and on replacement of summation by integration in the above equations, they form orthogonal polynomials suitable for linear least squares, as you can readily discover.

Exercise 6.6

(a) Show by generating the lowest-order orthogonal polynomials having $L = 0,1,2$, that if the weight factors are all unity and if the orthogonality summation over data values x_j is replaced by integration over x from -1 to $+1$, then $\phi_L(x)$ can be taken to be the Legendre polynomial of order L, $P_L(x)$, namely

$$\phi_0(x) = P_0(x) = 1$$
$$\phi_1(x) = P_1(x) = x$$
$$\phi_2(x) = P_2(x) = (3x^2 - 1)/2$$

(6.14)

The standard normalization of the Legendre polynomials is $P_L(1) = 1$.

(b) Verify the approximate correctness of the orthogonality condition (6.13) by using some Legendre polynomials in the numerical-integration programs in Section 4.6 to integrate products of the functions over the range -1 to $+1$. Notice that if you use the trapezoid rule for integration (Section 4.6), then the summation and integration are equivalent, if the x_j are equally spaced and if end-point corrections are negligible. ■

Orthogonal functions are much used in analysis and applications. In particular, if the orthogonal functions are polynomials (such as those of Chebyshev, Hermite, Laguerre, Legendre, and other ancient heroes), a general classification of their properties can be made. An extensive and accessible treatment of orthogonal polynomials, including applications to numerical analysis, partial differential equations, and probablity theory and random processes, is provided in the monograph by Beckmann.

Orthogonality and least squares

Suppose that all the parameters to be determined, a_L with $L = 1,2,...,N$ (if there are N parameters), appear linearly in the definition of the fitting function, Y, as

$$Y(x) = \sum_{L=1}^{N} a_L \phi_L(x) \qquad L = 1,2,...,N$$

(6.15)

We call the process by which the a_L in (6.15) are adjusted so as to minimize the objective function in (6.6) a *linear-least-squares fit*. This is to be distinguished from the more-restrictive least-squares fit to a straight line (Section 6.3), in which $Y(x) = a_1 + a_2 x$, so that $\phi_1(x) = 1$ and $\phi_2(x) = x$.

To find the fitting parameters that minimize O requires that the derivative of O with respect to each of the a_L be zero. In the situation that the x_j are precise, this requires from (6.6) that

$$0 = \sum_{j=1}^{n} \left[y_j - \phi(x_j) \right] \phi_L(x_j) w_j$$

(6.16)

By inserting the linear expansion (6.15) into this equation we get

$$\sum_{K=1}^{N} a_K \sum_{j=1}^{n} \phi_K(x_j) \phi_L(x_j) w_j = \sum_{j=1}^{n} y_j \phi_L(x_j) w_j \qquad L = 1,2,...,N$$

(6.17)

This is a set of N linear equations for the coefficients a_K that minimize the objective function. In general, these equations have to be solved by matrix methods. Further, adjustment of one coefficient propagates its influence to the values of all the other coefficients.

The use of orthogonal functions for the $\phi_L(x_j)$ greatly simplifies the solution of the equations (6.17), as you may immediately prove.

Exercise 6.7
Show that if the functions $\phi_L(x_j)$ satisfy the orthogonality condition (6.13), then the left-hand side of (6.17) collapses to a single nonzero term for each L, resulting in an immediate formula for the linear-least-squares coefficients a_L, namely

$$a_L = \sum_{j=1}^{n} y_j \phi_L(x_j) w_j \Big/ \sum_{j=1}^{n} \phi_L^2(x_j) w_j \qquad L = 1,2,...,N \qquad (6.18)$$

so that each coefficient is obtained by an independent calculation, but all coefficients are interrelated through the data values y_j and their weights w_j. ■

Thus, if appropriate orthogonal functions can be found, their use greatly simplifies the least-squares fitting, because each coefficient is found independently of the others. For example, if a different number of fitting functions is decided upon (for example if N is increased) there is no need to recalculate all the coefficients, because formula (6.18) does not depend upon the value of N. The quality of the fit, however, does depend on N. Because of this linear independence of the fitting coefficients, each of them can be uniquely associated with its corresponding function. For example, in Fourier expansions, if x is the time variable, then the a_L are the amplitudes of the successive harmonics of L times the fundamental frequency.

One must be careful with the application of orthogonal functions when weights are involved, for the following reason: A given type of function is orthogonal for a set of x_j only for a particular choice of weights. For example, the cosine and sine functions for Fourier expansions (Chapters 9 and 10) are orthogonal over the range 0 to 2π if the weight factors are unity (or constant). In data analysis such a choice of weights usually conflicts with the probability-derived weights discussed in Section 6.1, in which the weights are inversely proportional to standard deviations of measurements. A common compromise is to use (6.13) in formal work and in situations where the functions are orthogonal with weight factors of unity, but to use the general linear-least-squares formula when other weighting schemes or choices of x_j are made. Unpredictable results will be obtained if formulas derived from different weighting models are combined.

With the above general background, which also serves as the foundation for Fourier expansions in Chapters 9 and 10, we are ready to study the special case of straight-line least squares, which is of very common use in scientific and engineering applications.

6.3 ERRORS IN BOTH VARIABLES: STRAIGHT-LINE LEAST SQUARES

In least-squares fitting models there are two related components that are usually discussed in only a cursory way. The first component is a model for the errors in the variables, and the second component is a model for the weight that each datum has in the fitting. In Section 6.1 we indicated in the context of maximum likelihood the connection between a possible error model (Gaussian distributions of independent errors from point to point) and the weighting values in (6.6). The subject of least-squares fitting when both variables contain errors is perennially interesting, with many pitfalls and possibilities. They have been summarized in an article by Macdonald and Thompson.

In the following we discuss a variety of weighting models for straight-line least-squares fitting, we then particularize to the case in which there is a constant ratio of x-data weights to y-data weights from point to point, then we derive some interesting yet useful symmetry and scaling properties for this weighting model.

Weighting models

The ideas behind generalized weighting models can be illustrated by a mechanical analogy that will be very familiar to readers with a background in physics or engineering. Suppose that we are making a straight-line least-squares fit, so that in (6.6) the X_j and Y_j define the best-fit straight line. If we literally hung weights w_{xj} and w_{yj} at the data points (x_j, y_j), then the objective function (6.6) would be proportional to the moment of inertia of the distribution of mass about the line defined by the (X_j, Y_j). The best-fit straight line would then be that which minimizes the moment of inertia about this line.

As an example of weighting of both x and y data, we choose the four (x, y) data pairs in Table 6.1, which apparently have large errors, so that the distinction between the various weighting models is accentuated.

TABLE 6.1 Data and weights for straight-line least squares with weighting model IDWMC, which has a constant ratio of x-data weights to y-data weights, λ, from point to point, as shown in Figure 6.2. Here $\lambda = 0.5$.

x data	y data	x and y weights for the IDWMC	
x_j	y_j	w_{xj}	w_{yj}
1	1	2	4
2	5	2.5	5
4	3	3	6
6	5	1	2

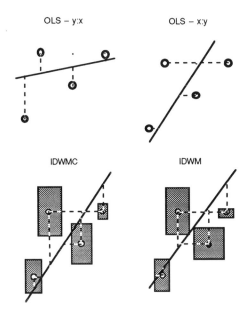

FIGURE 6.2 Various weighting schemes for least-squares fitting, shown for fits to a straight line. The conventional fitting method is OLS – y:x with y weights (dashed verticals) only. For x weights only (dashed horizontals) the fitting method is OLS – x:y. For weights on both x and y data that are in a constant ratio from point to point the fitting method is IDWMC. If the weight ratio is not constant, then we have IDWM.

Figure 6.2 shows graphically the ideas behind different weighting schemes. In order to motivate your interest in proceeding with the tedious analysis that follows, we first explain this figure, then we tackle the algebra. The best-fit straight lines shown are obtained using Program 6.2 developed in Section 6.6. We now interpret the various weighting schemes shown.

In Figure 6.2 *ordinary least squares* (OLS – y:x) is conventional least squares, in which the x values are assumed to be known precisely ($x_j = X_j$), so that the weighting is applied in the y direction only. For the four data illustrated this produces a fairly flat line if all the y data have equal weights. Alternatively, if the y values are assumed to be precise, then the weighting is in the x direction only, and a much steeper line results, as shown by OLS – x:y in Figure 6.2.

The most-general possibility for a weighting model is that the weights from point to point are independent and not simply related: We call this the *independent diagonal weighting model* (IDWM), using the nomenclature in Macdonald and Thompsons' article. To illustrate it we make the length of the weighting rectangle at each datum proportional to the weight w_{xj} and the height of the rectangle proportional to the weight w_{yj}. A general method for solving such a least-squares problem approximately was given by Deming, and an exact method for straight-line fits was described by York and by Reed.

Constant ratio of weights

Within IDWM there is an especially simple weighting model for which the ratio of x to y weights is *constant* from point to point. We call this IDWMC, with the C denoting a constant weight ratio, as illustrated in Figure 6.2. We express this as follows:

$$\lambda = \sigma^2(y_j)/\sigma^2(x_j) = w_{xj}/w_{yj} \tag{6.19}$$

Thus, for the IDWMC data in Table 6.1, $\lambda = 0.5$, because the x weights are half the y weights and this ratio of weights is constant for all four data points. We now derive the formula and some interesting results for straight-line fits in the IDWMC.

Write for the straight-line least-squares fit the model function as

$$Y = a_1 + a_2 X \tag{6.20}$$

and consider in (6.6) the contribution to the objective function from uncertainies in both the x_j and the y_j. There are two contributions to the error in the observed y_j. The first is from the uncertainty in measuring the y value, $\sigma(y_j)$, and the second is from the uncertainty in the x value, $\sigma(x_j)$. For independent errors the total uncertainty at the jth point has a standard deviation

$$\sigma_j^2 = \sigma^2(y_j) + a_2^2 \sigma^2(x_j) \tag{6.21}$$

It is therefore appropriate to use the inverse of this quantity for the weight w_{yj} in (6.6). With this method of setting up the objective function, O, the $(x_j - X_j)^2$ terms in (6.6) contribute just a constant, $\sigma(x_j)^2$, to the objective function, and do not affect its minimization. Thus, the quantity to be minimized is the y-data contribution to the objective function, O_y, given by

$$O_y = \sum_{j=1}^{n} O_{j\perp}^2 \tag{6.22}$$

where the contribution from the jth point is

$$O_{j\perp}^2 = \frac{O_j^2}{1 + a_2^2 \sigma^2(x_j)/\sigma^2(y_j)} \tag{6.23}$$

and the squared and weighted difference between data and prediction at this point is

$$O_j^2 = (y_j - Y_j)^2 w_j \tag{6.24}$$

in terms of the predicted value at the jth point

$$Y_j = a_1 + a_2 x_j \qquad (6.25)$$

and the weight coming from the variance in the y data at the jth datum

$$w_j = 1/\sigma^2(y_j) \qquad (6.26)$$

In this weighting model the contribution to the objective function from the data that are not modeled (here the x_j) is just the total number of data points, n.

Exercise 6.8
Verify each of the steps in the derivation of (6.22) for the objective function in the IDWMC model (Figure 6.3), starting with the objective function (6.6) and going through the steps from (6.19) onwards. ∎

All this algebra has a purpose, which becomes clearer when you look at Figure 6.3. Namely, the least-squares minimization that we are performing corresponds to minimizing the sums of the squares of the *shortest* distances (perpendiculars) from the data points to the best-fit straight line, as we now demonstrate.

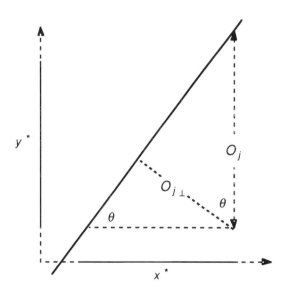

FIGURE 6.3 Geometry of the IDWMC model. The local coordinates are the x and y data divided by their errors. The perpendicular distance from the data values (x_j, y_j) to the fit that is minimized is $O_{j\perp}$, in terms of the scaled coordinates $x^* = x/\sigma(x_j)$ and $y^* = y/\sigma(y_j)$.

In order to justify the claim that minimization of O_y defined by (6.6) is equivalent to minimizing sums of squares of perpendiculars, we need to consider the dimensions and units of the variables x and y, and of the slope parameter a_2. If we just plotted y against x and discussed perpendiculars, the best fit would depend on the units used for x and y. In particular, the value of the slope (a_2) changes if their units change. An appropriate way around this difficulty for our weighting model is to define local dimensionless coordinates, the data divided by their standard deviations, as shown in Figure 6.3. Note that the slope in terms of $[x/\sigma(x_j), y/\sigma(y)]$ variables becomes

$$\tan\theta = a_2 \sigma(x_j)/\sigma(y_j) \qquad (6.27)$$

Exercise 6.9
Use the geometry of the triangles in Figure 6.3 to show that (6.23) through (6.27) are equivalent to the relation between the vertical and perpendicular distances in the figure. *Hint:* Use the identity $1/(1 + \tan^2\theta) = \cos^2\theta$. ■

Note that this result does not depend on the assumption of constant weight ratios, (the IDWMC model),.but it holds for the more general independent diagonal weighting model (IDWM in Figure 6.2).

We have obtained a visually appealing and intuitive criterion for fitting. Also, it is apparent from this geometric construction that the same straight line should be obtained if we plot x on y rather than y on x, because the perpendicular is invariant under this change of axes. This claim is validated in the following algebraic derivation of the slope and intercept formulas.

In the following let

$$S_w = \sum_{j=1}^{n} w_j \qquad S_x = \sum_{j=1}^{n} x_j w_j \qquad S_y = \sum_{j=1}^{n} y_j w_j \qquad (6.28)$$

By setting the derivative of O_y in (6.22) with respect to a_1 to zero, you will readily find that that the intercept on the y axis is given by

$$a_1 = \overline{y} - a_2 \overline{x} \qquad (6.29)$$

where the weighted averages of x and y values are

$$\overline{x} = S_x/S_w \qquad \overline{y} = S_y/S_w \qquad (6.30)$$

This result for a_1 is not immediately usable, since it contains the unknown slope, a_2. We may use it, however, to calculate the differences between data, y_j, and fit, Y_j, in (6.24) as

$$y_j - Y_j = y_j - \overline{y} - a_2(x_j - \overline{x}) \qquad (6.31)$$

The calculation in terms of differences from the mean also has the advantage of minimizing subtractive-cancellation effects, as discussed in detail in Exercise 4.8 in Section 4.3. Now compute the sums of products

$$S_{xy} = \sum_{j=1}^{n} (x_j - \bar{x})(y_j - \bar{y}) w_j \qquad (6.32)$$

$$S_{xx} = \sum_{j=1}^{n} (x_j - \bar{x})^2 w \qquad (6.33)$$

$$S_{yy} = \sum_{j=1}^{n} (y_j - \bar{y})^2 w_j \qquad (6.34)$$

These are the key quantities used to calculate interesting properties of the straight-line fit.

Exercise 6.10

(a) Derive the least-squares equation for the slope of y on x, a_2, by substituting (6.31) in the formula (6.16) for the derivative of the objective function O_y with respect to a_2 to show that

$$a_2 S_{xy} + (S_{xx} \lambda - S_{yy}) - (1/a_2) S_{xy} \lambda = 0 \qquad (6.35)$$

where, λ, the ratio of x weights to y weights, is given by (6.19).

(b) Show that this equation predicts that the best-fit slope for x on y is just $1/a_2$. To do this, demonstrate the symmetry of this equation under interchange of x and y, recalling that under such an interchange λ is replaced by $1/\lambda$. ∎

The quadratic equation for the slope, (6.35), has a solution that is insensitive to subtractive cancellation (as discussed fully in Section 4.3) when computed in the form

$$a_2(\lambda) = \frac{2 S_{xy} \lambda}{\sqrt{\left[(S_{xx}\lambda - S_{yy})^2 + 4 S_{xy}^2 \lambda\right]} + (S_{xx}\lambda - S_{yy})} \qquad (6.36)$$

for $S_{xx}\lambda > S_{yy}$, for example when x errors are relatively negligible. The slope should be computed in the form

$$a_2(\lambda) = \frac{-(S_{xx}\lambda - S_{yy}) + \sqrt{(S_{xx}\lambda - S_{yy})^2 + 4 S_{xy}^2 \lambda}}{2 S_{xy}} \qquad (6.37)$$

for $S_{xx}\lambda < S_{yy}$, as when y errors are relatively negligble ($\lambda \to 0$).

Exercise 6.11

Use the results in Exercise 4.9 and the discussion below it to show that the appropriate formulas have been selected for solution (6.36) or (6.37) of the slope equation (6.35). ∎

To summarize our derivation for a constant ratio of weights, IDWMC: The slope is given by (6.36) or (6.37), then the intercept is simply obtained from (6.29). We next consider some properties of the straight-line least-squares slopes, then we derive a compact expression for the minimized objective function.

Properties of the least-squares slopes

When both variables have errors, even if these are in the constant ratio given by (6.19), there is an additional parameter in the straight-line least-squares analysis, namely the ratio of x to y weights, λ. It is therefore important to understand how λ influences the fit, as we investigate in the following.

An interesting limit of the result (6.36) for the slope is for negligible x errors, $\lambda \to \infty$ (OLS – y:x in Figure 6.2), for which (6.36) has as its limit

$$a_2(\infty) = S_{xy}/S_{xx} \tag{6.38}$$

Similarly, for negligible y errors $\lambda = 0$ (OLS – x:y in Figure 6.2) formula (6.37) has the limit

$$a_2(0) = S_{yy}/S_{xy} \tag{6.39}$$

Now to verify the result suggested by Figure 6.3, that is, either

$$|a_2(0)| \le |a_2(\lambda)| \le |a_2(\infty)| \tag{6.40}$$

or the slopes are limited by

$$|a_2(0)| \ge |a_2(\lambda)| \ge |a_2(\infty)| \tag{6.41}$$

so that the slope for any value of λ lies between that for y on x, $a_2(\infty)$, and that for x on y, $a_2(0)$.

Exercise 6.12

To derive the inequalities (6.40) or (6.41), consider the quantity under the square-root sign in the denominator of (6.36), then apply the Schwartz inequality $S_{xy} \le \sqrt{S_{xx}S_{yy}}$ to show that $a_2 \ge S_{xy}/S_{xx} = a_2(0)$ if $S_{xy} > 0$. Then, since $a_2(0)$ and $a_2(\infty)$ are reciprocal, the second inequality follows immediately. The inequalities are reversed if $S_{xy} < 0$, which is (6.41). ∎

Since a value of the slope intermediate between the two extremes is attained for any ratio of y to x weights, λ, you might guess that the geometric mean of the extreme values of λ would often give an appropriate value for the slope a_2.

Exercise 6.13

(a) Consider the slope $a_2(\lambda)$ as a function of the ratio of weights, λ. By using the results in (6.39) and (6.38) for $a_2(0)$ and $a_2(\infty)$, show that their geometric mean is $a_2(\lambda_g) = \sqrt{S_{yy}/S_{xx}}$ with sign that of S_{xy}.

(b) Insert this result for the slope into the defining equation (6.35) for a_2 to show that this choice of the geometric mean corresponds to a specific choice of the relative weights, namely

$$\lambda_g = S_{yy}/S_{xx} \tag{6.42}$$

This value is determined by the data and not by any estimates of the relative errors. So the geometric-mean choice of slopes is usually *not* appropriate. ∎

The weighting-model parameter λ defined in (6.19) should be estimated, for example, by repeated measurements of some x and y values in order to estimate their standard deviations, which should then be used in (6.19). Because this is a tedious process, which scientists in the heat of discovery are seldom willing to spend much time on, it is of interest to investigate the dependence of the slope on λ. Within the IDWMC model (Figure 6.2), this can be done without reference to particular data by appropriately scaling the variables. To this end, we introduce the *weighted correlation coefficient* between x and y variables, ρ, defined by

$$\rho = \frac{S_{xy}}{\sqrt{S_{xx}S_{yy}}} \tag{6.43}$$

From the Schwartz inequality we have that $|\rho| \le 1$. Further, introduce the ratio between weights and the geometric weight, v, defined by

$$v \equiv \lambda_g/\lambda \tag{6.44}$$

By using the dimension-free variables ρ and v just defined, we can derive from the slope equation (6.35) the more-general form, which is again free of all dimensional considerations:

$$\frac{a_2(v)}{a_2(1)} = \frac{2\rho}{\sqrt{(1-v)^2 + 4v\rho^2} + (1-v)} \tag{6.45}$$

Exercise 6.14

(*a*) In (6.35) eliminate the S variables in favor of the variables appearing in (6.43) and (6.44) in order to derive (6.45).

(*b*) Show algebraically that if the factor λ greatly exceeds its geometric-mean estimate λ_g ($\nu \to 0$), then the slope ratio in (6.45) tends to ρ.

(*c*) Show that if λ is much less than its geometric-mean estimate λ_g, then the slope ratio in (6.45) tends to $1/\rho$. ∎

From this exercise we see that the ratio of slopes exhibits complete reflection symmetry with respect to the weighting ratio ν about the value $\nu = 1$. The relationship (6.45) for interesting values of the correlation coefficient ρ is illustrated in Figure 6.4. This figure and the formula (6.45) have a simple explanation, as follows. For $\rho = 0.1$, x and y are only weakly correlated, so the slope is strongly dependent on the weights. For $\rho = 0.9$, they are strongly correlated, so the slope becomes almost independent of the weight ratio.

As a final topic in straight-line least squares with errors in both variables, we give the first form of the minimum value obtained by the y contribution to the objective function, O_y in (6.22), namely

$$O_y \geq O_{y\,min} = \frac{S_{yy} - 2a_2 S_{xy} + a_2^2 S_{xx}}{1 + a_2^2/\lambda} \qquad (6.46)$$

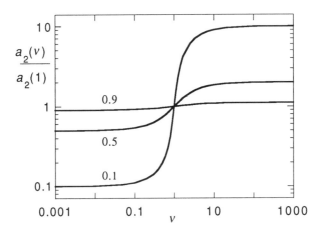

FIGURE 6.4 Dependence of the least-squares slope on the ratio of x weights to y weights in the IDWMC weighting model (Figure 6.2) according to (6.45). The slope dependence is shown for three values of the weighted correlation coefficient $\rho = 0.1$ (weakly correlated data), $\rho = 0.5$ (moderate correlation), and $\rho = 0.9$ (strong correlation).

Actually, this formula provides an unstable method (in the sense of Section 4.3) for computing $O_{y\,min}$, for the following reason. We know that the limit $\lambda \to 0$ is just that for negligible y errors (OLS – x:y in Figure 6.2) and is obtained by using the second slope formula (6.37), which is well-behaved as $\lambda \to 0$. Therefore we are not dealing with an unstable problem in the sense discussed in Section 4.3. In (6.46), however, the denominator is divergent in the limit $\lambda \to 0$. Part (b) of Exercise 6.15 shows you how to remove this difficulty.

Exercise 6.15

(a) Substitute in expressions (6.22) for the objective function and (6.37) for the best-fit slope a_2 in order to derive (6.46).

(b) Since, as just discussed, the $\lambda \to 0$ behavior in (6.46) cannot hold analytically (as contrasted to numerical behavior), the best way to avoid this unstable method is to eliminate λ from this expression by solving for it from (6.35), then substituting its solution in (6.46). By performing some algebraic simplification, show that the y contribution to the minimum objective function can be rewritten simply as

$$O_{y\,min} = S_{yy} - a_2 S_{xy} \qquad (6.47)$$

(c) Show that if $\lambda \to 0$ (negligible y errors), then, by using (6.37) for a_2, this contribution to the objective function vanishes and only the part from any x errors remains. Is this result expected? ∎

Formula (6.47) is used in the straight-line least-squares-fit function in Program 6.2 (Section 6.6). Note that the minimum value of the objective function obtained by using (6.47) depends upon the overall scale factor for the weights. Even if the weight sum S_w given in (6.28) is divided out, one still usually does not have the statistical chi-squared, because of the weighting scheme used. Therefore, interpretations of the probability content of such an objective function (such as physicists' favorite claim for a good fit that "the chi-squared per point is less than one") are often not meaningful.

Estimating the uncertainties in the slope and the intercept when both variables have errors involves some advanced concepts from statistics. Derivations of such estimation formulas are given in the article by Isobe et al., and the formulas are compared with Monte Carlo simulations in the article by Babu and Feigelson.

6.4 LEAST-SQUARES NORMALIZATION FACTORS

A common problem when analyzing data is to obtain a realistic estimate of best-fit normalization factors between data and fitting-function values. Such a factor may be applied to the fit values as an overall normalization, N_f, that best matches the data or it may be applied to the data, as N_d, to best match the fitting values. Under most best-fit criteria N_d is not the reciprocal of N_f. If the normalization is assumed to be independent of other parameters in the fitting (as when data are scaled to a model to calibrate an instrument), the procedures that we derive here will decrease the number of fitting parameters by one. Indeed, I worked up this section of the text after realizing while writing Section 7.2 on world-record sprints that one of the fitting parameters (the acceleration A) was linearly related to the data. The formula is seldom derived in books on statistics or data analysis, and then only for the case of errors in the y variables.

In this section we derive simple expressions for N_f and N_d that can be used with weighted least squares when there are errors in both variables. The formulas are of general applicability and are not restricted to straight-line or linear least-squares fitting. Also, they do not depend strongly on other assumptions about the fitting procedure, such as the weighting models used. The formulas are not special cases of determining the slope in a straight-line least-squares fit of the Y_j to the y_j, or vice versa, because here we constrain the intercept to be zero, whereas in a straight-line fit the analysis determines the intercept. After the best-fit normalizations have been computed, the corresponding best-fit objective function can be obtained from a simple formula that we derive. We work in the context of bivariate problems (x and y independent variables), then at the end indicate the extension to multivariate analyses.

In the first subsection we set up the problem and the method of solution for finding N_f. The same procedure is then indicated for N_d. We next derive the formula for the resulting minimized value of the objective function. In the final subsection we develop `Least Squares Normalization` for computing the normalizations and objective functions, together with suggestions for testing and applying it.

Normalizing fitting-function values to data

As you will recall from (6.6) in Section 6.1, the objective function to be minimized in a least-squares fit is usually chosen as

$$O = \sum_{j=1}^{N} \left[w_{xj}(x_j - X_j)^2 + w_{yj}(y_j - Y_j)^2 \right] \tag{6.48}$$

in which the weights may be chosen in various ways, as discussed in Section 6.3. In (6.48) the (x, y) pairs are the data and the (X, Y) pairs are the corresponding fitting values. Suppose that for each of the data points the fitting function is renormalized by a common factor N_f. That is,

$$Y_j = N_f Y_j^{(1)} \tag{6.49}$$

where the Y value on the right-hand side is assumed to have normalization factor of unity, which explains the superscript (1) on it. We want a formula for a value of N_f, common to all the N data points, so that O is minimized. This requires using only straightforward calculus and algebra, so why not try it yourself?

Exercise 6.16

(a) Substitute (6.49) in (6.48), differentiate the resulting expression with respect to N_f, then equate this derivative to zero. Thus derive the expression for N_f that produces an extremum of O, namely

$$N_f = \sum_{j=1}^{N} w_{yj} y\, Y_j^{(1)} / \sum_{j=1}^{N} w_{yj} \left[Y_j^{(1)} \right]^2 \qquad (6.50)$$

(b) Take the second derivative of O with respect to N_f and show that this derivative is always positive (assuming, as usual, that all the weights are positive). Thus argue that the extremum found for O must be a minimum. ■

Formula (6.50) is a general-purpose result for normalizing the fitting function, Y, to best match the data, y. It is used by the program in Section 7.2 for fitting data on world-record sprints.

A simple check to suggest the correctness of (6.50) is that for a single data point, $N = 1$, it gives $N_f = y_1/Y_1^{(1)}$, the fit for the y variables is exact, and only differences in x values (if any) contribute to the objective function O. This seemingly trivial result is quite useful for checking out the program Least Squares Normalization that is developed below.

Normalizing data to fitting values

It is important to know that the overall best-fit normalization constant to be applied to data, N_d, is usually not simply obtained as the reciprocal of N_f obtained from (6.50). Rather, if data are to be normalized to the fitting function, we must interchange the roles of y_j and Y_j in describing the normalization. Therefore, we set

$$y = N_d y_j^{(1)} \qquad (6.51)$$

in which the y data on the right-hand side are assumed to have unity for normalization, as indicated by the superscript (1). The procedure for determining the optimum N_d value is very similar to that in the preceding subsection, so it's left for you to work out.

Exercise 6.17

(a) Substitute (6.51) in (6.48) and differentiate the expression with respect to N_d. By equating this derivative to zero, derive the expression for the normalization of the data, N_d, needed to produce an extremum of O, namely

$$N_d = \sum_{j=1}^{N} w_{yj} y^{(1)}\, Y_j / \sum_{j=1}^{N} w_{yj} \left[y^{(1)} \right]^2 \qquad (6.52)$$

(*b*) Differentiate O once more with respect to N_d to show that this derivative is always negative if all the weights are positive. Thus show that the extremum found for O is a minimum.

(*c*) Show by comparison of the formulas (6.52) and (6.50) that N_d and N_f are generally not reciprocal. Show that a sufficient condition for reciprocity is that the fitting-function values Y_j and the data values y_j are all proportional to each other with the same proportionality constant. Show that another sufficient condition for reciprocity is that there be only one data point, $N = 1$. ■

The first condition derived in part (*c*) of this exercise is not particularly relevant, since it implies that after normalization the Y fit and the y data coincide at all data points, which is an unlikely situation. If this were true, then the only contributions to the objective function (6.48) would come from discrepancies between the values of the x_j and X_j. Both conditions for reciprocity are helpful, however, in checking the correctness of programs for the optimum normalization values. This observation indicates how results that are trivial in mathematics are sometimes useful when computing for scientists and engineers.

In applications one uses data renormalization when, for a fixed instrumentation setup, one can make a set of measurements for which the model is reliable but an instrument needs to be calibrated in terms of an overall normalization. Then the only sources of uncertainty are systematic and random errors in the data. The latter errors should be taken into account by appropriately choosing weights in (6.48).

The alert reader will have noticed that I have slipped by a point that needs to be discussed. Namely, the weights depend upon the data, as we acknowledge by writing them as w_{yj} rather than w_j. How does this affect the analysis? Why not find out yourself?

Exercise 6.18

(*a*) Show that formula (6.50) for the fitting-function value normalization, N_f, is unaffected by the dependence of the weights on the data values.

(*b*) Prove that if the weights depend on the data as a single power of the data, $w_{yj} \propto y_j{}^p$, and if the definition of the objective function is modified so that it includes division by the sum of all the weights, then formula (6.52) for N_d is unaltered. ■

You can see intuitively why the data normalization formula will be different if division by total weight is not included, because if the weights depend on inverse powers of the data, then O is reduced partly by scaling up the data values as they appear in the weights. One good reason not to include the total weight as a divisor in the objective function is that when the weights are just inversely proportional to the data, $w_{yj} \propto 1/y_j$, then we have Poisson statistics and O is proportional to the statistical chi-squared, for which many properties and interpretations are known. I have discussed elsewhere (Thompson, 1992) the particularly common weighting model with Poisson statistics. For consistency with the rest of our treatment, and with the works of others, the slightly wrong (but convenient) formula (6.52) for N_d will be maintained as-is. Alternatively, the formula for N_d becomes correct at the expense of a different formula for O, as shown in Exercise 6.18 (b).

For the X and x variables that appear in the objective function (6.48) the procedures for finding best-fit normalizations require only replacement of y and Y by x and X in (6.50) or (6.52). Indeed, by use of such formulas one may have any pair of renormalization factors drawn from (x,X), (X,x), (y,Y), or (Y,y), where the notation means that the first of the pair is normalized to the second of the pair. These options are available in the program Least Squares Normalization that we develop below.

You may be wondering what has happened to the effects of correlations in the data. These are implicit in the weight factors, which certainly do have major effects on the values of normalization constants, as (6.50) and (6.52) indicate and as you can verify by using the program with different weight factors.

The best-fit objective function

The final analysis topic in this section on least-squares normalization factors is to derive an expression for the minimized objective function, O_{min}, that is obtained after renormalization of either the fitting-function values or the data values.

A brute-force numerical method would be to use the normalization factors to recompute the normalized values, then to insert these in expression (6.48) for the objective function. If we do some algebra ahead of time, we can be accumulating the necessary sums while performing the summations for the normalization factors, thus improving computational efficiency and reducing the amount of code needed. So why don't you try some more algebra?

Exercise 6.19
Consider the case of normalizing the fitting function Y to the data y, for which the normalization factor is given by (6.50). In the objective function (6.48) expand the square of the differences of y values and expand this sum into three separate sums. Next, replace each Y_j by the substitution (6.49) in terms of the unnormalized values. Finally, use the result for N_f given by (6.50) to eliminate one of the sums in order to obtain the minimum objective function

$$O_{min} = \sum_{j=1}^{N} \left[w_{xj}(x_j - X_j)^2 + w_{yj} y_j^2 \right] - N_f^2 \sum_{j=1}^{N} w_{yj} \left[Y_j^{(1)} \right]^2 \quad (6.53)$$

in which all the summations involve only values before normalization. ∎

By appropriately relabeling variables, this formula is applicable to normalization of y data, x data, or x fits. Indeed, by the use of function references in a program, even this is not necessary, as the next subsection shows.

From our example with the two independent variables x and y, you can see how to generalize the treatment from a bivariate system to a multivariate one. Those variables that are normalized, such as y or Y, get a reduced value of their contribution to the objective function. Those that are not normalized, such as x or X, make an unchanged contribution to O. This simple extension arises from the assumed independence of the variables and their weightings.

Program for normalizing factors

The program Least Squares Normalization is designed to illustrate using the formulas derived earlier in this section for computing the normalization factor (6.50) and the resulting minimum objective function (6.53). Although the function NormObj that computes these is written in terms of normalizing the fit to the data, the variable choice in the main program lets you use the same function to compute any of the other three combinations of x or y quantities.

PROGRAM 6.1 Least-squares normalization factors and their minimum objective functions.

```
#include <stdio.h>
#include <math.h>
#define MAX 101

main()
{
/*   Least Squares Normalization */
/*   to test function   NormObj   */
double xd[MAX],xw[MAX],xf[MAX];
double yd[MAX],yw[MAX],yf[MAX];
double norm[4];
double xdin,xwin,xfin,ydin,ywin,yfin,omin;
int N,j,choice;
double NormObj();

printf("Least Squares Normalization\n");
N = 1;
while ( N > 0 )
  {
  printf("\nHow many data, N (N=0 to end)? ");   scanf("%i",&N);
  if ( N == 0 )
    {
    printf("\nEnd Least Squares Normalization");   exit(0);
    }
  if ( N > MAX-1 )
    printf("!! N=%i > array size %i\n",N,MAX-1);
```

```
    else
      {
      printf("\nInput j=1 to %i sets as\n",N);
      printf("xdata,xweight,xfit,ydata,yweight,yfit");
      for ( j = 1; j <= N; j++ )
        {
        printf("\n%i: ",j);
        scanf("%lf%lf%lf",&xdin,&xwin,&xfin);
        scanf("%lf%lf%lf",&ydin,&ywin,&yfin);
        xd[j] = xdin;  xw[j] = xwin;  xf[j] = xfin;
        yd[j] = ydin;  yw[j] = ywin;  yf[j] = yfin;
        }
      choice = 1;
      while ( choice > 0 )
        {
        printf("\n\nNormalization choice: (0 for new data set)\n");
        printf("1: xdata to xfit    2: xfit to xdata\n");
        printf("3: ydata to yfit    4: yfit to ydata\n");
        scanf("%i",&choice);
        if ( choice != 0 )
          {
          if ( choice <0  ||  choice > 4 )
            printf("\n !!choice=%i is out of range (1-4)\n",choice);
          else
            {
            switch(choice)
              {
              case 1: norm[1] = NormObj(xw,xf,xd,yw,yd,yf,N,&omin);
                      break;
              case 2: norm[2] = NormObj(xw,xd,xf,yw,yd,yf,N,&omin);
                      break;
              case 3: norm[3] = NormObj(yw,yf,yd,xw,xd,xf,N,&omin);
                      break;
              case 4: norm[4] = NormObj(yw,yd,yf,xw,xd,xf,N,&omin);
                      break;
              }
            printf("Normalization[%i]=%le\n",choice,norm[choice]);
            printf("Minimum objective function=%le",omin);
            }
          }
        } /* end   while choice  loop */
      }
    } /* end  while N>0  loop */
}
```

```
double NormObj(yw,yd,yf,xw,xd,xf,N,omin)
/* Least-squares best fit for normalization
          and minimum objective function  */
/* Written as if normalizing yf(=yfit) to yd(=ydata) */
double yw[],yd[],yf[],xw[],xd[],xf[];
double *omin;
int N;
{
double norm,ywyf,num,den,xdif,obj1;
int j;

num = 0;   den = 0;   obj1 = 0;
for ( j = 1; j <= N; j++ )                /* loop over the N data */
   {
  ywyf = yw[j]*yf[j];
  num = num+ywyf*yd[j];
  den = den+ywyf*yf[j];
  xdif = xd[j]-xf[j];
  obj1 = obj1+xw[j]*xdif*xdif+yw[j]*yd[j]*yd[j];
   }
norm = num/den; /* is normalization of yfit to ydata */
*omin = obj1-norm*norm*den;/* minimum objective function */
return norm;
}
```

The overall structure of the program Least Squares Normalization is as follows. The outermost loop is controlled by the number of data points N, with program execution being terminated only if N is input as zero. If the usable array size, MAX − 1, where MAX is in a preprocessor definition, is exceeded by N, then a warning is issued and another value of N is requested. The subtraction of unity from MAX is necessary because, in keeping with general practice in this book, the arrays start at [1] (as is common in Fortran) rather than at [0] (as in C). Given that N is within limits, N sets of six data items each are to be input. The program is now ready to calculate normalization factors for various combinations of x and y data.

There are four choices of normalization, as the program output describes. If the value of input variable choice is zero, then the loop of choices is terminated, and if the value input is negative or greater than 4, a warning message (!!) is issued. For a given choice, the function NormObj makes the appropriate normalization calculation, as presented in the preceding subsection, then the normalization and minimum objective function, (6.53), are output. A new choice of normalization for the same data is then requested.

Function NormObj is a direct implementation of (6.50) and (6.53), coded as if the fitted y values are being normalized to the y data. The normalization factor, norm, is returned as the value of the function, and the minimum objective function, omin, is passed back through the argument list of the function.

The program Least Squares Normalization may be tested for correctness by running a single data point sample N = 1, as suggested below Exercise 6.17, and by checking that (for any number of data points) if the data and fit values are exactly proportional to each other then N_d and N_f are reciprocal, as derived in Exercise 6.17 (c). Further, under such proportionality there is no contribution to O_{min} from the data-pair values that are normalized.

As an example of using Least Squares Normalization, consider the six y data with their weights in Table 6.2, which are also shown in Figure 6.5 as data and errors.

TABLE 6.2   Data (y_j), weights (w_{yj}), and fit values (Y_j) for the least-squares normalization example discussed in the text.

j	y_j	$\pm\sigma(y_j)$	w_{yj}	Y_j
1	12	8	1.6	2
2	15	9	1.2	3
3	13	5	4.0	4
4	1	−9	1.2	5
5	17	5	4.0	6
6	9	−5	4.0	7

The fitting function used in Table 6.2 is $Y_j = 1 + j$. In order to control the analysis, the data are generated as $y_j = 2Y_j + \sigma(y_j)$ in terms of the errors $\sigma(y_j)$. I generated these errors by using the Chapel Hill telephone directory to choose random numbers as the least significant digit of six directory entries, while their signs were chosen from six other entries and assigned as + for digits from 0 to 4 and − for digits from 5 to 9. The data weights, w_{yj}, are 100 times the reciprocal squares of the errors.

If the data are normalized to the fit, Least Squares Normalization produces a normalization factor $N_d = 0.36364$ $(1/N_d = 2.7500)$ and a minimum objective function $O_{min} = 99.78$. Alternatively, if the fit is normalized to the data, then the least-squares normalization factor is $N_f = 2.1418$ with $O_{min} = 58.77$. In the absence of errors we should obtain exactly $N_d = 0.5$, $N_f = 2$, and zero for the objective functions. The data, errors, and fits are shown in Figure 6.5.

Strictly speaking, when displaying renormalized values in Figure 6.5 we should apply the factor N_d to the data, but it is easier to display the comparison of data normalized to fit versus fit normalized to data by dividing the fit by N_d, as shown in the figure. The important point is that the two methods of estimating normalization factors produce normalizations that differ from each other by about 25%. Note that the normalization factors would not change if we changed all the errors (and therefore all the weights) by the same factor.

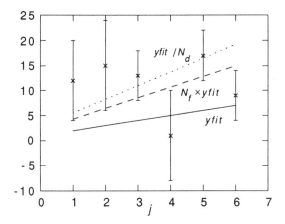

FIGURE 6.5 Least-squares normalization factors for the data in Table 6.2. The solid line is the fit before normalization. The dashed line is the best fit after the fit has been normalized to the data. The dotted line compares the scaled fit after the data have been normalized to the fit, but the normalization factor N_d has been divided into the data to simplify comparison.

Now that you see how the method is applied, how about trying it yourself?

Exercise 6.20

(*a*) Code program `Least Squares Normalization`, adapting the input and output functions to be convenient for your computing environment.

(*b*) Test all the program control options (by `N` and `choice`) for correct termination of all the `while` loops.

(*c*) Check the program using the tests for correctness suggested above.

(*d*) Use the data given in Table 6.2 to compare your normalization-factor values with those given above. Display the data and fits as in Figure 6.5. ∎

With the understanding that you have gained from this analysis of least-squares normalization factors, and with the function `NormObj` now being available for use in your other data-fitting programs, you have one more useful and versatile tool for comparing data and fits by least-squares methods.

6.5 LOGARITHMIC TRANSFORMATIONS AND PARAMETER BIASES

We now turn to another topic in least-squares analysis, namely an example of the effects of transforming variables. In order to obtain a tractable solution, we revert to the conventional approximation that the x data are known exactly but the y values are imprecise. The transformation that we consider is that of taking logarithms in order to linearize an exponential relationship.

Exponential growth and decay are ubiquitous in the natural sciences, but by the simple device of taking logarithms we can reduce a highly nonlinear problem to a linear one, from which estimates of the slope (exponent) and intercept (pre-exponential) can be readily obtained by using a linear-least-squares fitting program. This seemingly innocuous procedure of taking logs usually results in biased values of fitting parameters, but we show now that such biases can often be simply corrected. This problem is mentioned but not solved in, for example, Taylor's introduction to error analysis.

The origin of bias

A moment of reflection will show why biasing occurs in logarithmic transformations. Consider the example of data that range from $1/M$ through unity up to M, as shown in Figure 6.6.

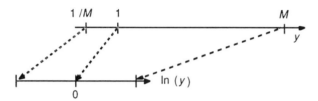

FIGURE 6.6 The origin of bias in logarithmic transformations. The top line shows the real line from 0 to M, with $M = 4$. Under logarithmic transformation the interval from $1/M$ to 1 maps into the same length as the interval from 1 to M.

For large M there are two subranges of data in Figure 6.6, having lengths roughly unity and M, respectively. After taking (natural) logs the subranges are each of length $\ln(M)$, so that the smallest values of the original data have been unnaturally extended relative to the larger values.

Exercise 6.21
Take any set of data (real or fictitious) in which the dependent variables, $y(x)$, are approximately exponentially related to the independent variables, x, and are roughly equally spaced. The range of the y values should span at least two decades. First make a linear plot of y against x, then plot $\ln(y)$ against x. Verify that the plots of the original and transformed independent variables differ as just described. ∎

The effect of the logarithmic transformation is therefore to make derived parameter estimates different from the true values. Although such parameter bias is evident, textbooks on statistical and data-analysis methods usually do not even mention it. Here we make an analysis that is realistic and that allows simple corrections for parameter bias to be made in many situations of interest. We also suggest a Monte Carlo simulation exercise by which you may confirm the analytical estimates.

Probability analysis for bias

Suppose that the fitting function, Y, is defined in terms of the independent variable, x, by the exponential relation

$$Y(x) = A \exp(Bx) \tag{6.54}$$

in which the fitting parameters are the pre-exponential A and the exponent B (positive for growth, negative for decay). Suppose that the data to be described by the function in (6.54) are $y(x_j)$; that is,

$$y(x_j) = Y(x_j) + e_j \tag{6.55}$$

in which e_j is the unknown random error for the jth datum. Only random errors, rather than systematic errors also, are assumed to be present. Under logarithmic transformation (6.54) and (6.55) result in

$$\ln(y_j) = \ln(A) + \ln[1 + e_j/Y(x_j)] + Bx_j \tag{6.56}$$

If the errors e_j were ignored, this would be a linear relation between the transformed data and the independent variable values x_j. If (6.54) is substituted into (6.56), A and B appear in a very complicated nonlinear way that prevents the use of linear least squares methods.

To procede requires an error model (Sections 6.1, 6.3) for the dependence of the distribution of the errors e_j upon the x_j. The only possibility in (6.56) that allows straightforward estimates of bias and that is independent of the x_j is to assume proportional random errors, that is,

$$e_j = \sigma Y(x_j) P(0,I_j) \tag{6.57}$$

in which σ is the same standard deviation of $e_j/Y(x_j)$ at each j. The notation $P(0,I_j)$ is to be interpreted as follows. In statistics $P(0,I)$ denotes an independent probability distribution, P, having zero mean and unity standard deviation at each j. Since I is a unit vector, $I_j = 1$ for all j, and $P(0,I_j)$ is a random choice from $P(0,I)$ for each j. For example, the Gaussian distribution in Figure 6.1 has

$$P(0,I_j) = P_G(0,I_j) = \exp(-v_j^2/2)/\sqrt{2\pi} \tag{6.58}$$

where v_j is a random variable parametrizing the distribution of errors at each data point. Proportional errors (constant percentage errors from point to point) are common in many scientific measurements by appropriate design of the experiment. Exceptions to proportional errors occur in radioactivity measurements in nuclear physics and photon counting in astronomy and other fields, both of which have Poisson statistics with square-root errors ($\sigma \propto \sqrt{y}$), unless counting intervals are steadily increased to compensate count rates that decrease with time. From the connection between error models and weighting models (Sections 6.1 and 6.3), note that proportional errors correspondingly imply inverse-squared proportional weights.

Before (6.56) and (6.57) can be used for fitting, we have to take expectation values, E in statistical nomenclature, on both sides, corresponding to many repeated measurements of each datum. This concept is clearly discussed in Chapter 6.3 of Snell's introductory-level book on probability   We assume that each x_j is precise, so that we obtain

$$E\{\ln(y_j)\} = 1(A_b) + Bx_j \tag{6.59}$$

in which the biased estimate of the intercept, A_b, is given by

$$A_b = A \exp(E\{\ln[1 + \sigma\,(0,I)]\}) \tag{6.60}$$

The use of I rather than I_j is a reminder that the expectation value is to be taken over all the data. Even when only a single set of observations is available, it is still most appropriate to correct the bias in the estimate of A by using (6.60) in the way described below. An estimate of the fractional standard deviation σ can be obtained experimentally by choosing a representative x_j and making repeated measurements of y_j,. Computationally, an error estimate can be obtained from the standard deviation of the least-squares fit.

Equation (6.60) shows that a straightforward least-squares fit of data transformed logarithmically gives us a biased estimate for A, namely A_b, and that the amount of bias depends both upon the size of the error (σ) and on its distribution (P) but, most important, not at all on the x_j. Note also that in this error model the exponent B (which is often of primary interest) is unbiased.

The bias in A can be estimated by expanding the logarithm in (6.60) in a Maclaurin series, then evaluating the expectation values term by term. The unbiased value, A, can be estimated from the extracted biased value A_b in (6.60) by solving for A to obtain

$$A = A_b \exp[L_b(P)] \tag{6.61}$$

where the bias term, $L_b(P)$, is given by

$$L_b() = \sigma^2/2 + S(P) \tag{6.62}$$

with the sum that depends upon the error distribution being given by

$$S(P) = \sum_{m=3}^{\infty} (-1)^m \sigma^m E_m(P)/m \tag{6.63}$$

where E_m denotes the mth moment of the distribution P. The first term in the Maclaurin series vanishes because P is to have zero mean, while the second term contributes $\sigma^2/2$, since P is to have unity standard deviation (second moment about the mean).

Exercise 6.22

Work through the steps in the derivation of the bias expressions (6.54) through (6.63). ■

Thus we have obtained for proportional errors a prescription for removing the bias induced by a logarithmic transformation of data. We now show that this bias is only weakly dependent on the probability distribution function satisfied by the errors.

Dependence of bias on error distribution

In the bias term, $L_b(P)$ in (6.62), the sum, $S(P)$, depends upon the error distribution, P. For many common probability distributions, only the even moments are nonzero, so then all the terms in the sum in (6.63) are strictly positive and higher moments must necessarily increase the bias.

Consider the commonly assumed Gaussian probability distribution $P = P_G$, discussed in Section 6.1, which is also called the normal distribution. Its third moment vanishes because of its symmetry about the mean value. The fourth moment of the Gaussian gives

$$L_b(P_G) \approx \sigma^2/2 + 3\sigma^4/4 \tag{6.64}$$

Another convenient and commonly used, but slightly unrealistic, distribution is the uniform probability distribution, $P = P_U$, shown in Figure 6.7.

Exercise 6.23

Show that the height and width of the standardized uniform probability distribution are as indicated in Figure 6.7. ■

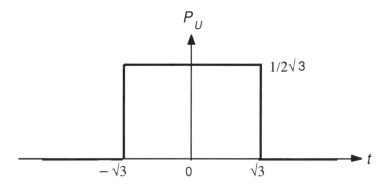

FIGURE 6.7 The uniform probability distribution, $P_U(0, I)$, also called the rectangular distribution. The distribution shown has zero mean, unity normalization, and unity standard deviation.

For the uniform distribution $P = P_U$, one obtains (similarly to the Gaussian distribution) the exponent of the bias factor given by

$$L_b(P_U) \approx \sigma^2/2 + 9\sigma^4/20 \tag{6.65}$$

through the fourth moment of the distribution. For a given standard deviation σ, the fourth moment contributes 30% (6/20) more for P_G than for P_U, primarily because the wings of the Gaussian distribution include larger values of v_j for a given σ. Other error models can be incorporated in the estimates of the pre-exponential for proportional errors if their probability distributions can be calculated.

Exercise 6.24

(a) Verify the correctness of the second terms of the bias factors L_b in (6.64) and (6.65) for the Gaussian and uniform distributions, respectively, by calculating the corresponding fourth moments of their probability distributions.

(b) As numerical examples of the bias induced in the pre-exponential A by a logarithmic transformation, show that A needs to be corrected upward by 0.5% for data with 10% standard deviation ($\sigma = 0.1$) and upward by about 5% for data with 30% random errors ($\sigma = 0.3$). ∎

Mathematically punctilious readers should object to the analysis of the Gaussian distribution, because the argument of the *ln* function in the error model may become negative, even though this is very improbable if σ is small. (For $\sigma = 0.2$ the probability is about 3×10^{-7}.) Therefore, in the analytical treatment for the Gaussian, (6.63) represents an asymptotic series which eventually diverges. In the Monte Carlo simulation suggested in Exercise 6.25 if the sample size is very large the chance of getting a negative argument increases, and the program should include a trap for this condition.

Formally, we can circumvent the problem with the logarithmic transformation for a Gaussian distribution of errors by defining suitably truncated distributions whose low-order moments are nearly the same as the complete distributions, so that there are no practical consequences for the bias estimates. For a uniform distribution, the problem arises only if $\sigma > 1/\sqrt{3} = 0.58$ (see Figure 6.7), which would usually be considered too large an error to justify anything more than a cursory fit to the data.

The simple corrections given by (6.61) and (6.62) are worth making if the assumption of proportional random errors, (6.57), is realistic. It is also reassuring that the exponent B is unbiased under this assumption. For any other error model the logarithmic transformation induces biases in both the exponent B and the pre-exponent A which cannot easily be corrected. As a final remark, data taken with a small number of samples, typically with fewer than fifty samples, will have errors dominated by the statistics of the data, which overwhelm any effects of the transformation bias that we have just investigated.

As a way of confirming the analysis of the bias in taking logarithms for analyzing exponential behavior, try a Monte Carlo simulation of a Gaussian random-error distribution, as suggested in the following exercise.

Exercise 6.25

(*a*) Use a computer random-number generator to provide a sample of say 10,000 values from a Gaussian distribution, and force this sample to have zero mean and unity standard deviation, as the above analysis uses. Then sort the sample into 100 bins.

(*b*) Choose $\sigma = 0.2$ (an average 20% error for each *y* datum), then make from the sampled Gaussian in (*a*) the distribution $\ln[1 + \sigma P(0, I)]$, as appears in (6.60). The Monte Carlo estimate of the bias is just the negative of the mean value (expectation value, E) of this distribution. How good is the agreement with our analytic estimate, $L_b(P_G)$ obtained from (6.64)? ■

In Sections 6.4 and 6.5 we explored two common and useful aspects of fitting imprecise data, to wit, normalization factors and logarithmic transformations. In the next section we return to straight-line least squares with errors in both variables, emphasizing the numerics of this problem.

6.6 PROJECT 6: PROGRAM FOR STRAIGHT-LINE LEAST-SQUARES FITS

Our aim in this section is to implement a program that makes a straight-line least-squares fit to data in which there are errors in both variables, following the algorithm developed in Section 6.3. We use the Independent Diagonal Weighting Model with Constant ratio of *x* to *y* weights from point to point, called IDWMC, as illustrated in Figure 6.2. This model for the weights is often quite realistic, and it also allows the simple formulas (6.36) or (6.37) to be used to estimate the slope of the best-fit line in terms of the data, the weights at each datum, and the ratio of *x* weights to *y* weights, λ.

Organization of Straight-Line Least Squares

The organization and implementation of `Straight-Line Least Squares` are quite straightforward, so we describe the features of the code, then we present it. The overall structure of `Straight-Line Least Squares` is very simple. A `while` loop over the number of data points, n, is used to input the number of data points. A value of n = 0 is used to terminate execution, while a value larger than the array bounds, NMAX – 1 (since the [0] element is not used), causes an error message (!!). A value of n < 2 is also diagnosed as an error because the problem is then undetermined, since one data point can't determine two parameters.

If the input   n value is acceptable, the parameter for the ratio of x weights to y weights, λ, is input, followed by n triples of x values, y values, and y weights. Then the program invokes function LeastSquares, which returns the intercept (a1), the slope (a2), the y-fit contribution to the minimum objective function (oymin), and the correlation coefficient between x and y values (rho).

The C function LeastSquares is a direct coding of formulas in Section 6.3. The discussion in Section 6.3 of the properties of the least-squares slopes and Figure 6.4 should convince you that once λ becomes small its value is not important. Try it in the program and see. The quadratic equation for the slope has two forms of solution, in order to avoid errors from subtractive cancellation, either (6.36) or (6.37).

PROGRAM 6.2 Straight-line least squares with errors in both variables, but with a constant ratio of x weights to y weights.

```c
#include <stdio.h>
#include <math.h>
#define MAX 101

main()
{
/* Straight-Line Least Squares */
/* Weighted fit with errors in both variables */
double xlist[MAX], ylist[MAX], wght[MAX]; /* data & weights */
double a1,a2,oymin,rho;/* intercept,slope,objective,correlation */
double lambda; /* ratio of x weights to y weights */
double xin, yin, win; /* temporaries for input */
int j, n; /* data counter, number of data sets */
void LeastSquares();

printf("Straight Line Least Squares\n");
n = 2;
while ( n > 0 )
   {
   printf("\nInput n (n=0 to end): ");   scanf("%i",&n);
   if ( n <= 0 )
     {
     printf("\nEnd Straight Line Least Squares"); exit(0);
     }
   if ( n > MAX-1 ) printf("\n!! More than %i data sets\n",MAX-1);
   else
```

```
    {
    if ( n < 2 )  printf("\n !!Fewer than 2 data sets\n");
    else
       {
       printf("Input ratio of x weights to y weights, lambda: ");
       scanf("%lf",&lambda);
       printf("Input %i triples of x, y, y weight:\n",n);
       for ( j = 1; j <= n; j++ )
          {
          scanf("%lf%lf%lf",&xin,&yin,&win);
          xlist[j] = xin;  ylist[j] = yin;  wght[j] = win;
          }
       LeastSquares(xlist,ylist,wght,lambda,n,&a1,&a2,&oymin,&rho);
       printf("Intercept a1 = %g and slope a2 = %g\n",a1,a2);
       printf("Minimized y objective function = %g\n", oymin);
       printf("Correlation coefficient = %g\n",rho);
       }
    }
  } /* end   while n  loop */
}


void LeastSquares(xlist,ylist,wght,lambda,n,a1,a2,oymin,rho)
/* Straight-line weighted least squares */
double xlist[],ylist[],wght[];
double lambda;
double *a1,*a2,*oymin,*rho;
int n;
{
double sw,sx,sy,xav,yav,xk,yk,sxx,sxy,syy,quad;
int j;

sw = 0;  sx = 0;  sy = 0;/* Weighted averages of x and y */
for ( j = 1; j <= n; j++ )
   {  /* sum weights, then sum weighted x and y values */
   sw = sw+wght[j];
   sx = sx+xlist[j]*wght[j];  sy = sy+ylist[j]*wght[j];
   }
xav = sx/sw;  yav = sy/sw;
/* Make weighted bilinear sums */
sxx = 0;  sxy = 0;  syy = 0;
for ( j = 1; j <= n; j++ )
   {
   xk = xlist[j]-xav;  yk = ylist[j]-yav;
   sxx = sxx+xk*xk*wght[j];
   sxy = sxy+xk*yk*wght[j];
```

```
    syy = syy+yk*yk*wght[j];
    }
quad = sxx*lambda-syy; /* Stable formulas for slope */
if ( quad > 0 )
    *a2 = 2*sxy*lambda/(sqrt(quad*quad+4*sxy*sxy*lambda)+quad);
else
    *a2 = (-quad+sqrt(quad*quad+4*sxy*sxy*lambda))/(2*sxy);
*a1 = yav-(*a2)*xav; /* intercept */
*oymin = syy-(*a2)*sxy; /* minimized y objective function */
*rho = sxy/sqrt(sxx*syy); /* correlation coefficient */
}
```

Testing and using the least-squares program

Now that you know from the preceding subsection how Straight-Line Least Squares is supposed to work, try adapting it and running it in your computing environment.

Exercise 6.26

(a) Implement the program Straight-Line Least Squares for your computer. For the first tests of program correctness check out the while loop in the main program by giving it the escape value, n = 0, and also the forbidden values n > 100 (for MAX = 101 as defined) and n = 1. Then try small data samples, say n = 2, for which the fit must be exact, so that oymin = 0.

(b) Check the reciprocity properties of x on y slopes and y on x slopes when the value of lambda is inverted, as discussed in Section 6.3. Also consider large, small, and intermediate values of lambda in order to verify (6.36) and (6.37).

(c) Improve your understanding of the dependence of the slopes on the ratio of x to y weights by generating some of the points on Figure 6.4. Analytically the dependence is given by (6.45). How closely do your numerical values of the slopes agree with this relation? ∎

Although graphical comparison of the fitted line with the data is not provided in Straight-Line Least Squares, your insight into least-squares analyses will be significantly increased if you make an appropriate program interface to make graphical output of data, weights, and best-fit lines, similarly to Figure 6.2. Since (as discussed in Section 1.3) the implementation of graphics is very dependent on the computing environment, I have not provided such an interface.

As an application of the methods for least-squares analyses that you have learned in this chapter, I suggest that you take from your own field of endeavor a small set of data for which the underlying model relating them predicts a linear relationship, and for which errors in x and y variables may realistically be considered to be in a constant ratio and of comparable size as a fraction of the data values. You will find it interesting to process these data using the program Straight-Line Least

Squares, varying λ to see the effects on the values of slope, intercept, and minimum objective function.

In this chapter we have emphasized using analytical formulas to estimate best-fit parameters, especially for straight-line fits (Sections 6.3, 6.6), normalization factors (Section 6.4), and for logarithmic transformation of exponentials (Section 6.5). It is also possible, and often desirable, to remove some of the constraints that we imposed in our least-squares fitting (such as linear parameterizations) by using Monte Carlo simulation techniques to model the distribution of errors, as is suggested in Exercise 6.25. With adequate computing power, such techniques can allow for arbitrary distributions of errors and correlations between fitting parameters. Introductions to such methods, including the "bootstrap" technique, are provided in the *Scientific American* article by Diaconis and Efron, in the article by Kinsella, and in Chapters 16 and 17 of Whitney's book. Application of the bootstrap technique to a very large database of 9000 measurements and their errors, in which there are highly nonlinear relations between parameters and data, is made in a review article by Varner et al.

Least-squares methods may be combined with spline methods (Chapter 5) at the expense of much more analysis and complexity. Smoothing of noisy data may thereby by achieved. Appropriate methods are described in the article by Woltring.

REFERENCES ON LEAST-SQUARES ANALYSIS

Babu, G. J., and E. D. Feigelson, "Analytical and Numerical Comparisons of Six Different Linear Least Squares Fits," Communications in Statistics – Simulation and Computation, **21**, 533 (1992).

Barlow, R. J., *Statistics: A Guide to the Use of Statistical Methods in the Physical Sciences*, Wiley, Chichester, England, 1989.

Beckmann, P., *Orthogonal Polynomials for Engineers and Physicists*, Golem Press, Boulder, Colorado, 1973.

Deming, W. E., *Statistical Adjustment of Data*, Wiley, New York 1943; reprinted by Dover, New York, 1964.

Diaconis, P., and B. Efron, "Computer-Intensive Methods in Statistics," Scientific American, **248**, May 1983, p. 116.

Draper, N. R., and H. Smith, *Applied Regression Analysis*, Wiley, New York, second edition, 1981.

Isobe, T., E. D. Feigelson, M. G. Akrita, and G. J. Babu, "Linear Regression in Astronomy.I," Astrophysical Journal, **364**, 104 (1990).

Jaffe, A. J., and H. F. Spirer, *Misused Statistics*, Marcel Dekker, New York, 1987.

Kinsella, A., "Numerical Methods for Error Evaluation," American Journal of Physics, **54**, 464 (1986).

Lichten, W., *Data and Error Analysis in the Introductory Physics Laboratory*, Allyn and Bacon, Boston, 1988.

Lyons, L., *A Practical Guide to Data Analysis for Physical Science Students*, Cambridge University Press, Cambridge, England, 1991.

Lyons, L., *Data Analysis for Nuclear and Particle Physics*, Cambridge University Press, New York, 1986.

Macdonald, J. R., and W. J. Thompson, "Least Squares Fitting When Both Variables Contain Errors: Pitfalls and Possibilities," American Journal of Physics, **60**, 66 (1992).

Reed, B. C., "Linear Least Squares When Both Variables Have Uncertainties," American Journal of Physics, **57**, 642 (1989); erratum *ibid*, **58**, 189 (1990).

Siegel, A. F., *Statistics and Data Analysis*, Wiley, New York, 1988.

Snell, J. L., *Introduction to Probability*, Random House, New York, 1987.

Solomon, F., *Probability and Stochastic Processes*, Prentice Hall, Englewood Cliffs, New Jersey, 1987.

Taylor, J. R., *An Introduction to Error Analysis*, University Science Books, Mill Valley, California, 1982, pp. 166, 167.

Thompson, W. J., and J. R. Macdonald, "Correcting Parameter Bias Caused by Taking Logs of Exponential Data," American Journal of Physics, **59**, 854 (1991).

Thompson, W. J., "Algorithms for Normalizing by Least Squares," Computers in Physics, July 1992.

Varner, R. L., W. J. Thompson, T. L. McAbee, E. J. Ludwig, and T. B. Clegg, Physics Reports, **201**, 57 (1991).

Whitney, C. A., *Random Processes in Physical Systems*, Wiley-Interscience, New York, 1990.

Woltring, H. J., "A Fortran Package for Generalized Cross-Validatory Spline Smoothing and Validation," Advances in Engineering Software, **8**, 104 (1986).

York, D., "Least-Squares Fitting of a Straight Line," Canadian Journal of Physics, **44**, 1079 (1966).

Chapter 7

INTRODUCTION TO
DIFFERENTIAL EQUATIONS

Differential equations are ubiquitous in science and engineering. The purpose of this chapter is to provide an introduction to setting up and solving such differential equations, predominantly as they describe physical systems. Both analytical and numerical techniques are developed and applied.

Our approach is to develop the methods of solution of differential equations from the particular to the general. That is, a scientifically interesting problem will be cast into differential equation form, then this particular equation will be solved as part of investigating general methods of solution. Then the more general method of solution will be discussed. This order of presentation will probably motivate you better than the reverse order that is typical (and perhaps desirable) in mathematics texts, but quite untypical in scientific applications. You may also recall that in the seventeenth century Isaac Newton invented the differential calculus in order to help him compute the motions of the planets. Only later was the mathematical edifice of differential calculus, including differential equations, developed extensively.

The applications in this chapter are broad; the kinematics of world-record sprints and the improving performance of women athletes provide an example of first-order linear equations (Section 7.2), while nonlinear differential equations are represented by those that describe logistic growth (Section 7.3). Numerical methods are introduced in Section 7.4, where we derive the algorithms. We present the programs for the first-order Euler method and of the Adams predictor as Project 7 in Section 7.5. Many examples and exercises on testing the numerical methods are then provided. References on first-order differential equations round out the chapter. Second-order differential equations are emphasized in Chapter 8, which contains further examples and develops the numerical methods introduced in this chapter.

7.1 DIFFERENTIAL EQUATIONS AND PHYSICAL SYSTEMS

We first survey how differential equations arise in formulating and solving problems in the applied sciences. In pure mathematics the emphasis in studying differential equations is often on the existence, limitations, and the general nature of solutions. For the field worker in applied science it is most important to recognize when a problem can be cast into differential equation form, to solve this differential equation, to use appropriate boundary conditions, to interpret the quantities that appear in it, and to relate the solutions to data and observations.

Why are there differential equations?

In physical systems, such as the motion of planets, the phenomena to which differential equations are applied are assumed to vary in a continuous way, since a differential equation relates rates of change for infinitesimal changes of independent variables. Nineteenth-century research using the differential calculus focused on mechanics and electromagnetism, while in the late twentieth century there have been the major developments of microelectronics and photonics. Both the principles and practical devices derived from these principles require for their design and understanding extensive use of differential equations.

We now recall some of the main ideas of differential equations. Formally speaking, a *differential equation* is an equation (or set of equations) involving one or more derivatives of a function, say $y(x)$, and an independent variable x. We say that we have solved the differential equation when we have produced a relation between y and x that is free of derivatives and that gives y explicitly in terms of x. This solution may be either a formula (analytical solution) or a table of numbers that relate y values to x values (numerical solution).

Exercise 7.1

(*a*) Given the differential equation

$$\frac{dy}{dx} = y(x) \tag{7.1}$$

verify that a solution of this differential equation is

$$y(x) = e^x \tag{7.2}$$

(*b*) Given a numerical table of x and corresponding y values that claim to represent solutions of a differential equation, what is necessarily incomplete about such a differential equation solution? ∎

Generally, there will be more than one solution for a given differential equation. For example, there is an infinite number of straight lines having the same slope, *a*, but different intercepts, *b*. They all satisfy the differential equation

$$\frac{dy}{dx} = a \qquad (7.3)$$

The choice of appropriate solution depends strongly on constraints on the solution, such as the value of the intercept in the example of a straight line. These constraints are called *boundary values,* especially when the independent variable refers to a spatial variable, such as the position coordinate x. The constraints may also be called *initial conditions*, which is appropriate if the independent variable is time. (For those readers who are into relativistic physics, the distinction is ambiguous.) Knowing the constraints for a particular problem is often a guide to solving the differential equation. This is illustrated frequently in the examples that follow.

Exercise 7.2

(*a*) Show that the differential equation (7.1) has an infinite number of solutions, differing from each other by different choices of overall scale factor for y.

(*b*) Given that a solution of the differential equation (7.3) is required to pass through the y origin when $x = 0$, show that the solution is then uniquely determined once a is specified.

(*c*) Make a logarithmic transformation from y in (7.1) to $z = \ln (y)$ in order to relate the solutions to parts (*a*) and (*b*). ■

Time is most often the independent variable appearing in the differential equation examples in this chapter and the next. This probably arises from the predictive capability obtained by solving differential equations in the time variable. Thus, tomorrow's weather and stock-market prices can probably be better predicted if differential equations for their dependence on time can be devised.

Notation and classification

We now briefly review notations and classifications for differential equations. We use two *notations* for derivatives. The first is the compact notation that makes differentiation look like an operation, which it is, by writing

$$D_x^n y \equiv \frac{d^n y}{dx^n} \qquad (7.4)$$

This notation we use most often within the text; for example $D_x y$ denotes the first derivative of y with respect to x, in which (by convention) we have dropped the superscript 1. The second notation for derivatives is dy/dx, suggestive of division of a change in y by a change in x. We use this notation if we want to emphasize the derivative as the limit of such a dividend, as in numerical solution of differential equations.

The *classification* of differential equations goes as follows; it is mercifully briefer than those used in biology or organic chemistry. Equations that involve total derivatives (such as $D_x y$) are called *ordinary differential equations*, whereas those involving partial derivatives (such as $\partial y / \partial x$) are termed *partial differential equations*. The aim of most methods of solution of partial differential equations is to change them into ordinary differential equations, especially in numerical methods of solution, so we use only the latter in this book. This greatly simplifies the analytical and numerical work. We note also that systems for doing mathematics by computer, such as *Mathematica*, are restricted to ordinary differential equations, both for analytical and numerical solutions. This limitation is discussed in Section 3.9 of Wolfram's book on *Mathematica*.

The *order* of a differential equation is described as follows. If the maximum number of times that the derivative is to be taken in a differential equation, that is, the maximum superscript in D_t^n is n, then the differential equation is said to be of *n*th *order*. For example, in mechanics if p is a momentum component, t is time, and F is the corresponding force component (assumed to depend on no time derivatives higher than first), then Newton's equation, $D_t p = F$, is first order in t. In terms of the mass, m, and the displacement component, x say, Newton's equation, $D_t^2 x = F$, is a second-order differential equation in variable t.

Exercise 7.3

(*a*) Write Newton's force equation for a single component direction as a pair of first-order ordinary differential equations.

(*b*) Show how any *n*th-order differential equation can be expressed in terms of n first-order differential equations. ∎

The result in (*b*) is very important in numerical methods of solving differential equations, as developed in Chapter 8. There remains one important item of terminology.

The *degree* of a differential equation is the highest power to which a derivative appears raised in that equation. Thus, if one identifies the highest value of m appearing in $(D_x y)^m$ in a differential equation, one has an *m*th-*degree differential equation*. If $m = 1$, one has a *linear differential equation*. Generally, it is preferable to be able to set up differential equations as linear equations, because then one isn't battling algebraic and differential equations simultaneously. The distinction between the order of a differential equation and its degree is important.

We consider mainly ordinary, first- and second-order, linear differential equations. Since there are also interesting systems that cannot be so described, we sometimes go beyond these, as for the logistic-growth equation (Section 7.3) and for catenaries (Section 8.2).

Homogeneous and linear equations

Two distinct definitions of "homogeneous" are used in the context of differential equations. The first definition is that a differential equation is homogeneous if it is invariant under multiplication of x and y by the same, nonzero, scale factor. For example, $D_x y = x \sin (y / x)$ is homogeneous according to this definition. Such equations are not very common in the sciences, because in order to have the invariance, x and y must have the same dimensions, which is uncommon if one is discussing dynamical variables, as opposed to geometrical or other self-similar objects.

The other definition of a homogeneous differential equation, usually applied to linear differential equations, is that an equation is homogeneous if constant multiples of solutions are also solutions. For example, $D_t^2 y = -ky$, the differential equation for simple harmonic motion, is a homogeneous differential equation. The power of such linear differential equations is that their solutions are additive, so we speak of the linear superposition of solutions of such equations. Almost all the fundamental equations of physics, chemistry, and engineering, such as Maxwell's, Schrödinger's, and Dirac's equations, are homogeneous linear differential equations.

Electronics devices are often designed to have a linear behavior, using the term in the same sense as a linear differential equation. Such a device is termed linear if its output is proportional to its input. For example, in signal detection the primary amplifiers are often designed to be linear so that weak signals are increased in the same proportion as strong signals. It is therefore not surprising that much of the theory of electronics circuits is in terms of linear differential equations. The differential equations for mechanical and electrical systems that we discuss in Section 8.1 are linear equations.

Nonlinear differential equations

Differential equations that are nonlinear in the dependent variable, y in the above discussions, are termed nonlinear, no matter what their order or degree. In general, such differential equations are difficult to solve, partly because there has not been extensive mathematical investigation of them. In science, however, such nonlinear equations are recently of much interest. This is because linear differential equations typically describe systems that respond only weakly to an external stimulus, which itself is not very strong. On the contrary, as an example of systems that should be described by nonlinear equations, many optical materials respond nonlinearly when strong laser beams are focused on them.

The example of a nonlinear differential equation that we explore in this chapter is the logistic-growth equation in Section 7.3. It illustrates how the methods of solving nonlinear differential equations depend quite strongly on the problem at hand.

7.2   FIRST-ORDER LINEAR EQUATIONS:
WORLD-RECORD SPRINTS

We motivate our study of first-order linear differential equations by studying a realistic problem from athletic competition, namely the kinematics involved in sprinting. Our analysis is phenomenological in that it is concerned only with the kinematics and not with the dynamics, physiology, and psychology that determine the outcome of athletic contests. So we need only a model differential equation and a tabulation of world-record times for various distances. We do not need treadmills for the dynamics, specimen bottles for the physiology, and post-race interviews for the psychology. In the language of circuit theory and mathematical modeling, we are making a "lumped-parameter" analysis.

Kinematics of world-record sprints

In 1973, J. B. Keller published an analysis of competitive running sprints in which he showed that for world-class runners in races up to nearly 300 m distance their speed, v, at time t into the race can be well described in terms of the acceleration formula for world-record sprints

$$\frac{dv}{dt} = A - v(t)/\tau \tag{7.5}$$

He found that for men's world records in 1972 the appropriate constants were an acceleration parameter $A = 12.2$ m s^{-2} and a "relaxation time" $\tau = 0.892$ s. Instead of A, Keller used F, which is easily confused with force. In this differential equation the net acceleration is the driving acceleration, A, minus a resistance proportional to the speed at time t. The physiological justifications for this equation were discussed by Keller and were subsequently debated by various experts.

The data on which Keller's analysis was based are given in Table 7.1 in the column "Men 1972." Also shown in the table are world records as of 1968 and 1991. The shortest world-record distance now recognized is 100 m, the British system of lengths is no longer used, and the timing is now to 0.01 s, being done electronically rather than manually as in 1968. I have also included the data for women sprinters, since these allow an interesting comparison of speeds and improvements.

TABLE 7.1 World-record sprint times (in seconds) for men and women. For the 200-m race the notation (s) denotes a straight track and (t) denotes a track with a turn. From *Guinness Book of World Records* (1968, 1991) and *Reader's Digest Almanac* (1972). The noninteger distances are converted from British units.

Distance (m)	Men 1968	Men 1972	Men 1991	Women 1968	Women 1991
45.7		5.1			
50.0		5.5			
54.9		5.9			
60.0		6.5		7.2	
91.4		9.1			
100.0	10.0	9.9	9.92	11.1	10.49
200 (s)	19.5	19.5			
200 (t)	20.0		19.72	22.7	21.34
400.0	44.9		43.29	51.9	47.6

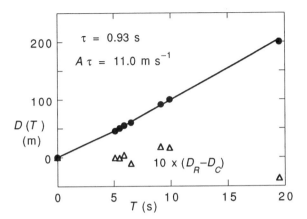

FIGURE 7.1 Distance vs time for world-record sprints. The line shows the best fit in the Keller model using the 1972 men's data (solid points). The triangles are the distance in the given time minus the calculated distance for that time, scaled up by a factor of 10.

Figure 7.1 shows the distance-time data for men sprinters in 1972. It may not be evident that its slope is increasing significantly for about the first 5 seconds. The aims of our analysis are first to solve the linear differential equation (7.5) for the speed and the distance covered, to repeat the analysis of Keller's data, then to analyze some of the other data in Table 7.1. This will be most efficient if we write a program to handle the arithmetic involved while we puzzle out the best choice of parameters A and τ. We will make weighted-least-squares fits of the data, making use of results from Section 6.4.

Warming up to the problem

We may use familiarity with athletics, and some mathematics, to build up the solution of (7.5). In track sprints the runners start from rest, so we must have the initial conditions that at $t = 0$ the speed $v = 0$. Therefore, near the start of the race we may neglect the second term on the right-hand side of (7.5), so that $v(t) \approx At$ at early times. On the other hand, if the A term in (7.5) were negligible, we would obtain the exponential function as the solution, namely $v(t) = C_e \exp(-t/\tau)$, where C_e is a constant. Is there a patched-up solution to the world-record sprint problem that contains both approximations?

Exercise 7.4

(a) Show by expanding the exponential to its linear term in t that the linear combination of the two linear solutions

$$v(t) = A + C_e e^{-t/\tau} \tag{7.6}$$

gives $v(t) = At$ for small t provided that $C_e = -A\tau$.

(*b*) By substituting in the original sprint differential equation (7.5), verify that

$$v(t) = A\tau\left(1 - e^{-t/\tau}\right) \tag{7.7}$$

is a complete solution of the equation that is also consistent with the initial conditions. ∎

Thus, by a combination of insight and algebra we have produced the solution to Keller's equation for modeling world-record sprints. You may recognize the differential equation (7.5) as an example of a linear first-order equation that can be solved by the use of *integrating factors*. For this method of solution, see (for example) the calculus text of Wylie and Barrett.

A simpler method of solution is to notice that (7.5) is also a separable differential equation, for which (as discussed in Wylie and Barrett) this equation can now be rearranged as the integral over a function of *v* and an integral over *t*.

Exercise 7.5
Convert (7.5) to integral form by writing it as

$$\int_0^{v(t)} \frac{D_t \prime v\prime}{A - v\prime/\tau} \, dt\prime = \int_0^t dt\prime \tag{7.8}$$

then recognizing the left-hand-side integrand as the derivative of a logarithm. Evaluate this logarithm for the given limits of integration, then take the exponential on both sides, show that the solution (7.7) for *v* (*t*) is obtained. ∎

Thus we have a second method for solving the sprint differential equation (7.5)

For our comparison with data we need an expression for the distance, *D*, run in a given time *T*. (Actually, we would prefer the relation in the other direction, but there is no closed expression for this.) It is also informative to calculate the runner's average speed during the race. We have that

$$D(T) = \int_0^T v(t) \, dt \tag{7.9}$$

where we have built-in the initial condition that $D = 0$ if $T = 0$; that is, the runners must toe the line.

Exercise 7.6
(*a*) Substitute (7.7) for *v* (*t*) then carry out the time integration in order to show

$$D(T) = A\tau\left[T - \tau\left(1 - e^{-T/\tau}\right)\right] \qquad (7.10)$$

which predicts D for given T.

(*b*) The average speed during the race is just the total distance divided by the elapsed time. Use (7.10) to show that this average speed, $v_{av}(T)$, is

$$v_{av}(T) = A\tau\left[1 - \frac{\tau}{T}\left(1 - e^{-T/\tau}\right)\right] \qquad (7.11)$$

(*c*) From (7.11) show by expanding the exponential to the third order in T/τ that near the start of the race

$$v_{av}(T) \approx \frac{1}{2}AT\left(1 - \frac{2T}{3\tau}\right) \qquad (7.12)$$

so that the average speed increases nearly linearly with time.

(*d*) Show that for long times, such that $T/t \gg 1$, the exponential term dies away and a uniform average speed is predicted, having value

$$v_{av}(T) \rightarrow A\tau \qquad (7.13)$$

which is just the solution of (7.5) as the left-hand side tends to zero, a steady speed that gradually comes to represent the average speed. ∎

With all these exercises you are probably getting more of a workout than an athlete, so it is time to apply Keller's mathematical model to the world-record data. A program is worthwhile for this purpose.

Program for analyzing sprint data

For analyzing the sprint times in Table 7.1 and for predicting average speeds and how they vary with improvements in athletic performance, some programming effort will allow the parameter space of A and τ to be explored thoroughly and efficiently.

Before outlining the program structure, examination of the equation for the distance covered, (7.10), is of interest. We notice that A appears only in the combination $A\tau$, which is a common normalization factor for all the data that are analyzed using a given value of τ. In Section 6.4 we show that in a weighted least-squares fit (not necessarily to straight lines) such a model normalization factor can be obtained directly from comparison between the model with the factor set to unity and the data. Equation (6.50) is the operative formula that we use to reduce our analysis from two parameters to a single parameter τ. This greatly decreases the time it takes to get an optimum fit, at the expense of only a half-dozen lines more code. The same arithmetic also produces the minimum value of the weighted objective function (6.53). Contrary to reality, but sufficient for our purposes, we assume that T is precise. We use weights that vary as $1/D^2$, as in Keller's analysis.

As suggested by (7.10), $A\tau$ is the relevant parameter for analysis, so this is to be calculated and output by the analysis program. It is also interesting to see the difference between the data and the calculated distances, $D_R - D_C$, in Figure 7.1. The average speeds, (7.11), are not of interest until we have an optimum τ value, so these can be readily be programmed on a pocket calculator or on your workstation.

The analysis program World Record Sprints has the following structure. The first section provides options for input of sprint data files, for example, those in Table 7.1. The next part readies matching output files, either to be written over or to be added to. The main work of the program is in the loop to optimize τ and A, which is controlled by the user. Only τ has to be input because A is derived automatically as a normalization of the calculated distances, as explained two paragraphs above. The output for each τ value chosen is sent to the console and to the output files for later graphical interpretation. Finally, the definition of the distance formula is provided in function DIST. If you decide that a different single-parameter model is appropriate, you need only modify DIST. The program listing is as follows.

PROGRAM 7.1 Analysis of world-record sprints using the Keller model (7.5).

```
#include <stdio.h>
#include <math.h>
#define MAX 8

main()
{
/* World Record Sprints */
FILE *fin,*fout;
FILE *fopen();
double DisR[MAX],Time[MAX],DisC[MAX];
double Din,Tin,A,Acalc,tau,Atau;
double obj,sum_yData,sum_yProd,sum_yFit,dif,weight;
int choice,ND,k;
char yn,wa;
double DIST();

printf("World Record Sprints\n");
choice = 1;
while ( choice > 0 )
  {
  printf("Choose data:\n0: Quit program\n1: Men 1968\n");
  printf("2: Women 1968\n3: Men 1991\n4: Women 1991\n");
  printf("5: Men 1972 (Keller)\n");
  scanf("%i",&choice);
  if ( choice == 0 )
    {
    printf("\nEnd World Record Sprints");  exit(0);
```

```
    }
if ( choice < 0 || choice > 5 )
   {
   printf("!! choice=%i only 1 - 5\n",choice); exit(1);
   }
printf("How many distances?\n");
scanf("%i",&ND);
if ( ND > MAX )
   {
   printf("!! # of Distances %i > MAX-1=%i",ND,MAX-1);
   exit(1);
   }
printf("Prepare input file? (y or n):\n");
scanf("%s",&yn);
if ( yn == 'y' )
   {
   printf("Input Distances & Times:\n");
   switch (choice) /* to ready input files for data */
      {
      case 1: fout = fopen("MEN68in","w"); break;
      case 2: fout = fopen("WOMEN68in","w"); break;
      case 3: fout = fopen("MEN91in","w"); break;
      case 4: fout = fopen("WOMEN91in","w"); break;
      case 5: fout = fopen("MEN72in","w"); break;
      }
   for ( k = 1; k <= ND; k++ ) /* input Distance & Time */
      {
      printf("\n%i:  ",k);
      scanf("%lf %lf",&Din,&Tin);
      DisR[k] = Din;   Time[k] = Tin;
      fprintf(fout,"%lf %lf\n",Din,Tin);
      }
   fclose(fout); rewind(fout); /* Ready for reuse */
   } /* end data-preparation loop */
printf("\nWrite over output (w) or Add on (a):\n");
scanf("%s",&wa);
switch (choice)  /* to ready input/output files */
      {
      case 1: fin = fopen("MEN68in","r");
              fout = fopen("MEN68out",&wa); break;
      case 2: fin = fopen("WOMEN68in","r");
              fout = fopen("WOMEN68out",&wa); break;
      case 3: fin = fopen("MEN91in","r");
              fout = fopen("MEN91out",&wa); break;
      case 4: fin = fopen("WOMEN91in","r");
```

```
            fout = fopen("WOMEN91out",&wa); break;
      case 5: fin = fopen("MEN72in","r");
            fout = fopen("MEN72out",&wa); break;
      }
  for ( k = 1; k <= ND; k++ )    /* Data input */
    {
    fscanf(fin,"%lf %lf\n",&Din,&Tin);
    DisR[k] = Din;  Time[k] = Tin;
    }
  printf("\nLoop to optimize  tau  &  A\n");
  tau = 1;
  while ( tau > 0 )
    {
    printf("\nInput  tau: (tau=0 to leave loop)\n");
    scanf("%le",&tau);
    if ( tau > 0 )
    {
    sum_yData = 0;  sum_yProd = 0;  sum_yFit = 0;
    A = 1; /* for clarity in definition of  DIST  */
    for ( k = 1; k <= ND; k++ ) /* loop over data */
    {
    DisC[k] = DIST(Time,k,A,tau);
    weight = 1/(DisR[k]*DisR[k]);
    sum_yData = sum_yData + weight*DisR[k]*DisR[k];
    sum_yProd = sum_yProd + weight*DisR[k]*DisC[k];
    sum_yFit = sum_yFit + weight*DisC[k]*DisC[k];
    }
    Acalc = sum_yProd/sum_yFit; /* best-fit */
    Atau = Acalc*tau;
    obj =  sum_yData - Acalc*Acalc*sum_yFit;
    printf("tau,A,A*tau,obj: "); /* obj is per data point */
    printf("%8.2lf %8.2lf %8.2lf %8.4le\n",
           tau,Acalc,Atau,obj/ND);
    for (k=1; k<=ND; k++) /* loop for normed calculation */
    {
    DisC[k] = Acalc*DisC[k];
    dif = DisR[k]-DisC[k];
   printf("%i %8.2lf %8.2lf %8.2lf %8.2lf\n",
          k,Time[k],DisR[k],DisC[k],dif);
  fprintf(fout,"%i %8.2lf %8.2lf %8.2lf %8.2lf\n",
          k,Time[k],DisR[k],DisC[k],dif);
    } /* end  k  loop */
    } /* end  tau>0  loop */
    } /* end  while (tau>0)  loop */
  } /* end  choice  loop */
```

```
}

double DIST(Time,k,A,tau)
/* Distance vs Time formula for world-record sprints */
double Time[],A,tau;
int k;
{
double length;
length = A*tau*(Time[k]-tau*(1-exp(-Time[k]/tau)));
return length;
}
```

As a check on the correctness of your formulas, it is a good idea to try the parameters that Keller used with the men's 1972 records up to 200 m. These are $\tau = 0.93$ s and $A = 12.2$ m s^{-2}, thus $A\tau = 11.0$ m s^{-1}. These are the parameters used to generate the fit shown in Figure 7.1. By running World Record Sprints with the men's 1972 data you may verify that there is a marginally better set of parameters for these data.

Because the model is not expected to hold past about 200 m, and because the only official race distances are now 100 m and 200 m, one should not attempt to find a best-fit τ from only these two distances. Also, there is about a 0.5-s penalty in a 200-m race if it is held on a track with a turn. In all the analysis results quoted here 0.5 s was subtracted from the 200-m times for races on turning tracks for both men and women to approximate the 200-m time on a straight track, a correction justified by looking at the 200-m records for 1968.

Exercise 7.7

(a) Code, check, and run World Record Sprints for the men's 1972 data given in Table 7.1. Search on τ in order to verify that the best-fit parameters are $\tau = 0.93 \pm 0.02$ s, $A = 11.8$ m s^{-2}, since $A\tau = 11.0$ m s^{-1}.

(b) Input the men's 1991 records for the 100-m and 200-m sprints, adjusting the times of the latter by subtracting 0.5 s, as justified above. Use the new value of τ in order to show that the best value of A is unchanged, since still $A\tau = 11.0$ m s^{-1}, within the validity of the model and the uncertainty in τ.

(c) Calculate the average speeds for the mens' 1972 records and compare them graphically with the Keller-model prediction (7.11). Show that the men's average speed is about 10.5 m s^{-1} for a 200-m dash.

(d) While I was preparing this book, Carl Lewis of the USA lowered the record time for the 100-m sprint from 9.92 s to 9.86 s. Use this time with the other data for men in Table 7.1 in order to see what effect this has on the best-fit values of τ and A. ∎

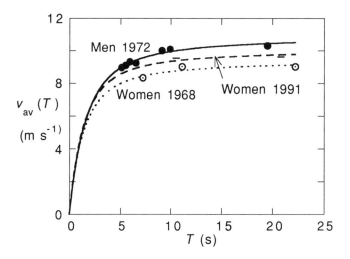

FIGURE 7.2 Average speeds for world-record sprints as a function of time for men in 1972 (solid points and solid line), women in 1968 (crossed circles and dotted line), and for women in 1991 (asterisks and dashed line). The lines are computed using the best fits to the Keller model.

The average speeds for races that are run in a time T are shown in Figure 7.2. Note the approximately linear increase in speed (constant acceleration) for the first second of the race, then the smooth approach to the steady speed (no acceleration) as the duration of the race increases.

Now that we have examined the records for the performance of male athletes, it is of interest to complement them with an analysis of the data for top female athletes.

Women sprinters are getting faster

Although Keller did not analyse the data for women sprinters, there are adequate records for several of the distances, both in 1968 and 1991, as given in Table 7.1. Therefore we can make a similar analysis of the women's records as for the men's records. By contrast with the men's records, which Exercise 7.7 showed have not resulted in improvements within the validity of the model, the women athletes have improved by about 5% over 20 years, as we now demonstrate.

Exercise 7.8

(*a*) Input to the program World Record Sprints the women's 1968 records given in Table 7.1, subtracting 0.5 s from the time for the 200-m race to correct for the track curvature. Show that the best value of $\tau = 0.78$ s and of A is 12.1 m s^{-2}, since $A\tau = 9.45$ m s^{-1}, which is about 14% slower than the male sprinters.

(*b*) Input the women's 1991 records from Table 7.1, again correcting for the curvature of the track. Using $\tau = 0.78$ s, derive the best-fit value of A as 12.9 m s^{-2}, since the speed has increased to $A\tau = 10.1$ m s^{-1}, a 7% improvement in 23 years, or about 6% if scaled uniformly for 1972 to 1991.

(*c*) Calculate the average speeds for the women's 1968 and 1991 records, then compare them graphically with the Keller-model prediction (7.11), as in Figure 7.2. Show that the women's average speed has increased from about 9.1 m s^{-1} to about 9.75 m s^{-1} over a 200-m dash, a 7% improvement.

(*d*) Estimate by linear extrapolation the decade in which women's sprint performances are predicted to match those of men sprinters. Predict that times for 100 m will match by about 2020 and for 200 m by about 2040. ∎

Athletic improvements for women compared to men are discussed by Whipp and Ward. The pitfalls of the linear extrapolations they used have been emphasized by Lichtenberg. The relatively slow increase in athletic performance of male sprinters compared with that of female sprinters over the past two decades might suggest that the speed limits of human sprinting have been essentially reached by males. This question and other constraints to improving athletic performance are discussed in the article by Diamond.

7.3   NONLINEAR DIFFERENTIAL EQUATIONS: LOGISTIC GROWTH

In this section we set up a nonlinear differential equation that is often used to model the growth of biological populations or other self-limiting system, such as a chemical system without feedback of chemical species. We first set up the differential equation, then we explore its properties both analytically and numerically, before showing how the equation may be generalized.

The logistic-growth curve

Although exponential-growth equations appear often in science, such growth cannot continue indefinitely. The corresponding differential equations must be modified so that increasing inhibitory effects take place as time goes by. For example, in an electronic circuit this damping could be obtained by having negative feedback proportional to the output. In ecology, the population dynamics of natural species often has self-generated or external constraints that limit the rate of growth of the species, as described in the introductory-level article on mathematical modeling by Tuchinsky. Chapter 3 of the classical treatise by Thompson provides a wealth of material on growth equations.

One interesting differential equation that models such inhibitory effects is the nonlinear *logistic-growth* differential equation

$$D_t N(t) = \lambda N(t) - \lambda_1 N^2(t) \tag{7.14}$$

in which both λ and λ_1 are positive. The first term on the right-hand side gives the familiar rate of increase proportional to the number present at time t, $N(t)$, and thus to exponential increase of N with time, since $\lambda > 0$. The second term on the right-hand side is an inhibitory effect since $\lambda_1 > 0$, and at any time t it increases as the square of the current value, $N(t)$. The competition between increase and decrease leads to a long-time equilibrium behavior. In the biological sciences (7.14) is often called the Verhulst equation, after the nineteenth-century mathematician.

Exercise 7.9

(a) Show that the equilibrium condition $D_t N = 0$ is satisfied by setting the right-hand side of (7.14) to zero to obtain the solution for the equilibrium N value as $N_e = \lambda/\lambda_1$, so that (7.14) can be rewritten as

$$D_t N(t) = \lambda N(t)\left[1 - N(t)/N_e\right] \tag{7.15}$$

(b) Noting that at early times such that $N(t)/N_e << 1$ we have the equation for exponential growth, and that for long times the slope tends to zero, assume a solution of the logistic-growth equation (7.14) of the form

$$N(t) = A/\left[1 + B\,e^{-\lambda t}\right] \tag{7.16}$$

By using the immediately preceding arguments, show that $A = N_e$.

(c) Show that the assumed form of the solution satisfies the original differential equation (7.14) for any value of B.

(d) If the number present at time zero is $N(0)$, show that this requires $B = N_e/N(0) - 1$. ∎

If we assemble the results from this exercise, we have that the solution of the differential equation for a quadratically self-inhibiting system has the number present at time t, $N(t)$, is given by

$$N(t) = N_e/\left[1 + (N_e/N(0) - 1)e^{-\lambda t}\right] \tag{7.17}$$

where N_e is the equilibrium population and $N(0)$ is the initial population. Equation (7.17) describes *logistic growth*.

The function $N(t)$ comes in various guises in various sciences. In applied mathematics, and ecology it is called the logistic, sigmoid, or Verhulst function. In statistics it is the logistic distribution, and in this context t is a random variable. As discussed in the book edited by Balakrishnan, its first derivative resembles the Gaussian distribution introduced in Section 6.1. In statistical mechanics, t usually represents energy, and $N(t)$ is called the Fermi distribution. It is called this (or the Woods-Saxon function) also in nuclear physics, where t is proportional to the distance from the center of the nucleus and the function measures the density of nuclear matter. In this context, many of the properties of $N(t)$ are summarized in the monograph by Hasse and Myers.

In order to explore the properties of logistic growth, it is convenient to recast (7.17) so that only dimensionless variables appear.

Exercise 7.10

(a) Define the number as a fraction of the equilibrium number as

$$n(t) \equiv N(t)/N_e \qquad (7.18)$$

and change the time unit to

$$t' \equiv \lambda t \qquad (7.19)$$

Show that the logistic differential equation can now be written in dimensionless form as

$$D_{t'} n(t') = n(t') - n^2(t') \qquad (7.20)$$

in which there are no parameters.

(b) Show that the solution to this equation can be written as

$$n(t') = \frac{1}{1 + [1/n(0) - 1] e^{-t'}} \qquad (7.21)$$

in which $n(0)$ is obtained from (7.18). ■

Graphs of $n(t')$ against t' are shown in Figure 7.3 for several values of $n(0)$. If the system is started at less than the equilibrium population, that is if $n(0) < 1$, then it grows towards equilibrium, and the opposite occurs for starting above equilibrium, as for $n(0) = 1.25$ in Figure 7.3. By contrast, if there is no feedback then the system population increases exponentially for all times, as shown in Figure 7.3 for an initial population of 0.125.

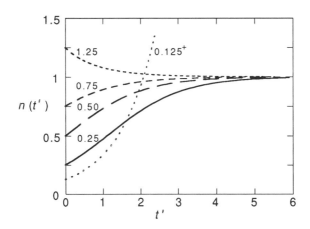

FIGURE 7.3 Solutions to the logistic differential equation (7.20) for various initial numbers as a fraction of the equilibrium number $n(0)$. The dotted curve shows exponential growth without feedback for an initial number of 0.125.

Now that we have understood how to model feedback by a nonlinear differential equation, it is interesting to explore logistic growth in more detail.

Exploring logistic-growth curves

Curves of logistic growth have many interesting properties, many of which are not self-evident because of the nonlinear nature of the differential equations (7.14) and (7.20). Several of the properties that we derive are important in studying ecological population dynamics, electronic feedback, and the stability of systems under disturbances. In order to simplify matters in the following, we use the scaled population, n, and we relabel the dimensionless time variable as t rather than t'.

Exercise 7.11
Show that if $n(0) < 1$, the logistic curve $n(t)$ has a point of inflexion (where $n'' = 0$) at the scaled time $t = \ln\ 1/n(0) - 1]$. This behavior may be visible in Figure 7.3, and it justifies another name often given to this curve, the *S curve*. Show that if $n(0) > 1$, there is no point of inflection. ∎

Even though for the logistic-growth curve the growth is limited, the following exercise may convince you that prediction of the final population from observations made near the beginning of the growth period is not straightforward.

Exercise 7.12
(a) From the differential equation (7.14) show that N_e can be predicted from the populations N_1, N_2 at times t_1 and t_2 and the growth rates at these times N'_1 and N'_2 according to

$$N_e = \frac{N_1 N'_2/N_2 - N_2 N'_1/N_1}{N'_2/N_2 - N'_1/N_1} \tag{7.22}$$

(b) Explain why an accurate value of N_e is difficult to obtain using this equation applied to empirical data with their associated errors. Consider the discussion in Section 4.3 about unstable problems and unstable methods. ∎

This exercise highlights the difficulties of deciding whether a natural species needs to be protected in order to avoid extinction. The question becomes especially interesting in the ecology of depleting resources, especially when one considers the following modification and application of the logistic-growth equation (7.14) to examine the question of harvesting nature's bounty.

Exercise 7.13
Consider the harvesting of a biological resource, such as fish, which might follow the logistic-growth curve (7.14) if undisturbed. Suppose, however, that the resource is harvested at a rate proportional to its present numbers.

(*a*) Show that the logistic-growth differential equation is then modified to

$$D_t N(t) = \lambda N(t)\left[1 - N(t)/N_e\right] - H N(t) \qquad (7.23)$$

where H (positive) is the harvesting fraction rate.

(*b*) Show that the logistic equation is regained, but with a reduced constant determining the equilibrium time

$$\lambda_H = \lambda - H \qquad (7.24)$$

and a reduced equilibrium population

$$N_{He} = N_e(1 - H/\lambda) \qquad (7.25)$$

(*c*) Verify that the model predicts that the resource will become extinct if $H/\lambda > 1$. Explain this result in words. ■

The logistic-growth equation and its relation to the stability of iteration methods and to chaos is considered in Chapter 5 of Hubbard and West's book and software system, in Chapters 4 and 8 in Beltrami's book, and in the chapter "Life's Ups and Downs" in Gleick's book. A discussion of solutions of the logistic equation, their representation in a phase plane, and the effects of lags are given in Haberman's book on mathematical modeling. Quasi-chaotic behavior induced by discretizing the solution of (7.15) is examined in the article by Yee, Sweby, and Griffiths.

Generalized logistic growth

Suppose that instead of quadratic feedback to stabilize the population, as we have in (7.14), there is feedback proportional to some power of n, say the $(p+1)$ th power. That is, (7.14) becomes the *generalized logistic equation*

$$D_t n(t) = n(t) - n^{p+1}(t) \qquad (7.26)$$

In mathematics this is sometimes called Bernouilli's equation. By following some simple analytical steps, you can find directly the solution of this equation.

Exercise 7.14
Divide throughout (7.26) by n, then convert the derivative to a logarithmic derivative and introduce a new time variable

$$t_p \equiv p\, t \qquad (7.27)$$

Also define

$$n_p(t_p) \equiv n^p(t_p) \qquad (7.28)$$

Show that n_p satisfies the differential equation

$$\frac{d \ln[n_p(t_p)]}{dt_p} = 1 - n_p(t_p) \tag{7.29}$$

which is a differential equation independent of the feedback power p. Thus, argue that, using the power factors given in (7.27) and (7.28), *all* the solutions of the generalized logistic equation are self-similar and can be obtained from that for $p = 1$, namely (7.21). ∎

The solution of the generalized logistic equation (7.26) may therefore be written

$$n(t') = \frac{1}{\left\{1 + [1/n^p(0) - 1]e^{-pt'}\right\}^{1/p}} \tag{7.30}$$

where, as in (7.18), we have n as the fraction of the equilibrium population, and t' as the scaled time, λt. This behavior is quite remarkable in that it is the particular form of the first-order nonlinear differential equation (7.14) that allows such a generalized solution to be obtained.

The behavior of the generalized logistic curve as a function of the feedback power p in (7.30) is shown in Figure 7.4 for $n(0) = 0.5$; that is, the initial population is one-half the final equilibrium population. As you would guess, as p increases, the population approaches the equilibrium value $n = 1$ more rapidly.

Exercise 7.15
Set up a calculation to compute formula (7.30) for given $n(0)$ and p. For a system that starts off above equilibrium population, $n(0) > 1$, run this calculation for a range of t' similar to that shown in Figure 7.4. Check your results against the curve shown in Figure 7.3 for $p = 1$, $n(0) = 1.25$. ∎

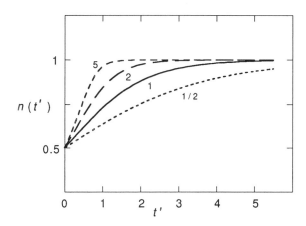

FIGURE 7.4 Generalized logistic equation solutions (7.30) for feedback proportional to the $(p + 1)$ power of the number present, shown for $n(0) = 0.5$ and four values of p.

The population growth as a function of time in units of $1/\lambda$ is well behaved even for a non-integer power, such as with $p = 1/2$ in Figure 7.4, for which the negative feedback in (7.26) is proportional to the slower $n^{3/2}$ power, which is more gradual than n^2 feedback ($p = 1$) in the conventional logistic-growth equation (7.14).

Our generalization of the logistic equation to (7.26), with its self-similar solution (7.30), does not seem to have been much investigated. The function (7.30) has been used as the Type I generalized logistic distribution, as discussed by Zeiterman and Balakrishnan in the statistics monograph edited by the latter author. For detailed applications in statistics, use of logistic distributions is described in the book by Hosmer and Lemeshow.

7.4   NUMERICAL METHODS FOR FIRST-ORDER EQUATIONS

As we learned in the preceding sections, if an analytical solution of a differential equation is possible, the method of solution is often specific to each equation. Numerical methods for differential equations, by contrast, can often be applied to a wide variety of problems. But, such methods are not foolproof and must be applied carefully and thoughtfully. (If numerical recipes are used by those who have not graduated from cooking school, then the dish may be quite unsavory.)

Our aim in this section is to emphasize basic principles and to develop general-purpose methods for solving first-order differential equations. In Project 7 in Section 7.5 we develop a program for the first-order Euler method. Numerical methods for differential equations are investigated further in Sections 8.3 – 8.6, where we emphasize second-order equations and include second-order Euler methods.

In numerical methods for differential equations, one often speaks of "integrating the differential equation." The reason for this terminology is that since

$$y(x) = \int^{x} \frac{dy}{dx'} dx' \tag{7.31}$$

we may consider solving a differential equation as an integration process. Indeed, this is often the formal basis for developing numerical methods.

In what follows we often abbreviate derivatives using the prime notation, in which $y'(x) = D_x y(x)$ for the first derivative, and the number of primes indicates the order of the derivative. Past the third derivative, numerals are used, (3), (4), etc.

Presenting error values

Errors in numerically estimated values can be presented in two ways. The first is to present the actual error in the y value at $x = x_k$, which we call e_k and define as

$$e_k \equiv y_k - y_k^{(\text{num})} \tag{7.32}$$

in which the first quantity is the exact value at x_k and the second is the numerically estimated value at this point. Unlike random errors of experimental data, but like their systematic errors, the value of e_k is usually reproduced when the calculation is repeated. (An exception is a Monte Carlo error estimate whenever different sampling of random numbers is made from one calculation to the next.) The sign of e_k is therefore significant, so we will be systematic in using definition (7.32).

The second way of presenting errors is as the relative error, r_k, defined by

$$r_k \equiv e_k/y_k \tag{7.33}$$

which we use only when y_k is not too close to zero. Sometimes we give $100 \times r_k$, the percentage error, or we show errors scaled by some power of 10. Clearly, we usually have only estimates of e_k or r_k, rather than exact values, as we indicate by using the "$\approx$" sign when appropriate. In the program `Numerical DE_1` in Section 7.5 the program presents the errors in both ways.

Euler predictor formulas

A predictor formula for the numerical solution of a differential equation is one in which previous and current values are used to predict y values at the next x values. The general first-order differential equation can be written

$$y'(x) = f(x,y) \tag{7.34}$$

where f is any well-behaved (usually continuous) function of x and y, but it should not contain any derivatives of y with respect to x past the first derivative. Suppose that y is known for a succession of x values (equally-spaced by an amount h, to make things simple) up to $x = x_k$. We now describe two basic ways of advancing the solution at x_k to that at $x_{k+1} = x_k + h$. For notational simplicity we abbreviate $f_k = f(x_k, y_k)$. We also use the abbreviation $y_k = y(x_k)$, with a similar notation for derivatives.

For a working example we choose a function for which the solution is well-known and simple, but which illustrates the various methods and their errors. We work with

$$f(x,y) = y \tag{7.35}$$

for which we already know a solution to (7.34), namely

$$y(x) = e^x \tag{7.36}$$

in which we have chosen the boundary condition $y(0) = 1$, so that actual errors and relative errors will coincide when x is near zero. This choice of f is also simple to deal with in estimating errors because all derivatives of y are just equal to y itself.

The first methods used for our first-order differential equation (7.34) are the so-called *Euler predictor formulas*. In the first version of this method we approximate the first derivative on the left-hand side of (7.34) by the forward-difference formula (4.25), namely

$$y'_k \approx (y_{k+1} - y_k)/h \qquad (7.37)$$

If h is small enough we may get a good approximation to y at the next x value by inserting this in (7.34) and solving for y_{k+1} as

$$y_{k+1} \approx y_k + hf(x_k, y_k) \qquad (7.38)$$

This forward-difference predictor formula we call the *forward Euler predictor* for a first-order differential equation.

An estimate of the error in each step of this predictor is the next term in the Taylor expansion of y_{k+1}. From (7.34) and the general Taylor series formula (3.6), this estimate is

$$e_k \approx \frac{h^2}{2} f'_k \qquad (7.39)$$

For the working example in which we use the exponential function (7.36) note that $y(x+h) = e^{x+h}$. Therefore

$$e_k = e^{x_k}\left[e^h - (1+h)\right] \qquad (7.40)$$

is the actual error, which agrees with the estimate (7.39) through h^2 terms.

Before turning to numerical comparisons, we consider a second Euler predictor formula. As shown in Section 4.4, the central-difference derivative is more accurate than the forward-difference derivative. We therefore expect that using the central differences in (7.34) will produce a more-accurate prediction of the next y value. Try it and see.

Exercise 7.16
Show that using central differences produces the predictor formula

$$y_{k+1} = y_{k-1} + 2hf(x_k, y_k) + e_k \qquad (7.41)$$

in which the error estimate is

$$e_k \approx \frac{h^3}{3} f_k \qquad (7.42)$$

obtained from the first neglected term in the Taylor expansion. ∎

Formula (7.41) we call the *central Euler predictor* for solving a first-order differential equation.

For our working example of a first-order equation, with $f(x, y) = y$, the error estimates are particularly simple, since the exact value, given our initial value of $y(0) = 1$, is just $y(x) = e^x$. You can therefore work out relative errors and compare the efficiencies of the two methods for yourself.

Exercise 7.17

(a) Consider the relative errors in the forward predictor formula (7.38), $r_k(f)$, and in the central predictor formula (7.41), $r_k(c)$, for the working example. Show that these error estimates for each step do not depend on x or y, and that for $h = 0.05$ are about 1.25×10^{-3} and 0.042×10^{-3}, respectively.

(b) Estimate the efficiency of the central-difference method compared to the forward-difference method for the working example as the ratio of step sizes h_c and h_f required for about the same errors in both methods. Show that the relative efficiency is about $\sqrt{3/(2h_c)}$, and show that this is at least a factor of 5 for $h_c = 0.05$. ∎

The above formulas advance the solution of the differential equation and provide estimates of the error incurred at each step.

Testing the Euler predictors

We now have two predictor formulas for solving first-order differential equations, formulas (7.38) and (7.41), together with error estimates for each step, (7.39) and (7.42). It is time to check these out numerically.

We take a look forward to the completed program for the first-order solutions that is given in Section 7.5, concentrating on the forward- and central-predictor methods that we have just analyzed. We use this with the exact analytical solution $y(x) = e^x$ for our working example, as explained above. It is probably easiest to first enter the definition, control and input-output parts of the program Numerical DE_1 from Section 7.5, and to write only the code cases for choice = 1 and choice = 2, the forward and central predictors, respectively. By doing this, you will be coding something that you understand and you will exercise your mind rather than just exercising your fingers.

Exercise 7.18

(a) Key in all parts of the program Numerical DE_1 given in Section 7.5, except case 3 and case 4, which we consider in Section 7.5. Note that the main program refers to f in (7.34) as FUNC. For general use, you will eventually substitute the coding for the function of interest to you. For our working example the function is y. Don't be tempted to anticipate the exact value by using exp (x) instead.

(b) Run this version of the program for an interesting range of x values. For simplicity xmin = 0, xmax = 4.0, and h = 0.05, are reasonable choices for getting started. As usual, there is a file (NUMDE_1) to record your output for use in a spreadsheet or graphics application.

(c) Add the sections of code that allow comparison of the numerical values just computed with analytical values for the solution of the differential equation. These values are computed by function ANALYT. Run Numerical DE_1 with compare = y (for 'yes'), and check your output against the values shown graphically in Figure 7.5. ∎

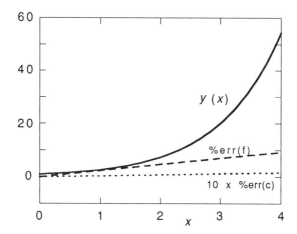

FIGURE 7.5 Euler predictor solutions of the first-order differential equations (7.34), (7.35) with boundary condition $y(0) = 1$. The solid line shows the analytical solution, $\exp(x)$, the dashed line shows the percentage error in the forward-difference numerical solution, and the dotted line shows 10 times the percentage error in the central-difference solution.

Sample output from `Numerical DE_1` for the Euler predictor methods (cases 1 and 2), using the parameters suggested in Exercise 7.18, is shown in Figure 7.5. The relative errors in the forward- and central-difference Euler predictors are also shown, as err(f) and err(c), respectively.

As you will notice from your output and from Figure 7.5, solution of the first-order equation by using the central-difference predictor has a much smaller accumulated relative error (a factor of 60 less at $x = 4$) than does the forward-difference predictor. From point to point the errors are in agreement with the estimates (7.39) and (7.42), respectively. The accumulated error after many steps is generally much more difficult to estimate, because errors may cancel in different regions of x. In our working example of the monotonically-increasing function $\exp(x)$ the error just accumulates linearly with the number of steps. This is clear from the linearity of the two error curves in Figure 7.5.

Adams predictor formulas

In the Euler predictor formulas for numerical solution of the first-order differential equation (7.34) that we developed and applied in the two preceding subsections, the methods rely on estimating derivatives. An alternative is to integrate (7.34) once with respect to x, then to approximate the integral of the right-hand side.

We now develop two examples of such formulas, beginning with the original linear differential equation

$$y'(x) = f(x, y) \tag{7.43}$$

We integrate this equation from x_k to x_{k+1}, to obtain

$$y_{k+1} = y_k + \int_{x_k}^{x_{k+1}} f(x,y)dx \qquad (7.44)$$

This equation is, like many beautiful things, true but not immediately useful. Hidden inside the integrand are all the unknown values of y in the interval between the k th and the $(k+1)$ th value, which is just what we are trying to find. Therefore, some practical means of approximating the integral is needed.

Our first approximation of the integral rule is simply to use the trapezoid rule from Section 4.6, namely

$$\int_{x_k}^{x_{k+1}} f(x,y)\,dx \approx [f(x_k,y_k) + f(x_{k+1},y_{k+1})]\,h/2 \qquad (7.45)$$

which has an error of order h^3. If this approximation to the integral is used in (7.44), then we have an implicit equation for y_{k+1}, since it then appears on both the right- and left-hand sides of (7.44). An iterative solution of this equation may be tried. That is, a trial value of y_{k+1} is used in f to predict its value on the left-hand side. We indicate this procedure by writing

$$y_{k+1}^{(n+1)} \approx y_k + \left[f(x_k,y_k) + f(x_{k+1},y_{k+1}^{(n)})\right]h/2 \qquad (7.46)$$

where the order of the iteration, $n \geq 0$. Note that iterative solution of this equation to a given limit does not guarantee a more-correct solution of the original differential equation than that provided by the trapezoid rule. In Section 7.5 we refer to (7.46) as the *Adams-trapezoid predictor*.

If the integral in (7.44) is estimated by a formula that is more accurate, then the solution of the differential equation will be more accurate for a given ste size h.

Exercise 7.19
Use Simpson's rule, (4.45), for the integral to derive the estimate

$$y_{k+1}^{(n+1)} \approx y_{k-1} + \left[f(x_{k-1},y_{k-1}) + 4f(x_k,y_k) + f(x_{k+1},y_{k+1}^{(n)})\right]h/3 \qquad (7.47)$$

which is to be solved by iterating on n until a desired convergence of y value is reached. ∎

The error in this formula, once iterated to convergence, is just the error in the Simpson integration rule, which is of order h^5. We refer to formula (7.47) as the *Adams-Simpson predictor*.

Formulas in which combinations of integration-iteration methods are used are called *Adams closed formulas* or *Adams-Moulton formulas*. We illustrate their use in the following project. It would be desirable to follow the extrapolation (predictor) step by a consistent corrector step in which the original differential equation is used to improve the solution accuracy. Such predictor-corrector methods are described in Chapter 8 of Vandergraft and in the book by Zill. Jain's treatise on numerical solution of differential equations contains much more than you will probably ever use to meet your needs.

In the next section we describe a program that you can use to explore numerically the four solution methods that we have developed.

7.5   PROJECT 7: PROGRAM FOR SOLVING FIRST-ORDER EQUATIONS

In the preceding section we worked out predictor formulas for numerical solution of first-order differential equations of the kind (7.34). We also gained experience with the Euler forward- and central-predictor methods, anticipating the program that is presented in this section. We now supplement these methods with the numerics and coding of the Adams integration formulas, (7.46) and (7.47), derived in the preceding subsection. We first describe the parts of the program that relate to the Adams methods, we describe how to test the methods, then we explore use of the program for solving differential equations numerically.

Programming the differential equation solver

The Adams predictor methods involve some extra steps in programming and coding than are needed for the Euler predictors, so we now describe these steps.

The realities of computing require us to be more specific about the iteration procedures (7.46) and (7.47), because we must have a stopping criterion. In the program `Numerical DE_1` for cases 3 and 4 (the Adams predictors) only one of two stopping criteria needs to be satisfied for the iteration on n to quit. The first criterion is that no more than `nmax` iterations are made, where `nmax` is an input parameter that forces a definite stopping of the iteration. The alternative criterion is that the fractional change in the y value since the last iteration should be less than an input value `epsilon`.

Exercise 7.20

(*a*) Explain why it is both inefficient and uninformative to use `nmax` as the only stopping criterion in the iterations (7.46) and (7.47).

(*b*) Explain why it is dangerous (for the computer) to use only `epsilon` as the iteration-stopping criterion. ■

Since either criterion used alone is ineffective, but their combination is informative and prudent, in `Numerical DE_1` if either criterion is satisfied, the iteration stops. This explains in `case` 3 and 4 the `while` (`&&`) statements for the iteration loops, where `&&` is C-language for the logical "and" operation.

The comparison with `epsilon` is straightforward for the Adams-trapezoid case (3), but for Adams-Simpson (case 4) it is more complicated because we are iterating on two adjacent k values. A reasonable compromise that is made in the program is to average the absolute fractional differences for these two values. Experience shows that if only one difference is used and the fractional difference for the other is ignored, then the error for the ignored k values tends to grow unreasonably quickly. Averaging forces a moderating compromise.

In either Adams method, if `epsilon` is made comparable to the number of significant figures that your computer is using, there will be a complicated and confusing interplay between convergence of the iteration and computer limitations from subtractive cancellation and other roundoff problems of the kinds examined in Section 4.3.

So, with all these warnings we should be ready to plunge into the maelstrom of numerical mathematics. Here is the complete program.

PROGRAM 7.2 Numerical solution of first-order differential equations by Euler and Adams predictor methods.

```
#include <stdio.h>
#include <math.h>
#define MAX 201

main()
{
/* Numerical DE_1; Euler & Adams Predictors */
/*     for First-Order Differential Equations */
FILE *fout;
FILE *fopen();
double y[MAX],err[MAX],relerr[MAX];
double xmin,xmax,h,epsilon,x,ylast1,ylast2,absdiff,analytical;
int choice,Nk,nmax,k,kmin,kstep,n;
char wa;
double FUNC(),ANALYT();

printf("Numerical DE; First Order\n");
printf("Write over output (w) or Add on (a):\n");
scanf("%s",&wa);    fout = fopen("NUMDE_1",&wa);
choice = 1;
while ( choice > 0 )
  {
  printf("\nChoose method of solution (zero to quit):\n");
```

```
printf("1 Euler-forward predictor;\n");
printf("2 Euler-central predictor;\n");
printf("3 Adams-trapezoid predictor;\n");
printf("4 Adams-Simpson predictor;\n");
scanf("%i",&choice);
if ( choice == 0 )
  {
   printf("\nEnd Numerical DE; First Order");  exit(0);
  }
if ( choice < 0 || choice > 4 )
  {
  printf("!! choice=%i only 1,2,3 or 4\n",choice);exit(1);
  }
printf("Input  xmin, xmax, h\n");
scanf("%le%le%le",&xmin,&xmax,&h);
Nk = (xmax-xmin)/h+1.1;
if ( Nk > MAX-1 )
  {
  printf("!! # of xsteps, %i, > %i\n",Nk,MAX-1); exit(1);
  }
switch (choice)      /* Boundary values */
  {
  case 1: /* Euler-forward predictor */
    {
    printf("1:  Input y(xmin);\n");
    scanf("%le",&y[1]);
    kmin = 1;  kstep = 1;  break;
    }
  case 2: /* Euler-central predictor */
    {
    printf("2:  Input y(xmin),y(xmin+h);\n");
    scanf("%le%le",&y[1],&y[2]);
    kmin = 2;  kstep = 2;  break;
    }
  case 3: /* Adams-trapezoid predictor */
    {
    printf("3:  Input y(xmin),nmax,epsilon;\n");
    scanf("%le%i%le",&y[1],&nmax,&epsilon);
    kmin = 1;  kstep = 1;  break;
    }
  case 4:  /* Adams-Simpson predictor */
    {
    printf("4:  Input y(xmin),y(xmin+h),nmax,epsilon;\n");
    scanf("%le%le%i%le",&y[1],&y[2],&nmax,&epsilon);
    kmin = 2;  kstep = 2;  break;
```

```
        }
    }

  x = xmin;                        /* Recurrence in  x  */
  for ( k = kmin; k < Nk; k = k+kstep )
    {
    switch (choice)
        {
        case 1: /* Euler-forward predictor */
           {
           y[k+1] = y[k]+h*FUNC(x,y,k);  break;
           }
        case 2: /* Euler-central predictor */
           {
           x = x+h;
           y[k+1] = y[k-1]+2*h*FUNC(x,y,k);
           y[k+2] = y[k]+2*h*FUNC(x+h,y,k+1);  break;
           }
        case 3: /* Adams-trapezoid predictor */
           { /* Start iteration */
           y[k+1] = y[k];  n = 1;  absdiff = 1;
           while ( n <= nmax && absdiff > epsilon ) /* iterate */
              {
              ylast1 = y[k+1];
              y[k+1] = y[k]+
                        (FUNC(x,y,k)+FUNC(x+h,y,k+1))*h/2;
              absdiff = fabs((y[k+1]-ylast1)/ylast1);
              n = n+1;
              }
           break;
           }
        case 4: /* Adams-Simpson predictor */
           {
           x = x+h; /* Start iteration */
           y[k+1] = y[k];  y[k+2] = y[k];  n = 1;  absdiff = 1;
           while ( n <= nmax && absdiff > epsilon ) /* iterate */
              {
              ylast1 = y[k+1];  ylast2 = y[k+2];
              y[k+1] = y[k-1]+
              (FUNC(x-h,y,k-1)+4*FUNC(x,y,k)+FUNC(x+h,y,k+1))*h/3;
              y[k+2] = y[k]+
              (FUNC(x,y,k)+4*FUNC(x+h,y,k+1)+FUNC(x+2*h,y,k+2))*h/3;
              /* For convergence average fractional diferences */
              absdiff = (fabs((y[k+1]-ylast1)/ylast1)+
                          fabs((y[k+2]-ylast2)/ylast2))/2;
```

```
        n = n+1;
        }
      break;
      }
    }
  x = x+h;
  } /* end  k   for  loop */

  printf("Comparison with analytical values\n");
  x = xmin;
  for ( k = 1; k <= Nk; k++ )
    {
    analytical = ANALYT(x);/* analytical and error */
    err[k] = analytical-y[k];
    relerr[k] = 100*err[k]/analytical;
    x = x+h;
    printf("%6.2le %10.6lg %10.6lg %10.6lg\n",
            x,y[k],err[k],relerr[k]);
    fprintf(fout,"%6.2le %10.6lg %10.6lg %10.6lg\n",
            x,y[k],err[k],relerr[k]);
    } /* end  k  output loop */
  } /* end  choice  while  loop */
}

double FUNC(x,y,k)
/*  dy/dx = FUNC  */
double y[];
double x;
int k;
{
double value;

/* Using working example of  f=y  for FUNC */
value = y[k];
return value;
}

double ANALYT(x)
/* Analytical solution of the differential equation */
double x;
{
double value;

/* Using working example of  y=exp(x)  for ANALYT */
value = exp(x);
```

```
return value;
}
```

Exercise 7.21

(*a*) Key in those parts of `Numerical DE_1` that refer to `case 3` and `case 4`, assuming that you did Exercise 7.18. (Otherwise, key in the complete program.) Note that the main program refers to *f* in (7.34) as `FUNC`. For use after testing you will substitute the coding for the function of interest to you. For our working example the function is *y*. Don't anticipate the exact value by using exp (*x*) instead.

(*b*) Run this version of the program for an interesting range of *x* values. For simplicity `xmin = 0, xmax = 4.0` and `h = 0.05`, as in Exercise 7.18, are reasonable choices. The file (`NUMDE_1`) will record your output for use in a spreadsheet or graphics application. The iteration-control values used in Figure 7.6 are `nmax = 5` and `epsilon = 10`$^{-3}$, with the latter convergence criterion usually being satisfied after *n* = 3 iterations.

(*c*) Compare the numerical values just computed with analytical values for the solution of the differential equation that are computed by function `ANALYT`. Run `Numerical DE_1` to check your output against the values shown graphically in Figure 7.6.   ■

The exponential solution of our working-example differential equation is shown in Figure 7.6 for the two Adams predictor methods, along with the relative errors for the two integration methods (trapezoid and Simpson) shown as err(trap) and err(Simp) respectively. Note the major error reduction, by about a factor of 100, for Adams-trapezoid method over Euler-forward method and for Adams-Simpson over Euler-central. This is not all gain without pain, because if *n* iterations are used in the Adams methods, the computing time is increased by about the same factor. Dependent on how complex it is to calculate `FUNC` *f*, it might sometimes be more efficient for the same numerical accuracy to use a smaller stepsize, *h*, in one of the Euler methods. Try it and compare.

Exploring numerical first-order equations

Now that we have gone to the trouble to develop a practical program for numerical solution of first-order differential equations, `Numerical DE_1`, it is interesting to explore other equations with it.

The first equation that I suggest trying is

$$y'(x) = f(x,y) = -\sin(x) \tag{7.48}$$

This has the well-known exact solution

$$y(x) = \cos(x) + y(0) - 1 \tag{7.49}$$

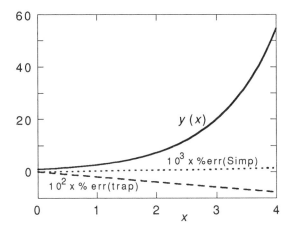

FIGURE 7.6 Adams-predictor solutions of the first-order differential equations (7.34), (7.35) with boundary condition $y(0) = 1$. The solid line shows the analytical solution, $\exp(x)$, the dashed line shows 10^2 times the percentage error in the Adams-trapezoid numerical solution, and the dotted line shows 10^3 times the percentage error in the Adams-Simpson solution.

Because of the oscillatory behavior and the initial decrease of the function as one moves away from $x = 0$, we can see whether the error buildup is sensitive to the oscillatory behavior, something that we could not do for the exponentially increasing function used as the working example.

For simplicity, starting at $x = 0$ and going to just over $x = 2\pi$ is a good choice. This allows us to see how nearly periodic is the numerical solution of (7.48). Because the exact solution becomes zero if x is an odd multiple of $\pi/2$, the output is less confusing if actual errors, e_k from (7.32), are used instead of the relative errors, r_k from (7.33), shown in the preceding computing exercises in this chapter.

Exercise 7.22

(a) In Numerical DE_1 modify function FUNC to assign $-\sin(x)$ to value for the first derivative, and also modify function ANALYT to assign $\cos(x)$ to value for the analytical solution of the differential equation (7.48). Change the file name to SINDE_1 so that the file output from this problem and from the exponential-function problem do not become jumbled.

(b) Run the modified program over an interesting range of x values, from say xmin $= 0$ to xmax $= 7$, with h $= 0.05$. Use each method of solution, choicefrom 1 to 4 to study the accuracy of the four predictor methods of solving the differential equation. In methods 3 and 4 iteration-control values of nmax $= 5$ and epsilon $= 10^{-3}$ are reasonable. For all methods y[1] $= 1$, while the program will request for methods 2 and 4 the second starting value (for h $= 0.05$) y[2] $= 0.9987503$ to seven decimals.

(c) Make a graphical comparison of the predicted y values over the range calculated. With the above accuracy parameters, these will probably be indistinguishable at the resolution of your graphics device. To point out the differences between the results from each method of solution, display the actual errors on a
convenient scale. If you use relative errors, there will be anomalous spikes near
the zeros of $\cos(x)$. Either believe me, or try it and see. ■

Figure 7.7 displays the numerical solution of the linear differential equation (7.48)
and the actual errors from the four methods that we have developed. Note carefully
the scale factors that have ben applied before plotting the errors Generally, the errors increase in magnitude as the function decreases, and the errors in the Euler
methods, which are based on derivatives (f and c), are greater than those in the
Adams methods (*trap* and *Simp*), which are based on integration-iteration.

For the Adams-Simpson method the errors indicated by $err(Simp)$ are actually
alternating errors, for the even and odd k values in y_k. The analysis of such errors
is difficult because the accuracy of the computer, of the computation of the sine
function used in function ANALYT, and of the starting value y_2 are all interacting
with the errors of the numerical method. This may not be of much practical consequence because all the errors are about 0.5×10^{-7} or less, which is about a factor
of 10 million less than the value of the function that is being computed numerically.

Now that you have some experience with numerical solution of first-order differential equations, you may like to try other equations, such as the world-record-
sprints equation in Section 7.2, the logistic equation in Section 7.3, or an interesting equation from your research and study.

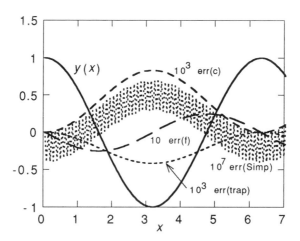

FIGURE 7.7 Solutions of the differential equation (7.48) with boundary condition $y(0) = 1$.
The solid curve shows the analytical solution $\cos(x)$. The curve of long dashes shows 10 times
the error in the forward-difference Euler method, the shorter dashes show 10^3 times the error in the
central-difference Euler predictor, the dotted curve shows 10^3 times the error in the Adams-trapezoid
predictor, and the dashed band shows 10^7 times the error in the Adams-Simpson predictor.

REFERENCES ON FIRST-ORDER EQUATIONS

Balakrishnan, N., Ed. *Handbook of the Logistic Distribution*, Dekker, New York, 1992.

Beltrami, E., *Mathematics for Dynamic Modeling*, Academic Press, Boston, 1987.

Diamond, J., "The Athlete's Dilemma," Discover, **12**, August 1991, p. 79.

Gleick, J., *Chaos: Making a New Science*, Penguin Books, New York, 1987.

Haberman, R., *Mathematical Models*, Prentice Hall, Englewood Cliffs, New Jersey, 1977.

Hasse, R. W., and W. D. Myers, *Geometrical Relationships of Macroscopic Nuclear Physics*, Springer-Verlag, Berlin, 1988.

Hosmer, D. W., and S. Lemeshow, *Applied Logistic Regression*, Wiley, New York, 1989.

Hubbard, J. H., and B. H. West, *Differential Equations*, Part 1, Springer-Verlag, New York, 1991.

Hubbard, J. H., and B. H. West, *MacMath: A Dynamical Systems Software Package for the Macintosh*, Springer-Verlag, New York, 1991.

Jain, M. K., *Numerical Solution of Differential Equations*, Wiley Eastern, New Delhi, second edition, 1984.

Keller, J. B., "A Theory of Competitive Running, " Physics Today, September 1973, p. 43.

Lichtenberg, D. B., A. L. Julin, and P. H. Sugden, Nature, **356**, 21 (1992).

Thompson, D'Arcy W., *On Growth and Form*, Cambridge University Press, New York, 1943.

Tuchinsky, P. M., "Least Squares, Fish Ecology, and the Chain Rule," UMAP Module 670, in *UMAP Modules Tools for Teaching*, COMAP, Arlington, Massachusetts, 1986, pp. 195 – 240.

Vandergraft, J. S., *Introduction to Numerical Computations*, Academic Press, New York, 1978.

Whipp, B. J., and S. A. Ward, "Will Women Soon Outrun Men?," Nature, **355**, 25 (1992).

Wolfram, S., *Mathematica: A System for Doing Mathematics by Computer*, Addison-Wesley, Redwood City, California, second edition, 1991.

Wylie, C. R., and L. C. Barrett, *Advanced Engineering Mathematics*, McGraw-Hill, New York, fifth edition, 1982.

Yee, H. C., P. K. Sweby, and D. F. Griffiths, "Dynamical Approach Study of Spurious Steady-State Numerical Solutions of Nonlinear Differential Equations," Journal of Computational Physics, **97**, 249 (1991).

Zill, D. G., *Differential Equations with Boundary-Value Problems*, Prindle, Weber & Schmidt, Boston, 1986.

Chapter 8

SECOND-ORDER DIFFERENTIAL EQUATIONS

Differential equations of second order, both linear and nonlinear, are very common in the natural sciences and in many applications. In the physical sciences and engineering Newton's force equations are ubiquitous, and they are second-order equations in the displacement as a function of the variable $x = t$, where t denotes time. In the quantum mechanics of molecules, atoms, and subatomic systems the Schrödinger equation for the wave function of a time-independent state is a second-order differential equation which is usually also linear. Thus, much of the mathematics (and therefore much of the computing) that describes dynamics in the natural sciences is encompassed by second-order differential equations.

This chapter builds on the foundation of first-order differential equations that we laid in Chapter 7, both analytically and computationally. For example, the Euler predictors developed in Sections 7.4 and 7.5 are now extended to second-order equations in Sections 8.3 and 8.4. Examples of second-order equations are drawn from both classical and quantum physics, such as the equations for resonance in Section 8.1, generalized catenaries in Section 8.2, and the quantum harmonic oscillator in Section 8.5. We introduce some new computing concepts, methods, and programs that are especially relevant to second-order differential equations, such as the notion of stiff differential equations (Section 8.6), the Noumerov method for linear second-order equations (Section 8.3), and the two programming projects for second-order Euler methods (Project 8A in Section 8.4) and for the Noumerov method (Project 8B in Section 8.5).

By understanding the materials in this chapter and the preceding, you will have good preparation for using a wide variety of analytical and computational techniques involving differential equations. The background that you have acquired should be sufficient for you to read and use the research literature on differential equations in applied mathematics and in the natural and applied sciences.

8.1   FORCES, SECOND-ORDER EQUATIONS, RESONANCES

In this section we review the relations between forces, second-order equations, and resonances. The background and more details are given in texts such as Part 1 of Pippard's treatise on the physics of vibration. The analogies between the differential equations for mechanical and electrical systems are summarized. Then second-order equations for free motion are presented, and their solutions are derived and interpreted. This analysis is preparatory to describing forced motion and resonant behavior. The latter topic is re-examined from the Fourier-transform viewpoint in Sections 10.2 and 10.3 when discussing the transform of a Lorentzian function that describes a resonance.

Forces and second-order equations

We summarize here the relation between forces and second-order differential equations in nonrelativistic classical mechanics. For simplicity of notation, we work implicitly with a single directional component of forces, momenta, and displacements.

Recall that Newton's equation for the momentum, p, and displacement, y, of a particle of mass M as a function of time, t, when subjected to a force F can be written as a single second-order equation

$$M \frac{d^2 y}{dt^2} = F \tag{8.1}$$

or, with more information content, as a pair of first-order equations, one for the momentum

$$\frac{dp}{dt} = F \tag{8.2}$$

and one for the displacement

$$M \frac{dy}{dt} = p \tag{8.3}$$

The distinction between a single second-order equation and a pair of first-order equations is also relevant when considering numerical methods for solving second-order differential equations. In principle, but usually not optimally in practice, any second-order differential equation can be treated by this two-step process. We illustrate the strengths and weaknesses of this procedure by adapting the first-order Euler methods from Sections 7.4 and 7.5 to solve second-order equations in Sections 8.3 and 8.4. The one-step solution of second-order equations using specially formulated methods is illustrated by the Noumerov method, which is derived in Section 8.3 then programmed and used in Section 8.5.

The greater information content in the pair of first-order equations is quite clear in the example of forces, momenta, and displacements. The momentum (or velocity) components can be measured, and they have dynamical interpretations. Although analytically the momentum can be obtained by differentiating the displacement with respect to time, in a numerical context this is usually not an accurate procedure, as is clear in the discussions of numerical noise and numerical differentiation Sections 4.3, 4.4, and 4.5.

Exercise 8.1

Make a list of basic equations in your scientific or engineering discipline. Which of the equations usually appear as second-order differential equations? For which of these equations is the interpretation clearer and more meaningful if the differential equation is written as a pair of first-order equations? ∎

With this overview and viewpoint of second-order differential equations, we are prepared to consider to consider specific equations for mechanical and analogous electrical systems.

Mechanical and electrical analogs

We consider first a mechanical system to model, for example the vertical displacement, y, of an automobile suspension system as the auto is driven over a bumpy road. For simplicity, the bumping forces as a function of time, t, are approximated as having a single angular frequency ω, thus varying as $F_0 \cos(\omega t)$. The inertial mass of the system we denote by M, and the system is modeled as having compressional forces from a Hooke's law spring (restoring force proportional to displacement) having spring constant K and damping forces proportional to speed with proportionality constant B.

Exercise 8.2

First write down Newton's force equation, (8.1), for the mechanical system just described, then rearrange it so that all the derivatives are on the left-hand side, in order to obtain the differential equation

$$M y''(t) + y'(t) + Ky(t) = F_0 \cos(\omega t) \tag{8.4}$$

for the mechanical model of the automobile suspension, where the primes denote derivatives with respect to time. ∎

Now consider an AC series circuit for which we will describe the free charge in the circuit at time t, $Q(t)$, responding to an external EMF of $E_0 \cos(\omega t)$. The inductive inertia of the system we denote by L, the system has a capacitance of C, and there is an Ohm's law resistance (damping) with value R. The electrical analog of relation (8.3) between mechanical displacement and momentum is given by

$$\frac{dQ}{dt} = I \tag{8.5}$$

relating the charge to the time-dependent current, I. Recall that the voltage drop across an inductor is $L \, dI / dt$, that the drop across a resistor is RI, while the drop across a capacitor is Q/C. This information can be used to derive a differential equation for the AC circuit containing the three elements in series.

Exercise 8.3

Write down Kirchhoff's second law (conservation of energy) to equate the sum of voltage drops across the three circuit elements to the source EMF, in order to obtain the differential equation for the electrical model of the series AC circuit,

$$LQ''(t) + RQ'(t) + \frac{1}{C}Q(t) = E_0 \cos(\omega t) \tag{8.6}$$

with the primes denoting derivatives with respect to time. ■

We see immediately that the mechanical and electrical systems are completely analogous and can be represented by a single second-order differential equation, namely

$$y''(x) + b\,y'(x) + ky(x) = S_0 \cos(\omega x) \tag{8.7}$$

in which the derivatives are with respect to control variable x. The correspondence between the variables in the last three differential equations is given in Table 8.1.

TABLE 8.1 Correspondence of mechanical, electrical, and general differential equation quantities.

Interpretation	Mechanical	Electrical	General
Control variable	t	t	x
Amplitude	$y(t)$	$Q(t)$	$y(x)$
Damping	B/M	R/L	b
Restoring	K/M	$1/LC$	k
Source term	$(F_0/M)\cos(\omega t)$	$(E_0/L)\cos(\omega t)$	$S_0\cos(\omega x)$
Natural frequency	$\sqrt{K/M}$	$1/\sqrt{LC}$	$\omega_0 = \sqrt{k}$
Damping parameter	$B/\sqrt{4KM}$	$R\sqrt{C/(4L)}$	$\beta = b/(2\omega_0)$

To understand how these analogies arise, apart from the correctness of the formal equations, try working the following exercise.

Exercise 8.4

In Table 8.1 of the analogies between mechanical and electrical systems, consider the intuitive explanations of the correspondences. To do this, discuss the frequency response of the amplitude, damping, and restoring terms in the differential equations (8.6) and (8.7). ■

With this preparation on setting up and interpreting the differential equations, we are prepared to solve and interpret the linear equations that arise when the source term is zero.

Solving and interpreting free-motion equations

The analysis of the general differential equation (8.7) is easiest to approach by first considering the situation with no driving term, $S_0 = 0$. Then we have a linear second-order equation. After solving this, it will be straightforward to solve the forced-motion equation in the next subsection. The differential equation to be solved is now

$$y''(x) + by'(x) + ky(x) = 0 \qquad (8.8)$$

We know from common experience (in the context that x represents time) that springs undulate and circuits oscillate, but that their motion eventually damps out if there is no driving term. This suggests that we look for solutions of the form

$$y(x) = y_0 \, e^{mx} \qquad (8.9)$$

where m is complex in order to allow both damping and oscillation. By substituting into the differential equation (8.8) one obtains directly that

$$\left(m^2 + bm + k\right)y(x) = 0 \qquad (8.10)$$

One solution of this equation is that $y(x) = 0$ for all x. In our examples the controlling variable x is the time, so this solution states that if the system is initially undisturbed it will remain undisturbed if there are no driving terms. A solution of (8.10) that is more interesting is to imagine that driving terms were present for some earlier x but were removed, so that the system then evolved according to (8.8). Then the appropriate solution of (8.10) is that the parentheses containing the m terms must be zero.

Exercise 8.5

(a) Equate the m-dependent expression in (8.10) to zero in order to find the two solutions for m, namely

$$m_{\pm} = \frac{-b}{2} \pm i \sqrt{D} \qquad (8.11)$$

where the *discriminant, D*, is given by

$$D = k - (b/2)^2 \tag{8.12}$$

(*b*) Show that the most general solution of (8.8) can therefore be expressed in terms of dimensionless variables as any linear superposition of the two basic solutions

$$y_{\pm}(\omega_0 x) = e^{-\beta \omega_0 x} e^{\pm i \sqrt{\delta} \, \omega_0 x} \tag{8.13}$$

in which the natural frequency is

$$\omega_0 = \sqrt{k} \tag{8.14}$$

the dimensionless *damping parameter* is

$$\beta = \frac{b}{2\omega_0} \tag{8.15}$$

and the dimensionless discriminant is

$$\delta = 1 - \beta^2 \tag{8.16}$$

so that the corresponding dimensionless solutions are

$$v_{\pm} = -\beta \pm \sqrt{1 - \beta^2} \tag{8.17}$$

(*c*) Show that the differential equation corresponding to (8.8) is

$$y''(\omega_0 x) + 2\beta y'(\omega_0 x) + y(\omega_0 x) = 0 \tag{8.18}$$

in which derivatives are taken with respect to dimensionless variable $\omega_0 x$. ∎

The expressions for the natural angular frequency of oscillation, ω_0, and the damping parameter, β, for the mechanical and electrical systems are given in Table 8.1. Plots of the relation (8.17) as a trajectory in the complex plane are explored in Section 2.3.

The solution (8.11) does not cover all bases, because we know that any second-order differential equation has two linearly-independent solutions, whereas (8.11) collapses to a single solution when $\delta = 0$, that is, when $\beta = 1$. This case requires special treatment.

Exercise 8.6

Verify by substitution into (8.18) that when $\beta = 1$ the expression

$$y_c(\omega_0 x) = (y_+ + y_- \omega_0 x) e^{-\omega_0 x} \tag{8.19}$$

solves the differential equation for any choice of the constants y_+ and y_-. ∎

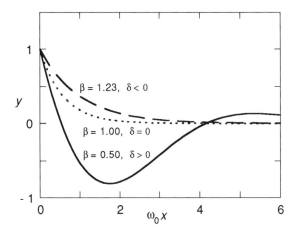

FIGURE 8.1 Amplitude of free motion, y, as a function of damping, using dimensionless variables. The solid curve ($\beta = 0.50$) is for damped oscillation, the dotted curve is for critical damping
($\beta = 1$), and the dashed curve ($\beta = 1.23$) is for overdamped motion.

The degenerate case (8.19) is called *critical damping* because it is sensitive to the
amount of damping, as indicated by b in the original differential equation (8.8).
Critical damping provides a dividing line between the oscillatory and nonoscillatory
solutions of (8.8) that we now consider.

Suppose that the modified discriminant, δ in (8.16), is positive, so that the
square root in (8.17) is real, then the solutions are clearly oscillatory, but damped.
If δ is negative, the solutions are always exponentially-damped as a function of x,
since $v_\pm$ is then real and negative. Such solutions are called *overdamped*.

Thus, the free-motion solutions to the damped harmonic oscillator equation are
of three types: damped oscillatory ($\delta > 0$), critically damped ($\delta = 0$), and overdamped ($\delta < 0$). They are illustrated in Figure 8.1.

Figure 8.1 shows that critical damping lies at a boundary between solutions that
oscillate with x, no matter how slowly, and solutions that die away uniformly without oscillation. It might therefore be expected that the transition from oscillatory to
overdamped motion is smooth. This is not so, as you can show by working the
following exercise.

Exercise 8.7

(a) To investigate the behavior of the amplitude y as a function of parameter β,
in (8.13) choose the y_+ solution and in (8.19) choose $y_- = 0$, $y_+ = 1$, so that
y values are the same at $\beta = 1$. Thus, continuity of values is assured. Next,
differentiate the expression for each y with respect to β. Show that as $\beta \to 1$
from below (oscillation side) the slope is continuous, but that as $\beta \to 1$ from
above (overdamped side) the slope $dy/d\beta \to -\infty$.

(b) Make graphs of y against β at fixed x values, say at $x = \pi/4$, $x = \pi/2$,
$x = \pi$, for β in the range 0 to 2. Show that there is always a cusp at $\beta = 1$. ∎

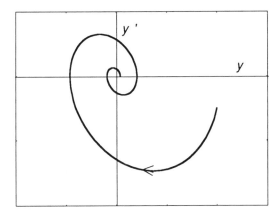

FIGURE 8.2 Phase-space plot (Poincaré map) for slightly damped motion ($\beta = 0.25$ in dimensionless units).

It is interesting to explore how $y' = dy/dx$, the analog of the speed in the mechanical system or the current in the electrical system, changes with y, the analog of displacement or charge, respectively. The graph of y' against y is called a "phase-space plot" or "Poincaré map." Figure 8.2 shows the phase-space plot for $\omega_0 = 1$ (dimensionless x units) and $\beta = 0.25$, which is not much damped.

Notice in Figure 8.2 how the motion is gradually damped as the "velocity" and "displacement" both tend to zero as x (or "time") increases. You may improve your understanding of the notion of phase space by making similar graphs yourself.

Exercise 8.8

(*a*) Prepare phase-space plots for the free-motion damped oscillator by choosing a value of β, then preparing tables of $y(x)$ against x and of $y'(x)$ against x, with x being in scaled units ($\omega_0 = 1$) and with $y'(0) = -y(0) = 1$. Then plot y' against y, as in Figure 8.2, which has $\beta = 1/4$. Other suitable values of β are 1/8 (very little damping) and $\beta = 1/2$ (strong damping).

(*b*) Show that for $\beta \geq 1$ the phase-space plot is a straight line with slope given by $-\beta + \sqrt{\beta^2 - 1}$. ∎

Extensive discussions of the phase plane and Poincaré maps are given in Pippard's omnibus text on response and stability, and also in Haberman's book on mathematical models.

Now that we have investigated the solutions for free motion, it is time to investigate the effects of a source term on the motion of an oscillator.

Forced motion and resonances

We return to the problem posed near the beginning of this section, namely motion with a source term, as given in general form by (8.7) and exemplified for the mechanical and electrical systems by (8.4) and (8.6) through the correspondences shown in Table 8.1. The equation with a driving term is

$$y''(x) + by'(x) + ky(x) = S_0 \cos(\omega x) \tag{8.20}$$

where the parameter S_0 indicates the strength of the source term, while parameter ω (angular frequency if x is time, but wavenumber, k, if x is distance) indicates the frequency of the forcing term. We assume, for simplicity, that only a single driving frequency is present. By superposing solutions for different ω, as indicated in Chapter 10, one can generalize the treatment given here.

It is most convenient to work with a complex-number representation of the variable, y, and the source term, S. As discussed in Sections 2.1 and 2.4, the modulus of each complex variable indicates its magnitude and the relation between the imaginary and real parts indicates its phase. So we write the source term in (8.20) as

$$S(x) = S_0 e^{i\omega x} \tag{8.21}$$

with the understanding that it is the real part of this expression that counts.

From everyday experience, such as pumping a swing, we know that a mechanical system under an oscillatory driving influence will settle into steady-state oscillations having the same frequency as the driver. Therefore, let's try a steady-state solution of (8.20) of the form

$$y(x) = y_0 e^{i\omega x} \tag{8.22}$$

The algebra and calculus of the solution are straightforward, so work them out for yourself.

Exercise 8.9

(a) Perform the indicated differentiations of (8.22) that are required for (8.20), substitute them in the latter, and notice that the complex-exponential factor can be factored out for all x, so that (8.22) is a solution of (8.20) under the condition that the complex amplitude y_0 is related to the source amplitude S_0 by

$$y_0 = -\frac{S_0}{\omega^2 - \omega_0^2 - i\omega b} \tag{8.23}$$

in which the frequency for free motion without damping appears as

$$\omega_0^2 = k \tag{8.24}$$

(b) Convert (8.23) into amplitude and phase form as

$$|y_0| = \frac{|S_0|}{\sqrt{\left(\omega^2 - \omega_0^2\right)^2 + \left(\omega b\right)^2}} \tag{8.25}$$

for the amplitude, and

$$\phi = \arctan\left[\frac{\omega b}{\omega^2 - \omega_0^2}\right] \tag{8.26}$$

which gives the phase angle by which the response, $y(x)$, lags the driving term, $S(x)$.

(c) Show that the amplitude relation (8.25) can be expressed in terms of the frequency ratio ω/ω_0 and the dimensionless damping variable β from (8.15), as the dimensionless ratio

$$A(\omega) \equiv \left|\frac{\omega_0^2 y_0}{S_0}\right| = \frac{1}{\sqrt{\left[(\omega/\omega_0)^2 - 1\right]^2 + \left(2\beta\omega/\omega_0\right)^2}} \tag{8.27}$$

while the phase can be expressed in terms of the same variables as

$$\phi(\omega) = \arctan\left[\frac{2\beta\omega/\omega_0}{(\omega/\omega_0)^2 - 1}\right] \tag{8.28}$$

which is usually displayed in the range 0 to π with a continuous distribution as a function of ω. ∎

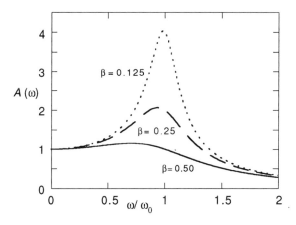

FIGURE 8.3 Resonance amplitude (dimensionless units) as a function of frequency relative to resonance frequency for three values of the damping parameter β.

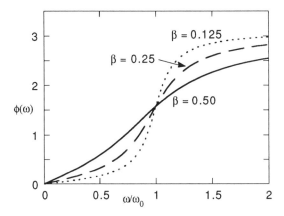

FIGURE 8.4 Resonance phase as a function of the ratio of frequency to resonance frequency for three values of damping parameter β, from small damping ($\beta = 0.125$) to moderate damping ($\beta = 0.50$).

The amplitude and phase are shown in Figures 8.3 and 8.4 for the same values of the damping parameter β as in Figure 8.2, which shows the response of the system under free motion for different amounts of damping. Notice that the amplitude, $A(\omega)$, is asymmetric about $\omega/\omega_0 = 1$, with the asymmetry decreasing as the damping decreases and the response becomes more resonant. The phase, $\phi(\omega)$, is nearly symmetric about $\omega/\omega_0 = 1$, and it varies more rapidly in the neighborhood of the resonance point as the damping decreases.

Resonant oscillations are important in practical applications; for example, one usually wants to avoid resonant oscillations of mechanical structures, but one often wants to enhance resonant behavior in an electrical circuit. It is therefore important to know for what driving frequency, ω_R, the amplitude is a maximum.

Exercise 8.10
Differentiate expression (8.25) for the amplitude with respect to ω and thereby show that the only physically meaningful solution for a maximum of $A(\omega)$ is at $\omega = \omega_R$ given by

$$\omega_R = \sqrt{\omega_0^2 - b^2/2} \qquad (8.29)$$

assuming that the quantity under the square root is positive, otherwise there is no relevant maximum. ∎

Notice that the amplitude has a maximum at a frequency slightly below the natural oscillation frequency for free motion, ω_0. The maximum rate of change of the phase, however, occurs right at this frequency for any value of the damping parameter β, as is clear from Figure 8.4.

For the typical resonances that are of interest to scientists and engineers the damping is relatively weak, that is, $b \ll \omega_0$. The amplitude formulas can then, with only mild approximations, be made analytically much more tractable. One sets

$$\omega_R \approx \omega_0 \tag{8.30}$$

and approximates the amplitude expression (8.25) by

$$A(\omega) \approx A_L(\omega) \equiv \frac{1}{2} \frac{1}{\sqrt{(\omega/\omega_0 - 1)^2 + \beta^2}} \tag{8.31}$$

and the phase expression (8.26) by

$$\phi(\omega) \approx \phi_L(\omega) \equiv \arctan\left[\frac{2\beta}{(\omega/\omega_0)^2 - 1}\right] \tag{8.32}$$

Exercise 8.11
Show the steps in the approximations that are necessary to produce (8.31) and (8.32) from the original expressions (8.27) and (8.28). ■

The function A_L will be revisited in Chapter 10. Within overall normalization, it is the square root of the Lorentzian function, with the damping parameter b being proportional to the width, γ, of the Lorentzian. We show in Figures 8.5 and 8.6 comparisons of the exact and approximate amplitude and phase expressions for $\beta = 0.125$.

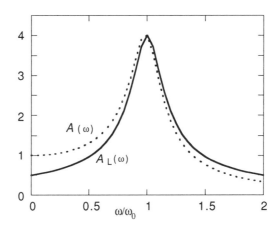

FIGURE 8.5 Resonance amplitude, $A(\omega)$, and its Lorentzian approximation, $A_L(\omega)$, as a function of the ratio of the frequency to the resonance frequency. Shown for damping parameter $\beta = 0.125$.

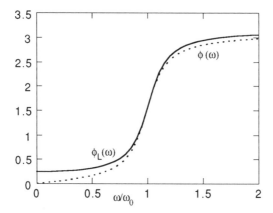

FIGURE 8.6 Resonance phase, $\phi(\omega)$, and its Lorentzian approximation, $\phi_L(\omega)$, as a function of the ratio of the frequency to the resonance frequency. Shown for damping parameter $\beta = 0.125$.

Notice that the Lorentzian function gives distributions that are symmetric about the resonance frequency, and that the approximation is particularly good for the phases. In applications of resonance, a value of β of order 10^{-6} is typical, so then the Lorentzian approximations (8.31) and (8.32) become very accurate. We explore Lorentzian functions and their properties extensively in Sections 10.2 – 10.4, where their relations to Fourier transforms and convolutions are emphasized. The relations between resonances, response, stability, and chaos are discussed extensively in Pippard's text on the physics of vibration.

The study of vibrations and waves, which we introduced in the context of complex variables in Chapter 2.4, is discussed extensively in the context of differential equations in the texts by Braun et al. and by Ingard. The musical connections are well described in the book by Backus and that by White and White.

8.2   CATENARIES, CATHEDRALS, AND NUPTIAL ARCHES

We now turn to the problem of modeling a static mechanical system, hanging chains of various density distributions. This provides us opportunities to develop our understanding of applications of differential equations. We also bring together many interesting and insightful results on general catenaries that have appeared in scattered form in the research and teaching literature of the past three centuries but which are not much appreciated by modern pedagogues.

The famous problem of the equilibrium shape of a cable or chain hanging under gravity provides insight into setting up and solving differential equations of second order. Also, the differential equation is nonlinear in that the quantity of interest, the height, y, as a function of horizontal distance from the center of the cable, x, is not linearly related to its derivatives. Once the equilibrium shapes have been determined — they're called "catenaries" from Latin for "chains" — the same equations can be applied with minor changes to the optimal shapes of cathedral archways and to the shape assumed by an arch of string uniformly levitated by balloons.

The equation of the catenary

The classical mechanics problem of finding the equilibrium shape of a flexible chain provides an introduction to the mechanics of continuous media ("continuum mechanics") and to nonlinear second-order differential equations. If we observe a chain hanging in a curve of given shape, what is its associated density distribution? Also, from an engineering viewpoint, what density distribution provides a constant strength (density-to-tension ratio), and what is the equilibrium shape of the chain?

In the following we derive a general formula for the density distribution in terms of first and second derivatives of the shape, then we introduce dimensionless variables (scaled units) to allow general characterization of catenaries. Examples for various shapes are then given and we examine the system under the condition of uniform strength.

In setting up the equations for the equilibrium of a chain, we allow the weight per unit length of the chain, $w(X)$, to depend on position along the chain, (X,Y). We assume that w is an even function of X, so that the chain hangs symmetrically about its midpoint $X = 0$. For the vertical coordinate, we choose $Y = 0$ at the midpoint. Referring to Fig. 8.7, we see that at $P = (X,Y)$ the conditions for equilibrium are that the tension T (tangential to the chain since the chain is flexible), the weight W, and the horizontal tension H (the same constant value on each side of the origin) are related by

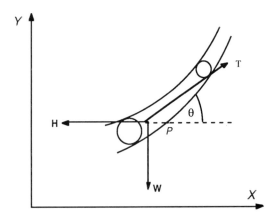

FIGURE 8.7   Forces acting at the point $P = (X, Y)$ of a chain hanging under gravity.

$$T \sin \theta = W \tag{8.33}$$

$$T \cos \theta = H \tag{8.34}$$

In terms of the arc length along the chain, S, we have

$$\frac{dW}{dS} = w(X) \tag{8.35}$$

From geometry we know that

$$\frac{dY}{dX} = \tan \theta \tag{8.36}$$

and also that

$$\frac{dS}{dX} = \sqrt{1 + \left(\frac{dY}{dX}\right)^2} \tag{8.37}$$

Our aim is to find an expression for the weight distribution, $w(X)$, given the shape, $Y(X)$. By eliminating T between (8.33) and (8.34) then differentiating the resulting expression with respect to X and using (8.35), we find immediately that

$$\frac{dW}{dX} = H \frac{d^2Y}{dX^2} \tag{8.38}$$

Use of (8.35) and (8.36) now gives an expression for the weight distribution in terms of the shape

$$\frac{w(X)}{H} = \frac{\dfrac{d^2Y}{dX^2}}{\sqrt{1 + \left(\dfrac{dY}{dX}\right)^2}} \tag{8.39}$$

We may verify this result for the uniform-density chain by substituting the standard equation for the catenary

$$Y(X) = L\left[\cosh\left(\frac{X}{L}\right) - 1\right] \tag{8.40}$$

to verify directly by substituting in (8.39) that

$$\rho \equiv \frac{Lw(X)}{H} = 1 \tag{8.41}$$

where the dimensionless variable ρ is a constant for the uniform-density catenary.

Exercise 8.12
Verify the result for the uniform-density chain by taking the first and second derivatives of (8.40) for the catenary then substituting into (8.39) to produce (8.41). ∎

Dimensionless variables are helpful when comparing catenaries with various weight distributions, because even for a uniform chain the shape depends upon the horizontal force (H) exerted. For the remainder of our analysis we therefore change to the following dimensionless variables in terms of an appropriate characteristic length, L, and the horizontal force, H :

$$x \equiv \frac{X}{L} \qquad y \equiv \frac{Y}{L}$$

$$\rho(x) \equiv \frac{Lw}{H} \qquad \tau(x) \equiv \frac{T}{H} \qquad \omega(x) \equiv \frac{W}{H} \qquad \sigma(x) \equiv \frac{Lw}{T} = \frac{\rho(x)}{\tau(x)}$$

(8.42)

in which the weight-to-tension ratio, σ, provides in dimensionless units the tensile strength required for the chain at each point.

The scaled density distribution is immediately found by inserting (8.39) for w into the definition in (8.42):

$$\rho(x) = \frac{\dfrac{d^2 y}{dx^2}}{\sqrt{1 + \left(\dfrac{dy}{dx}\right)^2}}$$

(8.43)

The tension at any point in the chain can readily be calculated by solving for T from (8.31), then using (8.36) to eliminate the angle variable. We thus find the tension in scaled units

$$\tau(x) = \sqrt{1 + \left(\frac{dy}{dx}\right)^2}$$

(8.44)

The weight of chain below x is then, from (8.42) with similar angle elimination, in scaled units,

$$\omega(x) = \frac{dy}{dx}$$

(8.45)

The strength of the chain is obtained by taking the ratio of ρ to τ. From (8.43) and (8.44), in scaled units this is

$$\sigma(x) = \frac{\dfrac{d^2 y}{dx^2}}{1 + \left(\dfrac{dy}{dx}\right)^2}$$

(8.46)

Exercise 8.13
Solve for T and W as suggested, then transform to the dimensionless variables of (8.42), to verify equations (8.43) through (8.46). ∎

With the above scaling, the characteristic dimensions (such as L for a uniform-density catenary) can be chosen such that *all* flexible chains will hang with the same shape near the bottom, where the slope of y on x is zero. Their scaled density, tension, weight below the bottom, and strength will satisfy

$$\rho(0) = 1 \qquad \tau(0) = 1 \qquad \omega(0) = 0 \qquad \sigma(0) = 1 \tag{8.47}$$

provided that at the origin $d^2y/dx^2 = 1$. In order to achieve this invariance, a given shape choice $y(x)$ should be expanded in a Maclaurin series in x, then constants in its formula adjusted so that the constant and first-derivative terms vanish and the second-derivative term has unity coefficient in order to make $\rho(0) = 1$. This procedure can be verified for the uniform-density catenary, (8.40), and is demonstrated for the parabolic shape in the next subsection.

Thus, we have solved a mechanics problem and have been able to express the significant quantities, (8.42), purely in terms of the geometry of the curve shape, $y(x)$. We now illustrate these scaled curves and corresponding scaled densities and forces.

Catenaries of various shapes and strengths

In the statics problem solved above we obtained the distribution of density, tension, weight below a point, and strength of a chain, given its shape, with the assumptions being that the shape, $y(x)$, is an even function of x and that its second derivative is positive, as required in (8.38). With these mild restrictions, a large variety of realistic catenaries can be investigated.

The *parabolic catenary* is the first shape that we examine. We write the shape equation as $Y(X) = X^2/L$, where L is some characteristic length. This example will explain why Galileo was nearly correct in his assertion in his *Two New Sciences* that a uniform-density chain hangs in a parabolic shape, and it will clarify use of the dimensionless variables introduced at the end of the preceding subsection. In terms of scaled lengths, we have $y(x) = \lambda x^2$, where the dimensionless quantity λ is determined by the condition $\rho(0) = 1$, that is, unity second derivative of y at $x = 0$, according to (8.43). Therefore $\lambda = 1/2$, and the scaled parabolic shape is

$$y(x) = \frac{x^2}{2} \tag{8.48}$$

It is now simple to take the first and second derivatives of y and by using (8.43) through (8.46) to compute the mechanical quantities of interest. Their formulas are given in Table 8.2, and the corresponding distributions with respect to x are shown as curves (1) in Figures 8.8 – 8.11.

TABLE 8.2 Examples of the relations between the shape of the equilibrium curve, the weight of chain below each point, the local linear density of the chain, the tension at a point (x, y), and the strength (ρ/τ) of the chain.

Curve	$y(x)$	$\omega(x)$	$\rho(x)$	$\tau(x)$	$\sigma(x)$
Parabola (1)	$x^2/2$	x	$1/\sqrt{1+x^2}$	$\sqrt{1+x^2}$	$1/(1+x^2)$
Uniform density (2)	$\cosh(x) - 1$	$\sinh(x)$	1	$\cosh(x)$	$1/\cosh(x)$
Uniform strength (3)	$\ln[\sec(x)]$	$\tan(x)$	$\sec(x)$	$\sec(x)$	1
Circle arc (4)	$1-\sqrt{1-x^2}$	$x/\sqrt{1-x^2}$	$1/(1-x^2)$	$1/\sqrt{1-x^2}$	$1/\sqrt{1-x^2}$

Exercise 8.14
For the parabolic catenary calculate the first and second derivatives, then substitute into (8.43) through (8.46) for the scaled density, tension, weight, and strength, thus verifying the results for line (1) in Table 8.2. ∎

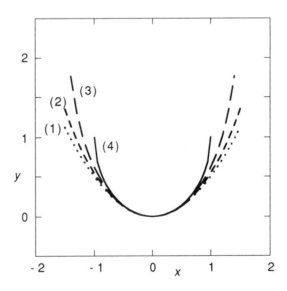

FIGURE 8.8 Equilibrium shapes of chains hanging under gravity: (1) for a parabola, (2) for a uniform-density catenary, (3) for uniform strength, and (4) for the arc of a circle. The formulas are given in Table 8.2.

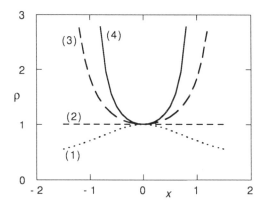

FIGURE 8.9 Scaled linear densities for the catenaries shown in Figure 8.8: (1) for a parabola, (2) for a uniform-density catenary, (3) for uniform strength, and (4) for the arc of a circle.

The *uniform-density catenary*, the usual example, can be solved by a similar analysis to that for the parabola to show directly that the scaled shape is given by

$$y(x) = \cosh(x) - 1 \tag{8.49}$$

which is just (8.40) with $L = 1$, as you might have guessed. This shape, and the density, tension, weight, and strength distributions that follow from application of (8.43) through (8.46), are given in Table 8.2 and are shown in Figures 8.8−8.11 as curves (2). Notice that by comparison with the parabola, the usual catenary becomes steeper as x increases. Since for this catenary the chain density is constant, correspondingly the flatter parabola must have a decreasing density, which agrees with the result in Table 8.2 and with curve (1) in Figure 8.9.

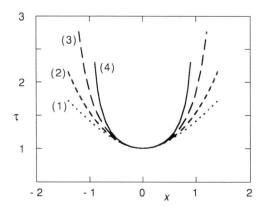

FIGURE 8.10 Tension as a function of x for the catenaries shown in Figure 8.8: (1) for a parabola, (2) for a uniform-density catenary, (3) for uniform strength, and (4) for the arc of a circle.

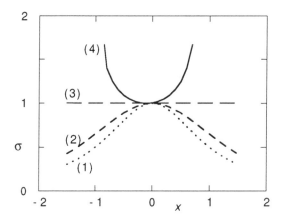

FIGURE 8.11 Strength as a function of x for the catenaries shown in Figure 8.8: (1) for a parabola, (2) for a uniform-density catenary, (3) for a uniform-strength catenary, and (4) for the arc of a circle. The formulas are given in Table 8.2.

Exercise 8.15

(*a*) For the uniform-density catenary calculate the first and second derivatives, then substitute into (8.43) through (8.46) for the scaled density, tension, weight, and strength, thus verifying the results for line (2) in Table 8.2.

(*b*) Write a small program then calculate and graph the x dependence of the regular catenary shape, tension, and strength, thus verifying the curves (2) in Figures 8.8 – 8.11. ∎

For a *circle-arc shape*, that is, a semicircle concave upward, what is the corresponding density needed to produce this shape? It's very simple, because the appropriate scaled circle, having zero value and unity second derivative at the x origin, is just

$$y(x) = 1 - \sqrt{1 - x^2} \qquad (8.50)$$

with $|x| < 1$ to ensure that the chain is under tension.

Exercise 8.16

(*a*) For the circle-arc catenary calculate the first and second derivatives, then substitute into (8.43) through (8.46) for the scaled density, tension, weight, and strength, thus verifying the results for line (4) of Table 8.2.

(*b*) Write a program, then calculate and graph the x dependence of this catenary shape, density, tension, and strength, thus verifying the curves (4) in Figures 8.8 – 8.11. ∎

The circular arc is even steeper than a catenary, so its density must increase as x increases, as Figure 8.9 shows. The tension also increases with x, but not as rapidly, so the strength of the chain steadily decreases with x. As $|x|$ tends to unity, the density and tension must become indefinitely large in order to produce the nearly-vertical shape. We therefore show the density and tension only up to $|x| = 0.8$. For real materials their elasticity would have to be considered for large densities and tensions, but this we have not included explicitly.

In two of the above examples the strength of the chain (ratio of cable weight per unit length at a given x to the tension in the cable at x) was found to decrease with x, so that for a large enough span the chain will exceed its elastic limit. For the third example, the circle arc, the strength increases with x, as shown in Table 8.2 and Figure 8.11, curve (4).

Consider a *constant-strength catenary*. If the strength is to be a constant with respect to x, what is its shape? Presumably, it will be intermediate between a uniform-density catenary and a circle, curves (2) and (4) in Figure 8.11. For the scaled units that we are using, (8.42) shows that the constant strength, σ, must be unity everywhere, since that is its value at $x = 0$. From (8.46) we deduce immediately the differential equation for the shape of the constant-strength catenary, namely

$$\frac{d^2y}{dx^2} = 1 + \left(\frac{dy}{dx}\right)^2 \tag{8.51}$$

It is straightforward to verify that integrating this once produces the slope condition

$$\frac{dy}{dx} = \tan(x) \tag{8.52}$$

in which the constant of integration is zero because the slope is to be zero at $x = 0$. Integrating (8.52) once more produces

$$y(x) = \ln[\sec(x)] \tag{8.53}$$

as can be verified by substitution in (8.52). Again, the constant of integration is zero, because in our coordinate system $y(0) = 0$.

Exercise 8.17
Carry out the steps indicated to produce the uniform-strength catenary formula (8.53). ∎

The result (8.53) for the chain of uniform strength is attributed to British engineer Davies Gilbert, who reported in 1826 his use of it in designing the first suspension bridge across the Menai Strait in north Wales. Because he did not use scaled variables, much of his article consists of very long tables of bridge-design parameters.

From (8.53) we see that, just as for the circular arc, the uniform-strength catenary is possible only for a limited range of the scaled x variable; here $|x| < \pi/2$. We therefore show in Figure 8.8 the uniform-strength catenary, curve (3), only up to $|x| = 1.4$. Notice that for uniform strength the shape is intermediate between that for the uniform-density catenary, curve (2), which has decreasing strength with increasing x, and the circle-arc catenary, curve (4).

For the uniform-strength catenary given by (8.53) it is straightforward to use (8.43) to (8.45) and to calculate the scaled density (which equals the scaled tension) and the scaled weight of chain below x, to produce the formulas in Table 8.2 and curves (3) in Figure 8.9 or 8.10. Generally, the uniform-strength catenary has mechanical properties intermediate between those for the uniform-density catenary and the circle-arc catenary.

Demonstrating arches

How can you demonstrate the relation between forces on a catenary and the catenary shape? A simple method is to follow up an idea described in an article by Honig and to use light helium balloons attached at equal horizontal intervals to a relatively light string. Thereby the weight distribution of the chain is replaced by the buoyancy of the balloons, and the analysis is identical to that above, except that the sign of y is reversed, so that inverted curves are obtained. For each shape, the size of the balloons will need to be adjusted for the appropriate buoyancy, the analog of $\rho(x)$.

We illustrate the setup for the circle-arc catenary because this has the most rapid variation of density with x, as curve (4) in Figure 8.9 shows. If the balloons are filled so that their volumes increase as $1/(1-x^2)$ and if the ends are fixed at horizontal points, then a curve which is almost a semicircle will be formed. The shape will be slightly different near the endpoints, where very large balloons would be needed to pull the endpoints straight up.

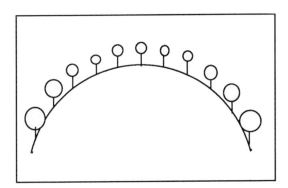

FIGURE 8.12 Arc of a circle made of a light string supported by helium balloons with diameters increasing to satisfy the weight relations, $\omega(x)$, for line (4) in Table 8.2.

A sketch of such an arrangement is shown in Figure 8.12, where the outermost balloons at $|x| = 0.9$ have diameters more than 1.7 times greater than that of the center balloon. When the system is in static equilibrium, the midpoint of the catenary should be above the base by half the separation of the ends. If helium is carefully released from the outer balloons until all the balloons have the same shape, then the graceful uniform-density "nuptial arch" referred to by Honig should form. If slightly more helium is released from the outer balloons, then a parabola will form.

Practical arches and catenaries

The history of the catenary and its relation to the stability of arches and domes in large buildings, such as cathedrals, is very interesting. Historically, the uniform-density catenary problem was pondered by Galileo, who explained (incorrectly) that "This [uniform] chain curves in a parabolic shape." The first correct solution was published (in Latin) by Johannis Bernoulli in the late seventeenth century, and he also indicated the general relation (8.43) between the shape and the density distribution. Robert Hooke, better known for his research on elasticity (Hooke's law) and in microscopy, solved the uniform-density catenary problem in 1676. Later, Newton about 1680, Huygens about 1691, and Taylor (of Taylor's theorem) in 1715 also provided proofs.

After the Great Fire of London in 1666, Hooke and Christopher Wren were architectural consultants for the rebuilding of St. Paul's cathedral, and Hooke probably told Wren about the stability of the catenary shape for arches and domes. For a solid arch, in the approximation that it is continuous and of uniform composition, the tension forces will all be tangential to the surface of the arch if it is of the regular catenary shape. This property will improve the stability of the arch. It is unlikely, however, that Wren used his colleague's advice when designing the cathedral dome.

8.3 NUMERICAL METHODS FOR SECOND-ORDER DIFFERENTIAL EQUATIONS

In the next four sections of this chapter we emphasize numerical methods for solving second-order differential equations, with several worked examples and applications of the methods. Beginning in this section, we first extend the Euler-type algorithms from the first-order equations developed in Sections 7.4 and 7.5 to second-order equations. Then we set the scene for the Noumerov method by showing how first derivatives can be removed from second-order linear equations, since the Noumerov algorithm assumes that first derivatives are absent. This algorithm is then derived.

Project 8A (Section 8.4) is on programming and testing the Euler-type algorithms, while Project 8B (Section 8.5) emphasizes programming and testing the Noumerov algorithm, then applying it to solve numerically the differential equation for the quantum harmonic oscillator. The problem of the numerics of stiff differential equations in Section 8.6 completes our study of numerical methods for second-order equations.

Euler-type algorithms for second-order equations

We now describe three Euler-type algorithms that are variants of the forward-difference predictor formula (7.38) used for first-order equations. As we see in Section 8.4, these algorithms are not very accurate for a given stepsize h. Their utility lies in their general applicability, because they make very few assumptions about the type of second-order differential equation that is to be solved numerically.

The Euler methods for second-order equations are based on the fact that the second-order differential equation

$$\frac{d^2y}{dx^2} = F(x,y) \tag{8.54}$$

is equivalent to a pair of first-order equations, namely

$$\frac{dy'}{dx} = F(x,y) \tag{8.55}$$

and, as an identity,

$$\frac{dy}{dx} = y' \tag{8.56}$$

Therefore, in principle, any second-order equation can be solved as a pair of first-order equations. In the practice of numerical methods for estimating the solutions of such equations, careful analysis and programming, as well as considerable skill and experience, are necessary if accurate results are to be obtained efficiently. The difficulties arise because with a finite stepsize h the numerical methods are necessarily approximate, as (7.38) indicates.

We now describe three variants of the forward Euler predictor that may be used for estimating the solutions of the differential-equation pair (8.55) and (8.56). These methods are generally of increasing accuracy, as indicated by the dependence of their error estimates on the stepsize h.

Method 1. Here we use the forward Euler predictors directly for advancing the solutions y and y' according to (7.38), namely

$$y'_{k+1} = y'_k + F_k h \tag{8.57}$$

and

$$y_{k+1} = y_k + y'_k h \tag{8.58}$$

The first equation neglects the change in F between steps k and $k+1$, while the second neglects the change of y' in this interval, so that it neglects a term of order Fh^2. The errors in both the function and its derivative are therefore of order h^2.

Method 2. In this variant we advance the derivative according to (8.57), but we then average the derivative just obtained with the derivative at the start of the interval in order to predict the new y value. That is,

$$y_{k+1} = y_k + (y'_k + y'_{k+1})h/2 \qquad (8.59)$$

The averaging of the two derivatives allows for what physicists would call an acceleration, since if x is time and y is displacement, then y' is speed and F is acceleration. The order of the error in y is h^3 in this method.

Method 3. Here the function is advanced by using terms up through its second derivative, that is, through F, so its error is of order h^3. Then the derivative is advanced using the average of the F values at the two ends, which gives an error in the derivative of order h^3. Thus

$$y_{k+1} = y_k + y'_k h + F_k h^2/2 \qquad (8.60)$$

then the derivative is advanced by

$$y'_{k+1} = y'_k + (F_k + F_{k+1})h/2 \qquad (8.61)$$

Notice that the order of evaluation is significant here if F is a function of y. Further, in (8.57) – (8.61) mathematically there should be "$\approx$" signs or computationally they should be assignment statements (":=" in Pascal).

The three Euler-type methods are summarized in Table 8.3. The programming of these algorithms is presented in Project 8A in Section 8.4.

TABLE 8.3 Euler-type algorithms for numerical solution of second-order differential equaions. In method 3 the formulas must be evaluated in the order shown. The dependence of the errors on stepsize, h, is also indicated.

Method	Algorithms (in sequence shown)	Order of error
1	$y'_{k+1} = y'_k + F_k h$	$O(h^2)$
	$y_{k+1} = y_k + y'_k h$	$O(h^2)$
2	$y'_{k+1} = y'_k + F_k h$	$O(h^2)$
	$y_{k+1} = y_k + (y'_k + y'_{k+1})h/2$	$O(h^3)$
3	$y_{k+1} = y_k + y'_k h + F_k h^2/2$	$O(h^3)$
	$y'_{k+1} = y'_k + (F_k + F_{k+1})h/2$	$O(h^3)$

Exercise 8.18

Make Taylor expansions of y and of y' about x_k in order to derive (8.57) through (8.61). Include enough terms in each expansion that the lowest neglected power of h is indicated. Use this to show that the error estimate for each formula has the approximate dependence on h that is claimed. ∎

A simple example will clarify these Euler-type algorithms for second-order equations. Suppose, for purposes of comparison, that we know the exact solution

$$y = x + x^2/2 + x^3/6 \tag{8.62}$$

Let us compare the exact solutions with those obtained from the three methods in Table 8.3 when they are used to advance the solutions from $x_1 = 1$ to $x_2 = 2$ in a single step with $h = 1$, assuming that all values are exact at $x = 1$. Working backwards, we have by differentiating (8.62) twice with respect to x that

$$F(x) = x \tag{8.63}$$

Starting with $y_1 = 1.6667$, $y'_1 = 2.5000$, and $F_1 = 2.0000$, the three methods produce the values given in Table 8.4 and displayed in Figures 8.13 and 8.14.

TABLE 8.4 Example of the Euler-type methods for advancing one step of the differential-equation solution. The second derivative is (8.63) and the exact solution is (8.62). The value of $x = 2$.

| Quantity | Exact value | Estimated value by method | | |
		1	2	3
y_2	5.3333	4.1667	5.1667	5.1667
y'_2	5.0000	4.5000	4.5000	5.0000

In order to check that you understand how each of the Euler-type methods works, try the following exercise.

Exercise 8.19

Use each of the algorithms in Table 8.3 in turn to predict the values of y_2 and y'_2, given the exact values in the text for y_1 and y'_1. Using that $h = 1$ and that the neglected terms can be calculated from the derivatives $F' = 1$ and $F'' = 0$, are the discrepancies between estimated and exact values correct? ∎

The behavior of the algorithms and their results are also shown in Figures 8.13 and 8.14 for y and its derivative y' respectively. Notice that the constancy of the derivative of F gives rise to estimates of y_2 that are the same (but not exact) for methods 2 and 3. For similar reasons, methods 1 and 2 agree (inexactly) on the value of y', while method 3 predicts this derivative exactly because only the second and higher derivatives of F (which are zero) have been neglected.

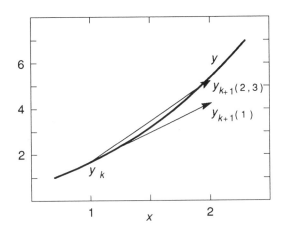

FIGURE 8.13 Numerical solutions by the three Euler methods for the exponential function, which is shown as the solid curve, $y(x)$.

This completes our derivation and preliminary discussion of the Euler-type algorithms for second-order differential equations. Numerical analyses that are much more extensive are part of Project 8A in Section 8.4.

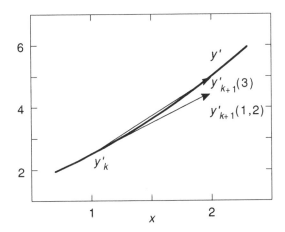

FIGURE 8.14 Numerical solutions by the three Euler methods for the derivative of the exponential function, whose derivative is shown as the solid curve, $y'(x)$.

Removing first derivatives from second-order linear equations

The Noumerov algorithm for second-order linear differential equations that we derive in the next subsection requires that the first derivative, y', be absent. This would seem to be a severe limitation of the algorithm, because it might not be able to handle such interesting differential equations as that for a damped harmonic oscillator (Section 8.1), where the damping is proportional to y'. The purpose of this subsection is to show how the first derivative in a second-order linear differential equation can always be transformed away.

Consider the general second-order linear differential equation

$$y''(x) + a(x) y'(x) + b(x) y(x) = 0 \tag{8.64}$$

The method of eliminating the first derivative is to find a transformation that makes the first two terms part of a single second derivative. Consider transforming $y(x)$ by multiplication by some, as yet unknown, function $g(x)$, to produce $z(x)$

$$z(x) \equiv g(x) y(x) \tag{8.65}$$

(Since unknowns are piling up at a great rate, this is clearly mathematics.) Also consider multiplication of the original equation (8.64) throughout by $g(x)$. Is there a choice of $g(x)$ that will remove the first derivative?

Exercise 8.20

(a) Show that

$$z'' = g''y + 2g'y' + gy'' \tag{8.66}$$

and therefore that if we force the first- and second-derivative terms in this equation and g times the original equation to coincide, we must have

$$2g' = ga \tag{8.67}$$

(b) Show that the solution of this equation is

$$g(x) = \exp\left[\frac{1}{2}\int^x a(x')dx'\right] \tag{8.68}$$

in which any constant of integration may be ignored because it just affects the overall scale of the differential equation solution.

(c) Thence show that the transformed variable z satisfies the following linear differential equation, in which its second derivative does not occur

$$z'' + (b - g''/g)z = 0 \tag{8.69}$$

(*d*) Express the second derivative g'' in this equation by using (8.67) in order to derive the final form of the transformed differential equation

$$z'' + \left[b - \frac{1}{2}\left(a' + \frac{a^2}{2} \right) \right] z = 0 \qquad (8.70)$$

in which all variables are implicitly functions of x. ∎

Thus, if the original equation (8.64) has a first-derivative term, then the function g calculated from (8.68) produces a differential equation without a first derivative, namely (8.70) for z. From the solution of this equation the original function, y, can be recovered by dividing z by g because of (8.65).

An example will make this formal analysis much clearer. Consider (8.64) for the damped harmonic oscillator, with the coefficient of the second derivative divided out for simplicity:

$$y''(x) + b y'(x) + k y(x) = 0 \qquad (8.71)$$

In the notation of (8.64) we have $a(x) = b$, a constant, so that $a' = 0$. Also, $b(x) = k$, again a constant. From (8.68) the function g is therefore just

$$g(x) = \exp(bx/2) \qquad (8.72)$$

The modified differential equation (8.70) becomes

$$z''(x) + K z(x) = 0 \qquad (8.73)$$

where $K = k - b^2/4$. By inspection, the solution for z is just a complex exponential (assuming that K is positive). Division by g produces the solution of the original damped-oscillator equation

$$y(x) = \exp\left(\pm i\, x \sqrt{k - b^2/4} - bx/2 \right) \qquad (8.74)$$

which is in agreement with (8.13). You will notice that elimination of the first derivative makes the differential equation much easier to solve analytically. As the Noumerov algorithm that we now derive shows, a similar result obtains for numerical solution of these differential equations.

Deriving the Noumerov algorithm for second-order equations

Boris V. Noumerov (1891 – 1943), a prominent Russian astronomer and geophysicist, derived the following algorithm as part of his numerical studies of the perturbations of planetary motions. The algorithm was first published in English in the reference given at the end of this chapter. It is accurate and efficient for the solution of second-order linear equations with no first-derivative term. As we showed in the preceding subsection, the latter restriction is not really an impediment.

We seek a recurrence relation relating y and its second derivative at a given x_k in terms of its values at the two preceding points, namely

$$y_{k+1} + A h^2 y''_{k+1} = B y_k + C h^2 y''_k + D y_{k-1} + E h^2 y''_{k-1} + R \qquad (8.75)$$

where A, B, C, D, and E are to be chosen to minimize the remainder R. The way to do this is indicated in the following exercise.

Exercise 8.21

(a) Expand each of y_{k+1} and y_{k-1} in Taylor expansions about y_k in order to show that choosing $D = -1$ in (8.75) eliminates all odd-order derivatives from it. Now add these two values and corresponding formulas for their second derivatives to obtain

$$y_{k+1} + y_{k-1} = 2\left(y_k + h^2 y''_k / 2 + \dots\right) \qquad (8.76)$$

and the recurrence for derivatives

$$y''_{k+1} + y''_{k-1} = 2\left(y''_k + h^2 y''''_k / 2 + \dots\right) \qquad (8.77)$$

(b) In (8.75), with the substitutions from the two preceding equations, equate corresponding derivatives on the left- and right-hand sides to show that $B = 2$, $A = -1/12$, $C = 5/6$, with remainder term $R \approx -h^6 y^{(6)}_n / 240$. ∎

The original differential equation can now be written

$$y_{k+1} - h^2 y''_{k+1}/12 + y''_{k-1} = 2y_k + 5h^2 y''_k/6$$
$$-y_{k-1} + h^2 y''_{k-1}/12 + R \qquad (8.78)$$

which holds for any ordinary differential equation. It becomes useful, in addition to being true, when we use the linearity condition on its right-hand side, as you may verify as follows.

Exercise 8.22

Define the iteration quantity

$$Y_k \equiv \left(1 - h^2 F_k/12\right) y_k \qquad (8.79)$$

and make a binomial expansion approximation for y_k in the terms of (8.78) that are of order h^2 and higher. Thus show that (8.78) can be written as

$$Y_{k+1} - Y_k = Y_k - Y_{k-1} + h^2 F_k \left(1 + h^2 F_k/12\right) Y_k + R' \qquad (8.80)$$

where the remainder R' is of order h^6. ∎

Formula (8.80), used with (8.79) for y_k and with neglect of the remainder, is called *Noumerov's algorithm* for advancing the solution of the differential equation from k to $k + 1$. In Section 8.5 we develop a program containing this algorithm, then we test its accuracy and efficiency before applying it to numerical calculations of quantum oscillator wave functions.

8.4   PROJECT 8A: PROGRAMMING SECOND-ORDER EULER METHODS

Our goal in this project is to convert the three Euler-type algorithms, derived in Section 8.3 and summarized in Table 8.3, into working programs, then to test and compare the accuracy of these algorithms in solving two simple differential equations, those for the exponential and cosine functions.

Programming the Euler algorithms

The Euler-type algorithms are straightforward to program, provided that care is taken to code the formulas in the order indicated for each method. Recall that this is important because the formulas are approximations for advancing the solution of the equation from x to $x + h$, rather than equations that determine the y values at these two values of the independent variable. In this sense, the = sign in the algorithms is like the assignment operator in programming languages, the = in C and Fortran and the := in Pascal.

For flexibility in testing and using the program, the second-derivative function, $F(x, y)$, is a separate function referred to by the main program that contains the three Euler-type algorithms as choices. The function name is FUNC. Similarly, the analytical solution to the second-order differential equation that has the same two boundary conditions that you use in your numerical solution is to be programmed in function ANALYT.

The overall structure of the program `Numerical DE_2; Euler-Type Methods` is quite conventional. It begins with the usual opening of a file, NUMDE_2xx, to be written over or added to in order to save the output for graphic display. The suffix "xx" in the file name is to be substituted by the function being calculated. For use with the exponential function the program listing has `xx = exp`. For the cosine function an appropriate suffix is `xx = cos`. In the program supplied you have to make new versions of the program for each new suffix. You should probably modify the file handling appropriately for your computing environment to avoid this awkwardness.

The three Euler-type algorithms appear adjacently rather than being hidden in their own functions. By this means, which is not generally good programming style, you easily see the similarities and differences between the coding of the three algorithms. If you wish to use them independently as functions, you will need care in moving the appropriate variables into the functions.

After the algorithm loop of the code comes the comparison with analytical values. Keeping the differential-equation solver loop separate from the comparison loop makes the program structure clearer. Notice that we calculate both actual errors (err) and percentage relative errors (relerr). The former are more appropriate for functions that oscillate in sign but are bounded, such as the cosine and sine functions. The second method of showing errors is better for functions with a large range of variation, such as the exponential function computed in the next subsection.

The complete program for solving second-order differential equations by the Euler-type methods summarized in Table 8.3 is as follows. Note that FUNC and ANALYT are appropriate for exponential functions, being coded as $F(x, y) = y$ and $y(x) = \exp(x)$, respectively.

PROGRAM 8.1 Euler-type algorithms for numerical solution of second-order differential equations. The three methods are summarized in Table 8.3.

```c
#include <stdio.h>
#include <math.h>
#define MAX 201

main()
{
/* Numerical DE_2;     Euler-Type Methods     */
/* for Second-Order Differential Equations */
FILE *fout;
FILE *fopen();
double y[MAX],yp[MAX],err[MAX],relerr[MAX];
double  xmin,xmax,h,x,Funkh;
double Funx,Funx12,dYkp1,Ykp1,analytical;
int   choice,Nk,k;
char wa;
double FUNC(),ANALYT();

printf("Numerical DE; Second-Order Euler-type Methods\n");
printf("Write over output (w) or Add on (a): ");
scanf("%s",&wa);    fout = fopen("NUMDE_2Eexp",&wa);
choice = 1;
while ( choice > 0 )
  {
  printf("\nChoose Euler method (zero to quit):\n");
  printf("1, 2, or 3; ");
  scanf("%i",&choice);
  if ( choice == 0 )
    {
    printf("\nEnd Numerical DE_2; Euler-Type Methods"); exit(0);
    }
```

```
if ( choice < 0 || choice > 3 )  /*  || is 'or' */
  {
  printf("!! choice=%i only 1,2,3\n",choice); exit(1);
  }
printf("Input  xmin, xmax, h\n");
scanf("%le%le%le",&xmin,&xmax,&h);
Nk = (xmax-xmin)/h+1.1;
if ( Nk > MAX-1 )
  {
  printf("!! # of steps, %i, > %i\n",Nk,MAX-1); exit(1);
  }
  printf("Input y(xmin), y'(xmin):\n");
scanf("%le%le",&y[1],&yp[1]);

/*       Euler-type algorithms;  Iteration in  x  */
x = xmin;
for ( k = 1; k <= Nk-1; k++ )
  {
  Funkh = FUNC(x,y,k)*h;
  yp[k+1] = yp[k]+Funkh;
  switch (choice)
    {
    case 1: /* First method */
            yp[k+1] = yp[k]+Funkh;
            y[k+1] = y[k]+yp[k]*h; break;
    case 2: /* Second method */
            yp[k+1] = yp[k]+Funkh;
            y[k+1] = y[k]+(yp[k]+yp[k+1])*h/2; break;
    case 3: /* Third method */
            y[k+1] = y[k]+yp[k]*h+Funkh*h/2;
            yp[k+1] = yp[k]+(Funkh+FUNC(x+h,y,k+1)*h)/2;
            break;
    }
  x = x+h;
  }/* end  k  for loop &  Euler-type algorithms  */

printf("\nComparison with analytical values\n");
x = xmin;
for ( k = 1; k <= Nk; k++ )
  {
  analytical = ANALYT(x); /* analytical and error */
  err[k] = analytical-y[k];
  relerr[k] = 100*err[k]/analytical;
  printf("%6.2le %10.6lg %10.6lg %10.6lg\n",
    x,y[k],err[k],relerr[k]);
```

```
    fprintf(fout,"%6.21e %10.61g %10.61g %10.61g\n",
      x,y[k],err[k],relerr[k]);
    x = x+h;  /* next  x  */
    } /* end  k  output loop */
  } /* end  choice  while loop */
}


double FUNC(x,y,k)
/* d2y/dx2 = FUNC(x,y) at y = y[k] */
double y[];
double x;
int k;
{
double value;

/* Using working example of  F = y  for FUNC */
value = y[k];
return value;
}


double ANALYT(x)
/* Analytical solution of the differential equation */
double x;
{
double value;

/* Using working example of y = exp(x) for ANALYT */
value = exp(x);
return value;
}
```

The following exercise may be used to check the correctness of coding the program before you start exploring use of Numerical DE_2; Euler-Type Methods for solving differential equations. The checking also illustrates a three-stage strategy for developing numerical algorithms and associated programs. The three stages are:

(1) Verify correctness of coding the algorithm by comparison against test examples where it should be exact.

(2) Check appropriateness and limitations of the coded algorithm against known functions.

(3) Use the program for solving analytically intractable problems numerically, guided by its limitations that have been investigated at stage (2).

Coding verification, stage (1), is illustrated in the following exercise.

Exercise 8.23

(*a*) Show that each of the three Euler-type algorithms for second-order differential equations is exact if the function $F(x, y)$ is identically zero. Similarly, show that the second and third algorithms are exact and identical if $F(x, y)$ is a constant. If you have physics background, you can use the correspondence to the equations of kinematics with constant acceleration (y is displacement, x is time, and F is acceleration) to establish these results.

(*b*) Modify the function FUNC so that it returns zero, and modify the function ANALYT so that it returns x. Run the program with xmin = 0 and initial conditions $y[1] = y(0) = 0$, yp $[1] = y'(0) = 1$. Verify that all three methods give results that are exact within computer roundoff errors, with the first derivative having the constant value of unity.

(*c*) Modify function FUNC so that it returns the constant value of 2, and modify function ANALYT so that it returns x^2. Run the program with xmin = 0 and initial conditions $y[1] = y(0) = 0$, yp $[1] = y'(0) = 0$. Verify that choice = 2 and choice = 3 methods give results that are exact within roundoff errors. Also verify that the first derivative has the value $2x$. ∎

With this program verification completed, we are ready for stage (2). That is, we investigate how well these algorithms handle typical differential equations whose solutions are already known.

Euler algorithms and the exponential function

The second-order differential equation $d^2y/dx^2 = y(x)$, with solution the increasing exponential function, $y(x) = \exp(x)$ if $y(0) = y'(0) = 1$, is a fairly demanding differential equation against which to test the three Euler-type algorithms in Table 8.3. The same solution arises for the first-order differential equation $dy/dx = y(x)$ with boundary condition $y(0) = 1$, as is investigated in Section 7.4, so a comparison of methods can be made. Indeed, method 1 for second-order equations is identical to the forward-predictor method for first-order equations, because second-derivative terms in advancing $y(x)$ are ignored in both algorithms.

The three algorithms for the second-order equations are compared in Figure 8.15 for a step size $h = 0.05$. Method 1 produces an error of about 10% by $x = 4$, that is after $4/0.05 = 80$ steps, or an error per step of about 1.25×10^{-3}. Notice that the line labeled %err(1) in this figure is identical with that labeled %err(f) in Figure 7.5 because they involve the same numerical methods for handling the same differential equation. Notice that the errors decrease by nearly an order of magnitude as we go from method 1 to method 2 to method 3. The numerical effort involved hardly increases, as you can see by comparing the coding for the three choices in Program 8.1, Numerical DE_2; Euler-Type Methods, which is listed in the preceding subsection.

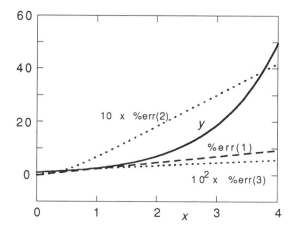

FIGURE 8.15 Errors in three Euler-type methods for the second-order exponential equation.

In this example of the increasing exponential function, all the derivatives are equal to the function value, so that once the function starts heading upward it continues uniformly upward. Even in method 3 derivatives beyond the second are ignored, so the error also increases uniformly, albeit slowly. Because, for all three methods, we see from Figure 8.15 that the errors increase proportionally to x (since the error lines are straight), the fractional error per step is constant. This result can be proved analytically for the exponential function. It is also of interest to investigate numerically the dependence of the errors in each method on the stepsize h.

Exercise 8.24

(*a*) Compile Numerical DE_2; Euler-Type Methods so that FUNC returns value y[k] and ANALYT returns value exp (x). Then run the program for each of the three Euler methods with xmin = 0, xmax = 4, and stepsize h = 0.05. Input y (xmin) = y' (xmin) = 1. Examine the values of the computed y[k] and the percentage relative errors in order to verify approximate agreement with the curves in Figure 8.15.

(*b*) Repeat the above calculations for step sizes of h = 0.1 and 0.025, going out to $x = 4$ each time. After taking into account that the number of steps to reach a given x is inversely proportional to h, do your estimates of the error per step agree with the estimates summarized in Table 8.3? ■

This exercise should convince you that the Euler-type algorithms for second-order differential equations are fairly inaccurate, because the error to propagate the solution to a given x decreases one power of h more slowly than the error per step that is indicated in Table 8.3. Probably only method 3 is suitable for typical applications. The merits of the three methods when used for oscillatory functions are explored in the next subsection.

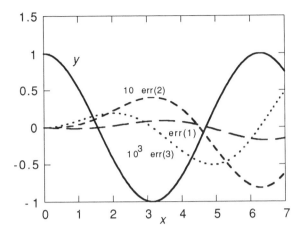

FIGURE 8.16 Errors in three Euler-type methods for the second-order harmonic equation with solution the cosine function.

Euler algorithms and the cosine function

For oscillatory functions it is interesting to examine the buildup and decay of the errors of the three Euler-type methods as the function varies. The function that exactly solves the second-order linear differential equation $d^2y/dx^2 = -y\,(x)$ with boundary conditions $y\,(0) = 1$ and $y'\,(0) = 0$, namely $y\,(x)$ $\cos\,(x)$, is the same as explored for the first-order differential equation solvers at the end of Section 7.5. There the equation was $dy/dx = -\sin\,(x)$ with $y\,(0) = 1$. In real applications it would not be sensible to go to the extra trouble of solving a second-order equation if there is a first-order equation available, but doing so here provides insight from the comparison. We shall see immediately why there is extra trouble.

The numerical solutions using the three Euler-type algorithms with a stepsize of $h = 0.05$ are shown in Figure 8.16 and may be compared directly with the first-order solutions shown in Figure 7.7. Notice that only one of the second-order methods we use is more accurate than any of our first-order methods, namely method 3. It is more accurate than the first-order forward-difference method by about a factor of 100 and it is of comparable accuracy to the first-order central-difference and trapezoid methods. It is worthwhile to convince yourself of the correctness of the numerical results for the second-order equation algorithms.

Exercise 8.25

(*a*) Rename and recompile program Numerical DE_2; Euler-Type Methods so that FUNC now returns value $-y\,[k]$ and ANALYT returns value $\cos\,(x)$. Run the modified program for each of the three Euler methods with xmin = 0, xmax = 7, and a stepsize h = 0.05. Input the values $y\,(\text{xmin}) = 1$, $y'\,(\text{xmin}) = 0$. Examine the values of the computed y[k] and the percentage relative errors in order to verify approximate agreement with the curves in Figure 8.16.

(*b*) Repeat these calculations for stepsizes of h = 0.1 and 0.025, going out to $x = 7$ each time. (The array sizes will need to be at least MAX = 282 for the latter calculation.) After taking into account that the number of steps to reach a given x is inversely proportional to h, do your estimates of the error per step agree with the estimates summarized in Table 8.3? ■

We may conclude from these numerical studies that numerical solutions of low-order differential equations are usually more accurate than the solutions obtained from higher-order equations. The only advantage of using a second-order equation over a first-order equation for the cosine function example is that one obtains numerical estimates of slopes as a by-product of estimating the values of the equation solution.

In the Euler-type methods investigated in this section and in Sections 7.4 and 7.5 there was no restriction to linear differential equations, although our examples do not illustrate this. If the restriction to linear equations can be made, so that $F(x,y) = F(x)y$ in (8.54), there remains a very large range of scientifically interesting differential equations, to be solved numerically. The more accurate methods, such as Noumerov's algorithm that we derived in Section 8.3, are then appropriate. The numerics of this algorithm is the topic of the next section.

8.5 PROJECT 8B: NOUMEROV METHOD FOR LINEAR SECOND-ORDER EQUATIONS

In this project our goal is to convert the Noumerov algorithm derived in Section 8.3 into a working program, to evaluate the method, then to apply it to a problem in quantum mechanics. We first outline the program structure and list the program, then provide two test functions (exponential and cosine) for evaluating the method. In the last two subsections we apply the method and adapt the program to solve numerically the quantum-mechanical Schrödinger equation for a harmonic oscillator in one dimension.

Programming the Noumerov method

Coding and testing of the Noumerov algorithm, (8.79) and (8.80), is quite straightforward. For ease of presentation it is coded in-line, but it is clearly distinguished from the main body of the program. The function that is iterated in-place is Y in (8.79), whereas it is the desired solution, y, that is stored in an array. The starting of the algorithm is assumed to be done by providing the first two values of y, in y[1] and y[2]. In some uses it might be more convenient to have the initial value of y and its first or second derivatives. These can be interconverted by use of a Taylor expansion and of the differential equation.

The overall program structure is by now familiar to you. The program `Numerical DE_2; Noumerov Method` begins with a file-control section for preparing file `NUMDE_2.N`, which receives the output for graphical or other subsequent processing. The file name should probably be made different for each application of the program, for example, the exponential, cosine, or harmonic-oscillator functions investigated below. The range of x values and the stepsize are requested, together with two initial values of y. Then the algorithm discussed above is executed, followed by comparison with analytic values made in terms of actual errors (`err`) and percentage relative errors (`relerr`).

The function `FUNC` is the x-dependent function that returns the second derivative, $F(x)$. The function `ANALYT` should be coded to return the analytical solution of the differential equation that has the same boundary conditions that the starting values of y indicate. It will be least confusing if you make new versions of the program whenever you change the file name or either of these function names. The version of `Numerical DE_2; Noumerov Method` that we now show is set up for testing the evolution of the cosine solution.

PROGRAM 8.2 Noumerov algorithm for second-order linear differential equations without first-derivative terms.

```
#include <stdio.h>
#include <math.h>
#define MAX 201

main()
{
/* Numerical DE_2;      Noumerov Method      */
/* for Second-Order Differential Equations */
FILE *fout;
FILE *fopen();
double y[MAX],err[MAX],relerr[MAX];
double   xmin,xmax,h,h2,Y1,Y2,dYk,Yk,x;
double Funx,Funx12,dYkp1,Ykp1,analytical;
int   Nk,k;
char wa;
double FUNC(),ANALYT();

printf("Numerical DE; Second Order: Noumerov Method\n");
printf("Write over output (w) or Add on (a): ");
scanf("%s",&wa);     fout = fopen("NUMDE_2.N",&wa);
printf("Input  xmin, xmax, h\n");
scanf("%le%le%le",&xmin,&xmax,&h);
Nk = (xmax-xmin)/h+1.1;
if ( Nk > MAX-1 )
```

```
  {
  printf("!! # of steps, %i, > %i\n",Nk,MAX-1); exit(1);
  }
printf("Input y(xmin), y(xmin+h): ");
scanf("%le%le",&y[1],&y[2]);

/*                 Noumerov algorithm;  Iteration in  x  */
h2 = h*h;
Y1 = (1-h2*FUNC(xmin)/12)*y[1]; /* starting Y values */
Y2 = (1-h2*FUNC(xmin+h)/12)*y[2];
dYk = Y2-Y1;    Yk = Y2;   x = xmin+2*h;
for ( k = 2; k <= Nk-1; k++ )
  {
  Funx = h2*FUNC(x);   Funx12 = (1+Funx/12);
  dYkp1 = dYk+Funx*y[k]; /* increment in  Y  */
  Ykp1 = Yk+dYkp1; /* new  Y  value */
  y[k+1] = Funx12*Ykp1; /* next  y  value */
  x = x+h;  dYk = dYkp1;  Yk = Ykp1; /* recycle values */
  }/* end  k  for loop   &   end  Noumerov algorithm  */

printf("\nComparison with analytical values\n");
x = xmin;
for ( k = 1; k <= Nk; k++ )
  {
  analytical = ANALYT(x); /* analytical and error */
  err[k] = analytical-y[k];
  relerr[k] = 100*err[k]/analytical;
  printf("%6.2le %10.6lg %10.6lg %10.6lg\n",
    x,y[k],err[k],relerr[k]);
  fprintf(fout,"%6.2le %10.6lg %10.6lg %10.6lg\n",
    x,y[k],err[k],relerr[k]);
  x = x+h; /* next  x  */
  } /* end  k  output loop */
printf("\n End of  Numerical DE_2; Noumerov Method");
}

double FUNC(x)
/* d2y/dx2 = FUNC(x)   */
/* Noumerov method is second-order and linear in y */
double x;
{
double value;

/* Using working example of  F = -1  for FUNC */
value = -1;
```

```
return value;
}

double ANALYT(x)
/* Analytical solution of differential equation */
double x;
{
double value;

/* Using working example of y = cos(x) for ANALYT */
value = cos(x);
return value;
}
```

With the program coding completed, we are ready to test it for correctness and accuracy in the next subsection, then apply it in the final subsection.

Testing Noumerov for exponentials and cosines

We test the accuracy and efficiency of the Noumerov algorithm for second-order linear differential equations by using the same test functions, exponentials and cosines, as in the previous tests in Chapters 7 and 8. We make the calculations over the same range of x as previously (0 to 4), but we anticipate that the Noumerov method will be much more accurate than both the Euler-type methods and the iteration-integration Adams methods for first-order equations. We therefore begin with twice as large a stepsize as used previously, that is h is increased from 0.05 to 0.1. The tests with the exponential-function solutions are straightforward to make, so I suggest that you make them youself.

Exercise 8.26

(*a*) Compile Numerical DE_2; Noumerov Method so that FUNC returns value = 1 and ANALYT returns value = exp (x). Then run the program for the Noumerov method with xmin = 0, xmax = 4, and stepsize h = 0.1. Input the initial values y[1] = y(xmin) = 1 and y[2] = y(xmin + h) = exp (0.1) = 1.105171. Examine the computed y[k] and prepare a graph to compare with the curves in Figure 8.15, which were computed using Euler-type methods. How much more accurate is the Noumerov method than the most-accurate (method 3) Euler method? (Take into account that you are using twice the stepsize used in the Euler methods.)

(*b*) Repeat the above calculations for step sizes of h = 0.05 and 0.025, going out to $x = 4$ each time. After taking into account that the number of steps to reach a given x is inversely proportional to h, do your estimates of the error per step agree in order of magnitude with the estimate derived in Section 8.3 for the Noumerov method? ∎

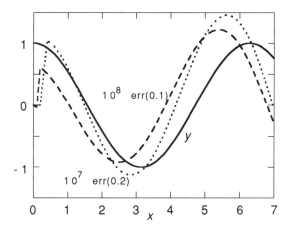

FIGURE 8.17 Noumerov-method solution of the second-order harmonic-oscillator differential equation with cosine solution. The solid curve is the analytical solution, the dashed curve is the scaled error in the numerical solution for stepsize of 0.1, and the dotted curve is the scaled error for stepsize of 0.2.

We now return to the cosine function, examined previously for first- and second-order differential equations by the Euler methods (Figures 7.7 and 8.16). We make calculations over the same range of x as previously (0 to 7), but we anticipate that the Noumerov method will be much more accurate than both the Euler-type methods and the iteration-integration Adams methods. Therefore we begin with twice as large a stepsize as used previously, that is, h is increased from 0.05 to 0.1. The resulting values of y [k] and the scaled errors from the numerical integration are shown for stepsizes of 0.1 and 0.2 in Figure 8.17.

Note the improvement by about five orders of magnitude in the accuracy of the computation, for very little increase in complexity of the algorithm, from Euler methods to Noumerov's method. To be fair in the general comparison, we should point out that the Noumerov algorithm is specially adapted to linear equations with the first derivative eliminated, whereas the Euler-type methods can also handle non-linear equations.

Just to show that all this improvement isn't just smoke and mirrors, why don't you verify for yourself the results shown in Figure 8.17?

Exercise 8.27

(*a*) Rename then recompile program Numerical DE_2; Noumerov Method so that FUNC now returns value −1 and ANALYT returns value cos (x). Run the modified program with xmin = 0, xmax = 7, and stepsize h = 0.1. Input the values y (xmin) = 0, y' (xmin) = 0. Examine the values of the computed y [k] and the percentage relative errors in order to verify approximate agreement with the curves in Figure 8.17.

(*b*) Repeat the calculations for stepsize $h = 0.2$, going out to $x = 7$. After taking into account that the number of steps to reach a given x is inversely proportional to h, show that your estimates of the error per step agree with the $O(h^6)$ estimates derived for the Noumerov method in Section 8.3. ∎

Now that we have a very accurate and efficient method for solving linear second-order differential equations, it is worthwhile to illustrate its use in quantum mechanics, where the Schrödinger equation is a linear second-order equation of wide applicability for describing molecular, atomic, and nuclear phenomena.

The quantum harmonic oscillator

The one-dimensional harmonic oscillator in quantum mechanics is of basic importance, analogously to the simple pendulum in classical mechanics, as discussed (for example) in Chapter 5 of the quantum-mechanics text by Cohen-Tannoudji et al. It is therefore interesting to apply the Noumerov algorithm to solve the Schrödinger equation numerically in order to verify the mathematical derivations of the quantized energy levels that are presented in quantum mechanics textbooks. Here we summarize the relevant mathematical results that are used for comparison with our numerical results. Two textbooks and software that are of interest for computer-based quantum mechanics are those by Brandt and Dahmen.

The time-independent Schrödinger equation for the wave function $\psi(x)$ in one dimension for an oscillator (Hooke's-law) potential is written in terms of the particle mass m, the position coordinate x, and the oscillator angular frequency ω, as

$$\frac{-h^2}{8\pi^2 m} \frac{d^2\psi(x)}{dx^2} + \frac{1}{2} m\omega^2 x^2 \, \psi(x) = E \, \psi(x) \tag{8.81}$$

where h is Planck's constant and E is the total energy. For time-independent states the energies are quantized according to

$$E = \left(n + \frac{1}{2}\right) h\nu \tag{8.82}$$

where n is a non-negative integer $(0,1,...)$ and $\omega = 2\pi\nu$. It is most convenient to rewrite this equation in terms of dimensionless variables, which also provides numbers that are conveniently scaled for computation.

Exercise 8.28

(*a*) Consider in the Schrödinger equation the change of variables

$$\frac{2\pi x}{\sqrt{h/(mv)}} \rightarrow x \tag{8.83}$$

(the denominator is 2π times the "oscillator length"), the scaling of the energy

$$\frac{2E}{hv} \rightarrow E \tag{8.84}$$

and renaming of the dependent variable from ψ to y. Show that (8.81) becomes

$$\frac{d^2y}{dx^2} = F(x)\, y(x) \tag{8.85}$$

where the function F that is appropriate for the Noumerov algorithm is given by

$$F(x) = x^2 - E \tag{8.86}$$

in which the variables x and E are dimension-free.

(b) Given the rule (8.82) for the quantized energies, show that in (8.86) the scaled energy is given by

$$E_n = 2n + 1 \tag{8.87}$$

with $n = 0,1,\dots$. ∎

In the quantum mechanics of bound systems the signature of a time-independent state, also called an "energy eigenstate," is that for large x the value of $y(x)$ tends to zero. Therefore the test that we will make is to use the Noumerov algorithm to solve (8.85) with (8.86) numerically for values of E near the E_n. Energy eigenstates should exhibit the claimed behavior for large x.

Analytic expressions for $y(x)$ can be obtained for the energy eigenstates. Appropriately scaled solutions of (8.85), $y_n(x)$, that are of order of magnitude unity, are given by

$$y_n(x) = \frac{1}{\sqrt{2^n n!}}\, H_n(x)\, e^{-x^2/2} \tag{8.88}$$

where the Hermite polynomials of order n, $H_n(x)$, may be obtained from

$$H_0(x) = 1 \qquad H_1(x) = 2x \tag{8.89}$$

$$H_k(x) = 2\left[H_{k-1}(x) - (k-1)\, H_{k-2}(x)\right] \qquad 2 \le k \le n \tag{8.90}$$

Since the solutions are polynomials times a Gaussian, they eventually die off for large x approximately as a Gaussian, with the falloff being slower for the larger n values associated with larger total energy.

Now that we have the analytic solutions of the differential equation (8.85) at energy eigenstates to check the accuracy of the Noumerov solutions, we are ready to adapt the program Numerical DE_2; Noumerov Method to solve (8.85).

Noumerov solution of the quantum oscillator

In order to solve the modified Schrödinger equation (8.85) for the quantum oscillator, our program for the Noumerov algorithm given two subsections above should be modified slightly in several places.

(1) First, change the output file name to NUM.SE or some distinguishing name.

(2) The next modification is to set xmin = 0 rather than making it an input variable, since the differential equation solution will start at zero. To avoid coding errors, the variable xmin should remain in the program.

(3) A convenient change is to enclose all the main program below the input of new variable n, xmax, and h within a while loop controlled by an input variable called energy for the scaled value of the variable E in (8.85). The program should stay in the loop as long as energy $\geq$ 0. The program can then be run for a range of energy values near the scaled energies E_n given by (8.87).

(4) Starting values should be obtained from the analytical values, rather than input. This is not essential, but it establishes a scale for the y values which cannot otherwise be determined because we are solving a linear equation. Therefore, set y[1] = ANALYT(xmin,n); y[2] = ANALYT(xmin+h,n);

(5) Modify the coding and reference of the function $F(x)$, called FUNC in the program, as follows:

```
double FUNC(x,energy)
/* d2y/dx2 = FUNC(x,energy) */
double x,energy;
{
double value;

/* Schroedinger equation second derivative */
value = x*x-energy;
return value;
}
```

(6) Modify the references to ANALYT, the analytical solution of the Schrödinger equation for quantum number n, from ANALYT(x) to ANALYT(x,n). Rewrite function ANALYT as follows:

```
double ANALYT(x,n)
/* Analytical solution of Schroedinger equation */
double x;
int n; /* energy level; n = 0,1,2,... */
{
double value,Hzero,Hone,Gauss,norm,Hn;
int k;
```

```
Hzero = 1;   Hone = 2*x;   Gauss = exp(-x*x/2);
if (n==0)   return Hzero*Gauss;
if (n==1)   return Hone*Gauss/sqrt(2);
norm = 2;
for (k=2; k<=n; k++) /* Hn by recurrence on k */
    {
    Hn = 2*(x*Hone-(k-1)*Hzero);
    Hzero = Hone;   Hone = Hn; /* update */
    norm = 2*k*norm;
    }
return Hn*Gauss/sqrt(norm);
}
```

The two lowest-order Hermite polynomials and the recurrence relations for the higher-order polynomials are as given by (8.89) and (8.90). The factor `norm` comes from (8.88).

Having made these changes, with appropriate declarations of new variables, we are ready to use the program to explore the numerical solutions of the quantum-oscillator wave functions.

Exercise 8.29

Run your modified program for the Noumerov method applied to the Schrödinger equation with input values $n = 0$, xmax $= 3$, h $= 0.02$, which are all dimensionless variables. Within the `energy` loop use values near the energy eigenvalue, $E_0 = 1$. Check that your results agree with Figure 8.18. ■

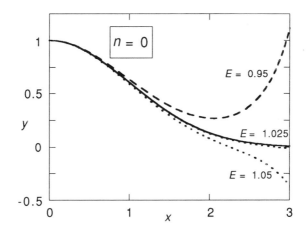

FIGURE 8.18 Numerical solution of the quantum harmonic oscillator Schrödinger equation for energies near the $n = 0$ eigenstate, using the Noumerov algorithm with stepsize $h = 0.02$.

Note that unusually small values of the stepsize h are needed to accurately track the decaying exponential or Gaussian function. Figure 8.18 displays the numerical wave function values y[k] that approximate $y(x)$. The three values of the energy E are close to the energy eigenstate value of unity for $n = 0$. Note that (to within ±0.002) the numerical solution that best fits the analytical solution differs from unity by 0.025. If you vary the stepsize h you will find that the discrepancy from unity decreases steadily as h decreases. Also, the decaying behavior of the numerical solution is extremely sensitive to the relative accuracy of the two starting values. Why not explore these two aspects of the numerical solution yourself?

Exercise 8.30

(a) Run your Schrödinger equation program as in Exercise 8.29, but now vary h from say 0.1 to 0.01 with (for minimum confusion) energy fixed at the energy eigenvalue of unity. Is the solution converging toward the analytical solution as h decreases?

(b) Modify the program so that y[2] is $1 + 10^{-6}$ times the analytical value. Repeat the calculations made in Exercise 8.29. Verify that for large x this part-per-million change in a starting value is amplified about 1000 times when $x = 3$ is reached. ■

It should now be clear to you that it is not good practice to use a numerical algorithm iteratively in a direction for which the function tends to decrease overall. For example, the *increasing* exponential function investigated by the Noumerov algorithm in Exercise 8.26 has part-per-million accuracy, whereas the accuracy at xmax for a *decreasing* exponential, as investigated in the two preceding exercises, is only part-per-cent or worse. Therefore, iteration by numerical algorithms should, whenever feasible, be carried out in the direction of increasing magnitude of y.

In our present example of the quantum oscillator one could begin with the Gaussian solutions at large x, then iterate inward. Unless you had selected an energy eigenstate, the solution would diverge near the origin because of the presence of that other solution of the second-order differential equation, the one irregular (divergent) near the origin. Such an iteration direction should give results that are much less sensitive to the stepsize than is the iteration direction we are using. An alternative scheme for numerical solution of such a "stiff" differential equation as we are trying to solve is presented in Section 8.6.

Finally, it is interesting to take a quantum leap and try the Noumerov method for the quantum oscillator on a state of higher energy, for which there is an oscillating region of the wave function followed by a decaying region for larger x if E is nearly an eigenstate value. Try it and see.

Exercise 8.31

Run the Noumerov method program for the Schrödinger equation with input values n = 3, xmax = 3, h = 0.02, which are all dimension-free variables. Within the energy loop use values near the energy eigenvalue, $E_3 = 7$, according to (8.87). Your results should be similar to those in Figure 8.19. ■

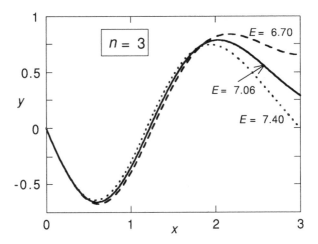

FIGURE 8.19 Numerical solution of the quantum harmonic oscillator Schrödinger equation for energies near the $n = 3$ eigenstate, using the Noumerov algorithm with stepsize $h = 0.02$.

In Figure 8.19 the value that best fits the analytical solution is at an energy $= E$ of 7.06, close to the analytical value of 7. The numerical solution is indistinguishable from the analytical solution within the resolution of the graph. Clearly the values of $E = 6.70$ or $E = 7.40$ do not give energy eigenstates.

Now that you understand how to solve the Schrödinger equation numerically to reasonable accuracy by using the Noumerov method, you will probably find it interesting (depending on your level of experience with quantum machanics) to explore other potentials both for bound states (as here) and for scattering states.

The subject of numerical solution of ordinary differential equations is vast. You can learn more about the analytical and numerical aspects in, for example, Chapters 9 and 10 of Nakamura's book and in Jain's compendium of numerical methods. Several programs in C are provided in Chapter 15 of the numerical recipes book by Press et al.

8.6 INTRODUCTION TO STIFF DIFFERENTIAL EQUATIONS

When applying the Noumerov algorithm to solve the differential equations for exponential-type functions in Section 8.5 we were struck by an unusual contrast. Namely, if we iterated the numerical solution in the direction that the function was generally increasing, as in Exercise 8.29, then the algorithm was remarkably accurate, typically at least part-per-million accurate for a stepsize $h = 0.1$. By comparison, if the solution was iterated so that the function was generally decreasing, as in Exercises 8.30 and 8.31, then even for $h = 0.02$ only part-per-cent accuracy was obtained.

Although we suggested at the end of Section 8.5 that one way around this problem is always to iterate in the direction of x such that the function is generally increasing in magnitude, this is often not possible or not practicable. Alternative schemes are presented in this section. We first describe what is meant by a "stiff" differential equation, then we give two prescriptions that may remove the stiffness, due to Riccati and to Madelung.

What is a stiff differential equation?

Suppose that in the second-order differential equation

$$y''(x) = F(x) y(x) \tag{8.91}$$

we have that $F(x) >> 0$ for a wide range of x. For any small range of x values, small enough that F does not change appreciably over this range, the solutions of (8.91) will be of the form

$$y(x) \approx A_+ \exp\left[\sqrt{F(x)}\, x\right] + A_- \exp\left[-\sqrt{F(x)}\, x\right] \tag{8.92}$$

in which A_+ and A_- are constants.

Exercise 8.32

(a) Verify the correctness of this approximate form of the solution of (8.91) by calculating the second derivative of this expression with respect to x, including the variation of F with respect to x. Show that if

$$\frac{F'}{F\sqrt{F}}\left(1 + \sqrt{F}\, x\right) << 1 \tag{8.93}$$

then the approximation is appropriate.

(b) Show that this condition can be weakened to

$$\frac{F'}{F\sqrt{F}} << 1 \tag{8.94}$$

by appropriately moving the x origin and adjusting A_+ and A_-. ∎

In many scientific and numerical applications the appropriate solution of (8.91) is the exponentially decaying one. For example, we may have the boundary condition that $y \to 0$ as $x \to \infty$, so that $A_+ = 0$ is required. If inaccuracies of the numerical algorithm or of roundoff error allow a small amount of exponentially-increasing solution to insinuate itself into the decaying solution, this increasing solution part will usually quickly increase. The differential equation is then "stiff" to solve, and its solution is an unstable problem in the sense discussed in Section 4.3.

One way of appreciating this insidious behavior is to consider the effect of a small error, ε, on the right-hand side of (8.91). The error in $y''(x)$ is then $F(x)\varepsilon$. Since $F \gg 0$, this amplifies the curve strongly upward or downward, depending on whether $\varepsilon > 0$ or $\varepsilon < 0$. This problem was introduced in Exercise 8.30 (b) and the subsequent discussion. There an ε as small as 10^{-6} was found to make a drastic change in the quantum-oscillator wave function. We now revisit this problem, but for the simpler example of the differential equation (8.91) having $F(x) = 1$ and initial values either increasing (stable behavior) or decreasing (unstable behavior). We use the Noumerov method, for which the increasing exponential was investigated numerically in Exercise 8.26.

Exercise 8.33

(a) Compile Numerical DE_2; Noumerov Method so that FUNC returns value $= -1$ and ANALYT returns value $= \exp(-x)$. Make an input option so that the second starting value can be multiplied by the factor $(1 + \varepsilon)$. Then run the program for the Noumerov method with xmin = 0, xmax = 1, using stepsize h = 0.05. Use as input values y[1] $= y(\text{xmin}) = 1$ and y[2] $= y(\text{xmin} + h) \times (1 + \varepsilon) = \exp(-0.05)(1 + \varepsilon) = 0.951229(1 + \varepsilon)$. Examine the computed y[k] and prepare a graph to compare the analytic and numerical-solution values for ε values in the range 10^{-4} to 10^{-2}.

(b) Repeat the above calculations for stepsizes of h = 0.025 and 0.01, going out to $x = 1$ each time. Show that the strong sensitivity to the accuracy of the second starting value persists for these smaller values of the stepsize. ∎

There are two remarks to be made about such an investigation of this very stiff second-order differential-equation solution. First, it doesn't matter at what x value the effect of a small error is introduced, since $\exp(x) = \exp(x_1)\exp(x - x_1)$ for any value of x_1, so only the distance the error has been propagated is significant. Second, this differential equation is particularly stiff because all the information about the initial slope is contained in the relative values of y[1] and y[2]. It would really be better, if possible, to have other ways of indicating whether it is the increasing or the decreasing solution that is to be found. For example, it might be possible to provide accurate first derivatives near the initial values.

There must be better ways to handle such stiff differential equations. The logarithmic transformation of the dependent variable that we now consider is such a way.

The Riccati transformation

Consider the logarithmic transformation of y in the differential equation (8.91), that is, let

$$R(x) \equiv \frac{d \ln(y)}{dx} = \frac{y'(x)}{y(x)} \qquad (8.95)$$

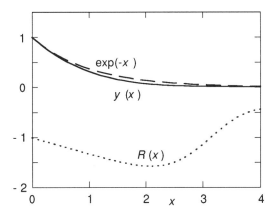

FIGURE 8.20   The function $y(x)$ in (8.96) is shown by the solid curve, the exponential function is indicated by the dashed curve, while the Riccati transformation of y, $R(x)$, is shown by the dotted curve.

The transformation (8.95) is called the Riccati transformation of $y(x)$. Note that the second relation is more general than the first, because it allows negative values of y. It is therefore to be preferred, although the first relation is the one most commonly encountered. The second relation has its own problems because y should not be identically zero anywhere it is used.

The motivation for this transformation is shown graphically in Figure 8.20 for the function

$$y(x) = e^{-ax}\left[1 + b\cos(x)\right]/(1 + b) \tag{8.96}$$

with $a = 1$ and $b = 0.5$. The choice of a and b in (8.96) is such that $y(0) = 1$ and $y'(0) = -1$. Note that on the linear scale the different slope of y from the exponentially decaying function is scarcely distinguishable, which is why a differential-equation solver will have trouble distinguishing between them. On the other hand, the "logarithmic derivative" $R(x) = y'/y$, shown in the lower part of Figure 8.20, clearly distinguishes the exponential decay of slope $-a$ from the modulating effect of the trigonometric factor.

Exercise 8.34

(*a*) Differentiate (8.96) with respect to x, then divide the derivative throughout by y in order to show that

$$R(x) = y'(x)/y(x) = -a - \frac{b\sin(x)}{1 + b\cos(x)} \tag{8.97}$$

so that in the logarithmic derivative there is no exponential decay.

(*b*) Prepare graphs of the logarithmic derivative (8.97) for *b* = 0.25, 0.5, and 0.75 with *a* = 1 over the range $0 < x < 4$. Compare with Figure 8.20 and justify whether it is *a* or *b* that is the more significant parameter in determining the behavior of the function $R(x)$. ∎

After this pictorial introduction to the Riccati transform we need some analysis to formulate the differential equation satisfied by $R(x)$. Suppose that *y* is a solution of

$$y''(x) + b\, y'(x) = F(x)y(x) \tag{8.98}$$

This can be rewritten in terms of *R* as

$$R'(x) = F(\ ) - [R(x) + b]R(x) \tag{8.99}$$

Therefore, solving the second-order differential equation (8.98) can be divided into the solution of two first-order equations. First, (8.99) is solved for *R*, then (8.95) is solved for *y* in terms of *R*.

Exercise 8.35
Derive the first-order Riccati form of the differential equation (8.99) for $R(x)$, starting from (8.98). ∎

Now that we have the mathematical analysis in good shape, we should try some numerics and programming to check out the Riccati transformation.

Programming the Riccati algorithm

The program Numerical DE; Second order by Riccati is based on the program Numerical DE_2; Euler-type Methods in Section 8.4. Here is the Riccati program.

PROGRAM 8.3 Solution of second-order differential equations by the Riccati transform.

```
#include <stdio.h>
#include <math.h>
#define MAX 201

main()
{
/* Numerical DE; Second order by Riccati */
FILE *fout;
FILE *fopen();
double Rb[MAX],errRb[MAX],y[MAX],erry[MAX];
double xmin,xmax,h,twoh,a,b,x,analRb,analy;
int choice,Nk,nmax,k;
```

```
char wa;
double FUNCRb(),FUNCy(),ANALYTRb(),ANALYTy();

printf("Numerical DE; Second Order by Riccati\n");
printf("Write over output (w) or Add on (a): ");
scanf("%s",&wa);    fout = fopen("NUMDE.NumRic.1",&wa);
printf("Input  xmin, xmax, h: ");
scanf("%le%le%le",&xmin,&xmax,&h);
Nk = (xmax-xmin)/h+1.1;
if ( Nk > MAX-1 )
  {
  printf("!! # of xsteps, %i, > %i\n",Nk,MAX-1); exit(1);
  }
twoh = 2*h;
printf("Input a,b: ");
scanf("%le%le",&a,&b);
Rb[1] = a+b;  Rb[2] = (a+b-h)/(1+(a+b)*h);
x = xmin;    /* Euler central for Rb */
for ( k = 2; k < Nk; k = k+2 )
  {
  x = x+h;
  Rb[k+1] = Rb[k-1]+twoh*FUNCRb(Rb,k);
  Rb[k+2] = Rb[k]+twoh*FUNCRb(Rb,k+1);
  x = x+h;
  }
y[1] = 1;  y[2] = 1+b*h;
x = xmin;     /* Euler central for y */
for ( k = 2; k < Nk; k = k+2 )
  {
  x = x+h;
  y[k+1] = y[k-1]+twoh*FUNCy(a,Rb,y,k);
  y[k+2] = y[k]+twoh*FUNCy(a,Rb,y,k+1);
  x = x+h;
  }
printf("Comparison with analytical values\n");
x = xmin;
for ( k = 1; k <= Nk; k++ )
  {
  analRb = ANALYTRb(x,a,b);/* Rb  analytical & error */
  errRb[k] = analRb-Rb[k];
  analy = ANALYTy(x,a,b);/* y  analytical & error */
  erry[k] = analy-y[k];
  printf("%5.2le %6.3lg %6.3lg %6.3lg %6.3lg %6.3lg\n",
  x,analRb,errRb[k],analy,y[k],erry[k]);
  fprintf(fout,"%5.2le %6.3lg %6.3lg %6.3lg %6.3lg %6.3lg\n",
```

```
  x,analRb,errRb[k],analy,y[k],erry[k]);
  x = x+h;
  }
printf("\nEnd  Numerical DE; Second Order by Riccati");
}


double FUNCRb(Rb,k)
/*  dRb/dx = FUNCRb  */
double Rb[];
int k;
{
double value;
value = Rb[k]*Rb[k]-1;
return value;
}


double FUNCy(a,Rb,y,k)
/*  dy/dx = FUNCy  */
double Rb[],y[];
double a;
int k;
{
double value;
value = (Rb[k]-a)*y[k];
return value;
}


double ANALYTRb(x,a,b)
/* Analytical solution for Rb=y'/y+a for oscillator */
double x,a,b;
{
double value;
value = (-sin(x)+(a+b)*cos(x))/(cos(x)+(a+b)*sin(x));
return value;
}


double ANALYTy(x,a,b)
/* Analytical solution for damped oscillator */
double x,a,b;
{
double value;
value = exp(-a*x)*(cos(x)+(a+b)*sin(x));
return value;
}
```

The structure of `Numerical DE; Second order by Riccati` is straightforward and the coding is very similar to that in the Euler-type and Noumerov methods, so you can proceed directly to implement, test, and use the program.

Exercise 8.36

(a) Code the program `Numerical DE; Second order by Riccati` to compute numerically the solution of (8.98). Test the program with $F(x) = F$ (a constant).

(b) Use your program to compute the numerical solution of the damped-exponential equation that has solution (8.96). Compare your numerical results for $a = 1$ and $b = 0.5$ with the analytical solution (8.96) and with the graphical output in Figure 8.20. ∎

We now have some idea how stiff differential equations of the exponentially decreasing type can be solved numerically, so it is interesting to consider briefly second-order differential equations with solutions that oscillate rapidly.

Madelung's transformation for stiff equations

Suppose that in the linear second-order differential equation

$$y''(x) = F(x)y(x) \tag{8.100}$$

we have $F(x) \ll 0$ for a wide range of x. The solutions of such a differential equation will be rapidly oscillating and therefore numerically troublesome, just as when F has the opposite sign and is large in magnitude we get a rapidly changing stiff differential equation that can be handled numerically by the Riccati transformation that we developed in the two preceding subsections.

The Madelung transformation is designed to reduce such problems by dividing y into an amplitude part, r, and a phase part, θ, according to

$$y(x) = r(x)e^{i\theta(x)} \tag{8.101}$$

where r is purely real. In order to identify real and imaginary quantities, we write

$$F(x) = F_R(x) + i F_I(x) \tag{8.102}$$

then insert this and (8.101) in the differential equation (8.100). The real quantities r and θ then satisfy

$$r''(x) - r(x)\big(\theta'(x)\big)^2 = r(x) F_R(x) \tag{8.103}$$

and

$$2r'(x)\,\theta'(x) + (x)\,\theta''(x) = r(x) F_I(x) \tag{8.104}$$

These equations can be simplified in the common case that F is purely real, as you may wish to prove.

Exercise 8.37

(*a*) Verify equations (8.103) and (8.104) by making the indicated substitutions.

(*b*) Show that if F is purely real, then (8.104) can be integrated to give

$$\theta'(x) = \frac{B}{r^2(x)} \qquad (8.105)$$

where B is a constant of integration.

(*c*) By substitution of this result into (8.103), show that, again only for real F,

$$r''(x) = r(x)F(x) + \frac{B}{r^3(x)} \qquad (8.106)$$

so that if this equation can be integrated the resulted can be substituted into the first-order differential equation (8.105) for θ.

(*d*) Verify that if $F(x) = -\alpha^2$, with α real, then r is a constant and, as expected, $\theta = \pm\alpha x$. ∎

Therefore the Madelung transformation for real and negative F in (8.100) can remove the stiffness of a linear differential-equation solution in much the same way as the Riccati transformation does for large real-positive values of F.

Our introduction to the mathematical and numerical analysis of differential equations should set you on the path to reading and understanding the extensive technical literature. A compendium of formulas is provided in the books by Jain and by Zill, while many formulas are listed in Abramowitz and Stegun. The problem of stiff differential equations is given particular attention in the book by Gear.

REFERENCES ON SECOND-ORDER EQUATIONS

Abramowitz, M., and I. A. Stegun, *Handbook of Mathematical Functions*, Dover, New York, 1964.

Backus, J., *The Acoustical Foundations of Music*, Norton, New York, second edition, 1977.

Bernoulli, J., *Opera Omnia*, Marc-Michel Bousquet, Laussanne, 1742, reprinted by George Olms, Hildesheim, Germany, 1968, edited by J. E. Hofmann; Vol. II pp. 232, 251, Vol. IV p. 234.

Brandt, S., and H. D. Dahmen, *Quantum Mechanics on the Personal Computer*, Springer-Verlag, Berlin, 1989.

Brandt, S., and H. D. Dahmen, *Quantum Mechanics on the Macintosh*, Springer-Verlag, Berlin, 1991.

Braun, M., C. S. Coleman, and D. A. Drew (eds.), *Differential Equation Models*, Springer-Verlag, New York, 1983.

Cohen-Tannoudji, C., B. Diu, and F. Laloë, *Quantum Mechanics*, Wiley-Interscience, New York, 1977.

Galilei, G., *Two New Sciences*, Elzevirs, Leyden, The Netherlands, 1638, translated by Stillman Drake, University of Wisconsin Press, Madison, 1974, p. 143.

Gear, C. W., *Applications and Algorithms in Computer Science*, Science Research Associates, Chicago, 1978.

Gilbert, D. "On the Mathematical Theory of Suspension Bridges, with Tables for Facilitating Their Construction," Philosophical Transactions of the Royal Society of London, **116**, 202 (1826).

Haberman, R., *Mathematical Models*, Prentice Hall, Englewood Cliffs, New Jersey, 1977.

Honig, E., "New Wine into Old Bottles: A Nuptial Arch," American Journal of Physics, **59**, 472 (1991).

Ingard, K. U., *Fundamentals of Waves and Oscillations*, Cambridge University Press, Cambridge, England, 1988.

Jain, M. K., *Numerical Solution of Differential Equations*, Wiley Eastern, New Delhi, second edition, 1984.

Nakamura, S., *Applied Numerical Methods with Software*, Prentice Hall, Englewood Cliffs, New Jersey, 1991.

Noumerov, B. V., Monthly Notices of the Royal Astronomical Society, **84**, 592 (1924).

Pippard, A. B., *The Physics of Vibration*, Cambridge University Press, Cambridge, England, 1988.

Press, W. H., B. P. Flannery, S. A. Teukolsky, and W. T. Vetterling, *Numerical Recipes in C,* Cambridge University Press, New York, 1988.

White, H. E., and D. H. White, *Physics and Music*, Saunders College, Philadelphia, 1980.

Zill, D. G., *Differential Equations with Boundary-Value Problems*, Prindle, Weber, & Schmidt, Boston, 1986.

Chapter 9

DISCRETE FOURIER TRANSFORMS
AND FOURIER SERIES

In this chapter and in the next we develop mathematical and computational techniques applicable to describing data and functions in terms of the orthogonal functions introduced in Chapter 6 for linear least-squares fitting. In particular, we use the complex-exponential or cosine and sine functions over appropriate ranges as the orthogonal functions. In pure mathematics the term *Fourier analysis* or *Fourier expansion* usually refers to the more general orthogonal functions, whereas in applied mathematics as used in science and engineering the trigonometric functions are often those of relevance because of their interpretation in terms of simple harmonic motions and harmonics of a fundamental frequency of oscillation.

The topics that we emphasize in this chapter are the following. In Section 9.1 we distinguish between the different kinds of Fourier expansions, the discrete transform, the series, and the integral transform. Section 9.2 is used to develop the discrete transform, whose practical computation is described by the Fast Fourier Transform (FFT) algorithm developed in Section 9.3. We then derive in Section 9.4 the common Fourier series from the discrete transform, and in Section 9.5 we give some obligatory exercises with series and show some of their novel applications. As a diversion, Section 9.6 discusses the pesky practical problems of the Wilbraham-Gibbs overshoot.

Turning to practical applications, we develop a program for the FFT in Section 9.7 as Project 9A. In Section 9.8 we implement as Project 9B a Fourier analysis of an electroencephalogram (EEG). We end the chapter with relevant references on Fourier expansions.

Chapter 10 we devote to Fourier integral transforms, a natural extension of the results in Chapter 9. Both mathematical and applied developments are made, with particular reference to powerful analytical results and their use in computing derivatives and integrals. We introduce convolutions and correlations, and their approximate computation by the FFT is discussed.

Although the results that we derive are mathematically correct, we do not examine rigorously such assumptions as the convergence of Fourier series. These are discussed with mathematical elegance in the monograph by Champeney on Fourier theorems, and in Chapter 6 of the book by Protter and Morrey.

9.1   OVERVIEW OF FOURIER EXPANSIONS

In this section we provide an overview of the uses of Fourier expansions, the different types of expansions, and the nomenclature used to describe them. The distinction between the different types is important both at the analysis level and in practical numerical applications.

The uses of Fourier expansions

There are several motivations for using Fourier expansions in mathematics, science, and engineering. A primary analytical motivation arises from the elegant mathematics of the complex-exponential function and its relation to cosine and sine functions, as reviewed in Section 2.3. A further motivation is the association of the successive terms in expansions, such as in successive Fourier-series terms, with harmonics of frequencies. This connection also relates vibrations and waves (Section 2.4) with resonances (Section 8.1).

At the level of numerics and applications, one motivation for using Fourier expansions arises from linear-least-squares fitting in terms of orthogonal functions, a topic developed in Section 6.2. There we showed that the fitting coefficients may be determined independently of each other if the fitting functions ϕ_k are orthogonal over appropriate regions of x and with suitable weight factors. As we derive in this chapter and the next, suitable orthogonal functions can be found for each type of Fourier expansion.

Types and nomenclature of Fourier expansions

The distinction between the various types of Fourier expansions and their nomenclature is confusing, and it is not always consistent between different authors. Our nomenclature is summarized here and is related to subsequent sections. The chart below illustrates schematically the features of the three types of Fourier expansions.

In each type of expansion the functions $\phi_k(x)$ are complex exponentials or are cosines and sines of kx. The left-hand side of the chart indicates the three types of Fourier expansions and their mathematical distinctions, while the right-hand side indicates the fast Fourier transform algorithm for computing the discrete transform, which is derived in Section 9.3 and programmed in Section 9.7. What are, generically, the similarities and differences between the three expansions?

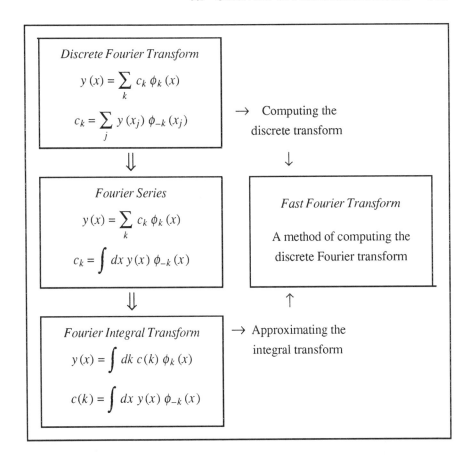

FIGURE 9.1 Schematic of the types and nomenclature of Fourier expansions.

- The *discrete Fourier transform* (DFT) uses summations in both stages It is therefore completely symmetric between the discrete data in the x domain, $y(x_j)$, and the coefficients c_k in the k domain, as we show in detail in Section 9.2. The DFT is most useful for analyzing data, especially because the Fast Fourier Transform (FFT) algorithm (Sections 9.3 and 9.7) allows very efficient computation of the discrete transform and its inverse.

- The *Fourier series* differs from the DFT in that the summation in computing the coefficients is replaced by integration. The y values must therefore be defined over a continuous range of x, which makes the Fourier series most applicable for $y(x)$ described by formulas, as we describe in Sections 9.4 and 9.5. Confusion between the discrete and series transforms is likely in practical applications because the integrals are often approximated by trapezoid-rule sums (Section 4.6).

- In the *Fourier integral transform* (FIT), also just called the Fourier transform, we go all the way, with $y(x)$ being given in terms of integrals over $\phi(k)$, and vice versa. The Fourier transform is often most useful for analytical work, especially because the symmetry between $y(x)$ and $\phi(k)$ has been regained. These analytical integral results are often approximated by summations in practical work, so there is often confusion between the FIT and the DFT, especially when approximating the integral transform numerically by using the FFT algorithm. The Fourier integral transform is so mathematically tractable and has so many applications that we devote Chapter 10 to exploring its properties and applications. For the complex-exponential function it is possible to calculate both its DFT and its FFT analytically, a fact apparently unknown to other authors. I therefore show in Sections 9.2 and 10.2 the calculations, then you have opportunity to compare the two types of Fourier expansion.

Our order of presenting the Fourier expansions indicated in the above chart is not inevitable. By beginning with discrete series, then taking limits to obtain integrals, we follow the Riemann and graphical approaches to the definition of integrals. If we began with the integral transforms and worked toward the discrete transforms, we would have to discretize the functions by introducing Dirac delta distributions, thus introducing additional mathematical pains without any gains. Various approaches are described in the treatises by Körner, Bôcher, and Champeney.

9.2   DISCRETE FOURIER TRANSFORMS

The first kind of Fourier expansion that we consider is direct use of summations in the expansion of a function or data in terms of the complex-exponential function for $\phi_k(x)$. Furthermore, we assume that the x values are available at equally-spaced points $x_j = jh$, with the origin of the x coordinates chosen so that $j = 1,2,...,N$.

Derivation of the discrete transform

Let us try a Fourier expansion of the form

$$y(x) = \sum_k c_k \phi_k(x) \tag{9.1}$$

where the expansion functions are the complex exponentials

$$\phi_k(x) = \frac{\exp(ik\alpha x)}{\sqrt{N}} \tag{9.2}$$

In (9.1), although the sum is over the index k, its range is not yet specified, nor has the quantity α in (9.2) been specified. Both the range and α will be chosen to get an orthogonal expansion, and the $\sqrt{N}$ factor in (9.2) will produce orthonormality, that is, a sum whose value is unity whenever it is nonzero. This property is derived below.

Consider the orthogonality sum, S, corresponding to (6.13) in the least-squares discussion. With the choice (9.2) for the expansion functions, S is given by

$$S = \sum_{j=1}^{N} \exp(ik\alpha x) \exp(il\alpha x)/N \tag{9.3}$$

Exercise 9.1

(a) To evaluate S, show that the substitution $r = \exp(i(k+l)\alpha h)$ produces the geometric-series formula (3.2), for $r \neq 1$,

$$S = \frac{r(1-r^N)}{N(1-r)} \tag{9.4}$$

(b) Thence show that for orthogonality ($S = 0$) one requires

$$\alpha = \frac{2\pi}{Nh} \tag{9.5}$$

(c) With this value of α, show for $r = 1$ ($l = -k$) that $S = 1$, which justifies the normalization of the function in (9.2). ∎

Thus, in the linear-least-squares formula (6.18), choosing weight factors w_j all unity, we have the coefficients for the discrete Fourier transform

$$c_k = \frac{\displaystyle\sum_{j=1}^{N} y_j \exp(-2\pi ijk/N)}{\sqrt{N}} \tag{9.6}$$

Here the coefficients have a range of k values $-N$ to N.

The complementary expansion of the functions in terms of the Fourier coefficients is

$$y(x) = \frac{\displaystyle\sum_{k=-N}^{N} c_k \exp(2\pi ikx/Nh)}{\sqrt{N}} \tag{9.7}$$

This expression may be used for any value of x, not just at the points where the y data exist, the y_j.

Note that the Fourier coefficients, c_k, are complex numbers even if the function $y(x)$ is purely real. This property and others will now be derived.

Properties of the discrete transform

It is worthwhile now to derive many properties of the discrete Fourier transform, since many of them extend to the other Fourier expansions, the series and the integral. The first property is called Parseval's theorem:

$$\sum_{k=-N}^{N} |c_k|^2 = \sum_{j=1}^{N} |y_j|^2 \tag{9.8}$$

In electrical engineering and physics if the k are frequencies, then the left-hand side is a measure of the total power in a system as computed in the frequency domain, while the right-hand side is the power computed in the time domain, so that $x = t$, the time. In quantum mechanics the k usually refer to momentum space and the x to configuration space, then the interpretation of the Parseval theorem is that total probabilities are the same in both representations.

Exercise 9.2
Use (9.8) and (9.6), with the orthonormality conditions on the $\phi_k(x)$, to prove Parseval's theorem. ∎

Parseval's theorem is a special case of a theorem on autocorrelations, usually attributed to Wiener and Khinchin, for which the autocorrelation may be computed in either k or x domains, giving the same result. Parseval's theorem is obtained from the Wiener-Khinchin theorem by setting the lag to zero. A special kind of orthogonality of the $\phi_k(x)$ produces the Parseval relation directly, as you may now prove.

Exercise 9.3
(a) Show that the Parseval theorem holds for any expansion of the form (9.1) provided that the expansion functions satisfy

$$\sum_{j=1}^{N} \phi_k(x_j) \, \phi_l^*(x_j) = \delta_{k,l} \tag{9.9}$$

where the complex-conjugate operation replaces the orthogonality defined in Section 6.2 and $\delta_{k,l}$ is the Kronecker delta, which is unity if $k = l$ and is zero otherwise.

(b) Show that the two definitions of orthogonality are equivalent when the $\phi_k(x)$ are the complex-exponential functions as in (9.2), where α is given by (9.5). ∎

Thus, the Parseval theorem is much more general than originally indicated, and has led us to discuss an alternative definition of orthogonality. The two definitions are confusing in the mathematics of this area, but they agree for the exponential function if the Kronecker delta in (9.9) is replaced by $\delta_{k,-l}$ and complex conjugation.

One aspect of the DFT may be puzzling you. How is it that there are $2N + 1$ complex-number coefficients c_k but only N (generally complex) y values? The answer is that the complex conjugate of each c_k is obtained just by taking complex conjugates on both sides of (9.6). Together with the constraint (9.8) provided by the Parseval relation, this reduces the number of real numbers needed to describe the c_k to $2N$, which is generally the number of real numbers describing the N values of y.

Since data that are processed by Fourier expansions are usually real numbers, it is worthwhile to check what simplifications arise in the DFT if the y_j are real.

Exercise 9.4

Suppose that all the y_j data values in the expression (9.6) for the DFT coefficients, c_k, are real.

(*a*) Show that for real data

$$c_{-k} = c_k^*$$ (9.10)

so that only the coefficients for $k \geq 0$ need to be computed.

(*b*) Thence show that c_0 must be purely real, given by

$$c_0 = \sqrt{N}\,\bar{y}$$ (9.11)

in terms of the average value of the y data, $\bar{y}$.

(*c*) Show that for real data the simplest expression for it in terms of the DFT coefficients is

$$y(x) = \bar{y} + \frac{2\,\mathrm{Re}\left[\displaystyle\sum_{k=1}^{N} c_k \exp\left(2\pi i k x / N\,h\right)\right]}{\sqrt{N}}$$ (9.12)

In applications the average value is often subtracted before the DFT is made. ∎

There are two major restrictions to use of the discrete Fourier transform:

1. The data values $y_j = y(x_j)$ must be obtained at equally spaced points x_j, with spacing h. If the y_j are experimental data, this is usually practicable. In mathematical analysis the discreteness of the data leads to relatively few interesting properties that can be investigated directly by the DFT, because the sums in (9.6) cannot usually be performed analytically.

2. The maximum frequency, k, for which coefficients, c_k, are obtained is related to the number of data points, N, by $k \leq N$. Although there may well be an underlying function with Fourier amplitudes at higher frequencies, one can never discover this by using the DFT. This fact, or variations of it, is called Shannon's sampling theorem or sometimes the Nyquist criterion.

The discrete Fourier transform is very convenient for computerized data analysis because the necessary sums can easily be summed into loop structures. Within a loop there appears to be, according to (9.6), a formidable amount of computation with complex exponentials. The evolution of the fast Fourier transform algorithm, together with the development of computers to handle the intricate logic involved in it, allows for very efficient computation of the DFT, as we describe in Section 9.3.

Exponential decay and harmonic oscillation

There are very few discrete Fourier transforms that can be calculated analytically, except for the rectangular pulse and for the train of spikes, which are not typical of real-world problems. However, the complex-exponential function, which is often encountered in applications (such as in Sections 6.5, 7.2, 8.1, and 8.6), can be transformed simply and analytically. This property does not seem to have been emphasized previously, since such authoritative sources as Brigham or Weaver do not mention it.

We first consider the general exponential, then specialize to exponential decay and to harmonic oscillation. The complex exponential for $x \geq 0$ is

$$y(x) = e^{-\alpha x} \qquad \text{Re } \alpha > 0 \qquad (9.13)$$

The condition on α ensures convergence for positive x when the number of points in the transform, N, is allowed to increase indefinitely, as is done to obtain the Fourier integral transform in Chapter 10. We first derive the transform of this exponential in closed form, then we specialize to pure exponential decay and to harmonic oscillation. Then we have several numerical examples to illustrate these analytical results. For convenience of analysis we modify our conventions for the transform over N points with step h by starting at the point $j = 0$, rather than at $j = 1$, since then we don't have to subtract an awkward expression for the average value of the function. Therefore we write the transform as

$$c_k = h \left[\sum_{j=0}^{N-1} e^{-\alpha j h} e^{-i \omega j h} - \frac{1}{2} \right] \qquad (9.14)$$

in which the frequency variable ω and the index k are related by

$$\omega = \frac{2\pi k}{Nh} \qquad (9.15)$$

In most uses of the transform the choice $k = 0, 1, ..., N - 1$ is made, but this is not necessary in what follows, so that nonintegral k values are allowed. Equation (9.14) also differs from our previous discrete transform by the overall factor h, which ensures convergence to the Fourier integral for large N as $h \to 0$, as shown in Section 10.1. The subtracted term in this equation ensures the correctness of the inverse transform by removing half the value at the point of discontinuity, $k = 0$.

The summation in (9.14) is just a geometric series (Section 3.1) with multiplier

$$r = e^{-(\alpha + i\omega)h} \tag{9.16}$$

and the series sum is given by

$$\sum_{j=0}^{N-1} r^j = \begin{vmatrix} \dfrac{1 - r^N}{1 - r} & r \neq 1 \\ \\ N & r = 1 \end{vmatrix} \tag{9.17}$$

The discrete Fourier transform of the complex exponential (9.13) is therefore

$$c_k = h \left[\frac{1 - e^{-\alpha N h}}{1 - e^{-\alpha h - 2\pi i k/N}} - \frac{1}{2} \right] \tag{9.18}$$

unless the product of the exponents in the denominator is unity, in which case

$$c_k = T \left[N - \frac{1}{2} \right] \tag{9.19}$$

which is also the value obtained by applying L'Hôpital's rule to (9.18). Thus, we have obtained directly a closed form for the discrete Fourier transform of the complex exponential (9.13). In (9.18) N may be any positive integer, so this exact and nontrivial expression may be used to check an FFT program, such as in Section 9.3. The symmetry property (9.10) is also readily verified for α real.

Exercise 9.5
Show that if α is real, then the coefficients for positive and negative k are related through (9.10). ∎

Exponential decay is described by choosing α in (9.13) to be real and positive. By appropriately choosing units for α and h, the time step h can be measured in units of $1/\alpha$, so we can then set $\alpha = 1$ and (9.18) becomes

$$c_k = h \left[\frac{1 - e^{-Nh}}{1 - e^{-h} e^{-2\pi i k/N}} - \frac{1}{2} \right] \tag{9.20}$$

which can be separated into real and imaginary parts for numerical computation.

Exercise 9.6
(a) Multiply numerator and denominator by the complex conjugate of the denominator. Then use Euler's theorem for the complex exponential to obtain sine and cosine expressions in the real and imaginary parts in (9.20).

(*b*) Write and run a simple program for this formula and compare your results with those in Figures 9.2 and 9.3, which have step $h = 1$ and use integer values of k from 0 up to $N - 1$ for $N = 32$, 64, and 128. These are powers of 2 for which the FFT in Section 9.3 may be used, and n is stopped at $N - 1$, as would be done in an FFT calculation. ■

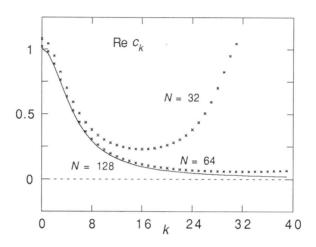

FIGURE 9.2 Real part of the discrete Fourier transform of the exponentially damped function (9.13), shown for 32, 64, and 128 points in the transform.

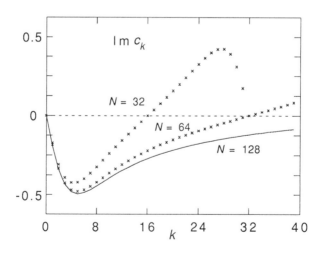

FIGURE 9.3 Imaginary part of the discrete Fourier transform of the exponentially damped function (9.13), shown for 32, 64, and 128 points in the transform.

The symmetry about $N/2$ of Re c_k and the antisymmetry of Im c_k are evident in Figures 9.2 and 9.3. This example of the discrete Fourier transform is discussed further in Section 9.3 when we consider an efficient numerical algorithm, the FFT, for computing the transform. It is also discussed further in Section 10.2 for the integral transform.

Harmonic oscillation is a second example of the discrete Fourier transform of the exponential (9.13). The mathematics of complex exponentials is discussed in Sections 2.3 and 2.4, while the science of resonances is presented in Section 8.1. The DFT of a harmonic oscillation is obtained by setting $\alpha = -i\omega_0$, a pure oscillator with a single frequency ω_0. (To treat this rigorously to produce convergence of the geometric series for large N, a small positive real part, ε, must be included in α. After convergence has been achieved, one can let $\varepsilon \to 0$.) The analytical result for the transform can be simplified by substituting in (9.13) $\alpha = -i\omega_0$, then expressing the complex exponentials in terms of half angles before converting to sines by using Euler's theorem on complex exponentials, (2.39). The complex amplitude may be expressed as

$$
c_k = \begin{vmatrix} h \left[\dfrac{\exp\left[i\pi \left(k_0 - \dfrac{k_0 - k}{N} \right) \right] \sin(\pi k_0)}{\sin\left[\pi \left(\dfrac{k_0 - k}{N} \right) \right]} - \dfrac{1}{2} \right] & k \neq k_0 \\[3em] h \left[N - \dfrac{1}{2} \right] & k = k_0 \end{vmatrix}
\tag{9.21}
$$

where the relation between frequency ω_0 and number k_0 is

$$
k_0 = \frac{N\, h\, n_0}{2\pi}
\tag{9.22}
$$

The second line in (9.21) is obtained either by applying L'Hôpital's rule from differential calculus to the first line or from (9.14) directly. Notice that this transform has no intrinsic dependence on the stepsize h used to compute it, except through the scale factor in (9.21), which was inserted in (9.14) only to produce the appropriate limit for the integral transform in Section 10.2, and through the conversion (9.15) between number and frequency. Therefore, when discussing the oscillator in the following we set $h = 1$.

Exercise 9.7

(a) Derive (9.21) from (9.14) by the route indicated.

(b) Apply L'Hôpital's rule to the first line of (9.21) to produce the second line in this equation. ∎

The symmetry of the discrete transform for real functions, (9.10), does not hold for the oscillator, but there is a related symmetry, namely

$$c_{N-k} (N - k_0) = c_k^* (k_0) \tag{9.23}$$

in which we have extended the notation so that the quantity in parentheses indicates the choice of "resonance" frequency. When k_0 is an integer or halfinteger the oscillator transform simplifies considerably, as can be shown by substituting the values of the circular functions at multiples of π or of $\pi/2$. For k_0 an integer

$$c_k (k_0) = \begin{vmatrix} -\dfrac{1}{2} & k \neq k_0 \\ N - \dfrac{1}{2} & k = k_0 \end{vmatrix} \tag{9.24}$$

so that the transform is discontinuous at k_0. For k_0 a half-integer, such as 1/2 or 15/2, we have

$$c_k (k_0) = \frac{1}{2} + i \cot \left[\frac{\pi (k_0 - k)}{N} \right] \tag{9.25}$$

which is singular across k_0. Thus, Re c_k is symmetric about k_0, while Im c_k is either zero or is antisymmetric about k_0.

Exercise 9.8

Derive the results (9.23) through (9.25) for the discrete Fourier transform of the harmonic oscillator. ∎

Notice that the Fourier transform from x domain at a single frequency produces in the k domain a dependence that is strongest near the frequency k_0. In Section 10.2 it is shown that the Fourier integral transform has a pole at $k = k_0$.

Formula (9.21) is essentially ready to compute numerically because the real and imaginary parts can be immediately identified in the complex exponential. The program Discrete Fourier Transform for Oscillator implements formula (9.21), allowing for any N value that is nonzero and including output to a file, DFToscr, that will record the real and imaginary transform values, Re_ck and Im_ck for each k. As coded, the discrete transform is computed for integer values of k from 0 to N-1, but this is easily modified if you want to look at the behavior of the transform near the discontinuities. When programming the formula the case $k = k_0$ must be handled specially according to (9.24).

Exercise 9.9

(a) Code and test program Discrete Fourier Transform for Oscillator, adapting the input/output to your computing environment. Spot check a few values against other calculations.

(b) Modify the program, or use the output in file DFToscr, to prepare graphics of the real and imaginary parts of c_k against k, as in Figures 9.4 and 9.5.

(c) Compare your results with those in the figure for the indicated values of the frequencies k_0. ∎

PROGRAM 9.1 Discrete Fourier transforms for harmonic oscillators.

```
#include <stdio.h>
#include <math.h>

main()
{
/* Discrete Fourier Transform for Oscillator */
FILE *fout;
double pi,kzero;
double pi_kzero,sinzero,w,wtop,cfact,Re_ck,Im_ck;
int N,k;
char wa;

pi = 4*atan(1.0);
printf("DFT for Oscillator\n");
printf("\nWrite over output (w) or Add on (a):\n");
scanf("%s",&wa);      fout = fopen("DFToscr",&wa);
N = 2;
while ( N != 0 )
  {
  printf("\n\nInput kzero,N (N=0 to end): ");
  scanf("%lf%i",&kzero,&N);
  if ( N == 0 )
    {
    printf("\nEnd  DFT for Oscillator");  exit(0);
    }
  /* Constants for  k  loop */
  pi_kzero = pi*kzero;
  sinzero = sin(pi_kzero);
  for ( k = 0; k <= N-1; k++ ) /* integers for k */
    {
    if ( k == kzero )
      {
      Re_ck = N-0.5;   Im_ck = 0;
      }
    else
      {
      w = (pi_kzero-pi*k)/N;
      wtop = pi_kzero-w;
      cfact = sinzero/sin(w);
      Re_ck = cos(wtop)*cfact-0.5;/* real part */
```

```
        Im_ck = sin(wtop)*cfact; /* imaginary part */
        }
    printf("\n%i  %g  %g",k,Re_ck,Im_ck);
    fprintf(fout,"\n%i  %g  %g",k,Re_ck,Im_ck);
    }
  }
}
```

In Figures 9.4 and 9.5 we choose $N = 16$ and connect the DFT values by line segments, except across the singularity in Im c_k for $k_0 = 7.5$. The discrete Fourier transform of the oscillator usually does not have a singularity at k_0, as the examples in the figure show.

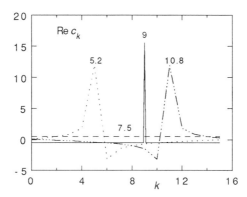

FIGURE 9.4 Real part of the discrete Fourier transform for harmonic oscillations at frequency k_0, for $k_0 = 5.2$ (dotted), $k_0 = 10.8$ (dash-dotted), $k_0 = 7.5$ (dashed, a constant value), and for $k_0 = 9$ (solid, with a spike at $k = 9$)..

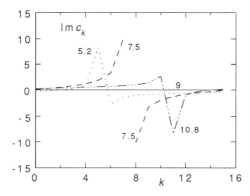

FIGURE 9.5 Imaginary part of the discrete Fourier transform for harmonic oscillations at frequency k_0, for $k_0 = 5.2$ (dotted), $k_0 = 10.8$ (dash-dotted), $k_0 = 7.5$ (dashed, discontinuous), and for $k_0 = 9$ (solid line at zero).

The symmetry (9.23) is made evident in Figures 9.4 and 9.5 by choosing two k_0 values whose sum is N, namely $k_0 = 5.2$ and $k_0 = 10.8$. Notice that for these k_0 values that are not integers or half integers there is a broad distribution of transform values about the oscillator frequency. These two examples give the extremes of bounded exponential behavior, namely, pure decay and pure oscillation. The more general case of the damped oscillator is just (9.18), and the transform will have a distribution of values about k_0, rather than a discontinuity there if $k = k_0$. The investigation of the damped oscillator DFT is an interesting project that you can easily develop from the general expression (9.18) and Program 9.1.

9.3   THE FAST FOURIER TRANSFORM ALGORITHM

Discrete Fourier transforms, whose theory is developed in Section 9.2, are widely used in data analysis, in imaging science, and in digital signal processing, because a very efficient algorithm, the fast Fourier transform (FFT), has been developed for their evaluation. The nomenclature of the FFT is misleading; the algorithm applies to the discrete transform in Section 9.2, *not* to the Fourier integral transform in Chapter 10. The distinction is indicated in Figure 9.1. By various approximations, however, one often estimates the integral transforms by the FFT.

Deriving the FFT algorithm

Here we derive a straightforward and complete derivation of the FFT algorithm appropriate for real data from which the average value of y (the DC level in electrical parlance) has been subtracted out. Equation (9.12) may then be written as

$$y(x) = 2 \, \mathrm{Re} \left[\sum_{k=1}^{N} C_k \, \exp \left(2\pi i k x /Nh \right) \right] /N \qquad (9.26)$$

with $C_k = c_k \sqrt{N}$. Then, from (9.6) we have

$$C_k = \sum_{j=1}^{N} y_j \, \exp \left(-2\pi i j k/N \right) \qquad (9.27)$$

Thus, the y values and the c values are treated symmetrically, the only difference in their treatment being that the exponent is complex-conjugated between their two formulas. Also, the summation range is between 1 and N for both, whereas for arbitrary data the summation range was slightly asymmetric, as shown in Section 9.2.

To continue the development, recall from Exercise 2.13 that the complex Nth roots of unity, E_N, are given by

$$E_N = \exp \left(-2\pi i/N \right) \qquad (9.28)$$

We can now write compactly

$$C_k = \sum_{j=1}^{N} y_j E_N^{jk} \tag{9.29}$$

Direct computation of the c_k requires a time that increases at least as fast as N^2 because each coefficient requires combining N values of the y_j and there are N coefficients. Let us call such a direct evaluation method a *conventional Fourier transform* (CFT) algorithm. If the calculation could be divided into two calculations each of length $N/2$, the time would be halved, since $2(N/2)^2 = N^2/2$. This divide-and-conquer strategy of breaking the calculation of the transform into smaller tasks is the genesis of various FFT algorithms. We now describe the simplest of these, the radix-2 FFT.

Assuming that N is even, one way of dividing the calculation is to combine odd values $j = 1,3,...,N-1$ (that is, $j = 2r-1$, with $r = 1,...,N/2$), then even values $j = 2,4,...,N-2,N$ ($j = 2r$, $r = ,1,...,N/2$). This pattern is illustrated in Figure 9.6 for $N = 8$. In this example the complex factor E_8 is the eighth root of unity, namely 8 values spaced by angles of $\pi/4$ in the complex plane, as shown. For real y_j data the location in the complex plane is appropriate for a DFT calculation of the fundamental frequency, $k = 1$. The suggested division is that between the outer level, labeled by $i = 1$, and the next level, $i = 2$.

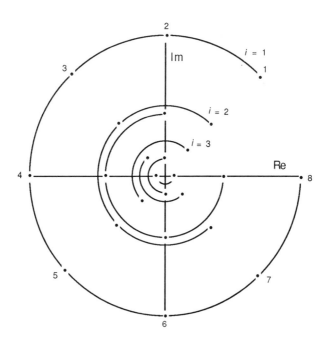

FIGURE 9.6 A mandala for conceptualizing the fast Fourier transform for $N = 8$ data. The positions in the complex plane of the 8 points for $k = 1$ are shown.

With this division the formula (9.29) for the modified Fourier amplitudes C_k can be written compactly as

$$C_k = \sum_{m=-1}^{0} E_N^{mk} \sum_{r=1}^{N/2} y_{2r+m} E_N^{rk} \tag{9.30}$$

In this equation we have used the distributive property of the complex exponential

$$E_N^{a+b} = E_N^a E_N^b \tag{9.31}$$

and the exponentiation property

$$E_N^{pa} = E_{(N/p)}^{a} \tag{9.32}$$

for $p \neq 0$. Note also that

$$E_N^0 = E_N^N = 1 \qquad E_N^{-1} = \exp(2\pi i/N) \tag{9.33}$$

Exercise 9.10
Prove in detail the immediately preceding properties of the complex exponential that are used for developing the FFT algorithm. (For complex exponentials see Section 2.3.) ■

By this level of division to the inner sum in (9.30), there are now two discrete Fourier transforms, each of half the previous length.

The process of halving each transform can be continued as long as the number of points to be divided in each group at each level is divisible by 2. For example, in Figure 9.6 at level $i = 3$ there are 4 groups, with only two points in each. These points are just combined with a sign difference, since they are exactly π out of phase with each other. Note that within each group along an arc the relative phase between points is the same as at the outer level, $i = 1$.

For simplicity, we restrict ourselves to cases in which the total number of data points, N, is a power of 2:

$$N = 2^v \tag{9.34}$$

where v is a positive integer. This is known as a *radix-2 FFT*. In our worked example in Figure 9.6 we have $v = 3$. Then the DFT can be halved v times, ending with $N/2$ subintervals, each of length 2, as in our example.

This essentially completes our outline of the radix-2 fast Fourier transform. After a little bookkeeping we will be ready to develop a computer program for the FFT, as we do in Section 9.7.

Bit reversal to reorder the FFT coefficients

At each step of the fast Fourier transform the values of the coefficients C_k become reordered because of the splitting of the transform into odd and even terms. This reordering is called *bit reversal*. We will now see how bit reversal can be efficiently done for our radix-2 FFT. Consider the example in Figure 9.6 for $N = 2^3 = 8$. The levels involved are $i = 1, 2, ..., v$, with $v = 3$. At the innermost level in this example, moving inward on the semicircular arcs, the subscript of the y_j values are $j = 1, 5, 3, 7, 2, 6, 4, 8$. What's the pattern here?

When we sort a sequence into odd and even terms, the odd terms all have the same least-significant bit in a binary representation, while the even terms all have the same (other) least-significant bit. Consider, for example, y_4, for which its index satisfies $4 - 1 = 3 = 2^2 0 + 2^1 1 + 2^0 1$, which is 011 in binary representation. After the transform is finished y_4 has been moved into the position occupied by y_8. Upon noting that $8 - 1 = 7 = 2^2 1 + 2^1 1 + 2^0 1$, which is 110 in binary, it seems that we have bit-reversed 011. We can formalize this example, as in the following exercise.

Exercise 9.11
Consider the binary representation of the integers between 0 and $2^v - 1$, the range of the index $k - 1$ for the FFT coefficients C_k. Such an integer can be represented by

$$j = j_0 + 2j_1 + ... + 2^{v-2}j_{v-2} + 2^{v-1}j_{v-1} \qquad (9.35)$$

in which each j_i is either 0 or 1. The bit-reversed integer is then j_R, where

$$j = 2^{v-1}j_0 + ... + j_{v-1} \qquad (9.36)$$

(a) Show that j_R can be generated similarly to Horner's algorithm (Section 4.1) as follows:

$j_T = j; \quad j_R = 0;$
Iterate the next steps in integer arithmetic
For $i = 1$ to v
$\quad \{$
$\quad\quad j_D = j_T / 2;$
$\quad\quad j_{i-1} = j_T - 2j_D;$
$\quad\quad j_R = 2j_R + j_{i-1};$
$\quad\quad j_T = j_D;$
$\quad \}$
End of for loop

(b) Verify this algorithm for the example $k = 4$, considered above. ∎

The final step in an FFT algorithm is therefore to bit-reverse the labels on the co-efficients according to this scheme. Note that it is $k - 1$, rather than k, that has to be reversed. This is because we used a labeling scheme for the frequencies in which they range from the first harmonic ($k = 1$) to the ($N - 1$) th harmonic, as in (9.26). Other derivations of the FFT may use a labeling from $k = 0$ to $k = N - 1$, thus avoiding this slight awkwardness. A related presentation of the DFT, lending itself to derivation of the FFT, is given in the article by Peters.

Having derived an FFT algorithm, was it worth our effort and will it be worth the effort to code it?

Efficiency of FFT and conventional transforms

We now discuss the efficiency of the FFT by comparing the times for calculating the discrete Fourier transform of N real data points using the fast (FFT) and conventional (CFT) algorithms. Our comparison is intended to estimate the dependence of the times on N, independent of details of the computer system used.

For the FFT the total number of arithmetic steps of the form (9.30) is roughly proportional to $Nv = N \log_2 N$. A more-detailed analysis, as presented by Brigham, shows that the time, $T(FFT)$, for the fast Fourier transform of N points in our radix-2 algorithm is about

$$T(FFT) = 5T_a N \log_2 N \qquad (9.37)$$

where T_a is the computer time for a typical operation in real arithmetic, such as multiplication or addition.

For the CFT algorithm the time, $T(CFT)$, to compute N coefficients involving N operations must scale as N^2. It is estimated by Brigham to be about

$$T(CFT) = 6T_a N^2 \qquad (9.38)$$

Graphical comparison of the two algorithms is made in Figure 9.7.

Exercise 9.12

(a) On my workstation I programmed and ran the FFT algorithm and program given in Section 9.7. I found that for $N = 128$ the running time was $T(FFT) = 2.1$ seconds, for $N = 256$ $T(FFT) = 4.4$ seconds, while for $N = 1024$ $T(FFT) = 20.1$ seconds. From these data, estimate a range for T_a, and show that the FFT time for a data sample of $N = 8192 = 2^{13}$ points is expected to be about 3.5 minutes.

(b) Use the estimates of T_a from (a) and (9.37) and (9.38) to show that a CFT is estimated to take the impractically long time of about 47 hours of continuous running to transform 8192 data points. Thus, the CFT would be about 800 times slower than the FFT. ∎

This exercise should help you to appreciate the practicality of the FFT for large N values, together with the impracticality of the CFT for the same values of N. For

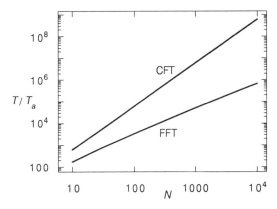

FIGURE 9.7 Comparative times of FFT and CFT according to (9.37) and (9.38).

small values of N, which we omit from the graph, there will be more startup time for the FFT, and some time (proportional to N) will be required for the bit reversal. Already for $N = 64$ the figure shows that the FFT is predicted to be one order of magnitude faster than the CFT, according to the estimates (9.37) and (9.38). You can use the program in Section 9.7 to compare the timing for the FFT on your own computer.

Further reading on the FFT algorithm, including its generalization to transforms in two dimensions, which are widely used in imaging science, is provided by the monograph of Brigham, in Oppenheim and Schafer, in the book on signal processing by Roberts and Mullis. Transformation into real sums of cosines and sines, which is particularly convenient for data, is called a Hartley transform, as discussed in Bracewell's book of this title. We apply the FFT algorithm in Section 9.7 to develop a program which is then applied in Section 9.8 to the Fourier analysis of an electroencephalogram.

9.4 FOURIER SERIES: HARMONIC APPROXIMATIONS

It is often convenient to obtain Fourier expansion coefficients by the procedure of integration, rather than by the summation used in the discrete Fourier transform considered in Sections 9.2 and 9.3. The main reason is that integrals are often easier to evaluate analytically than are sums. Integration requires, however, that the function be defined at all points in the range of integration. Therefore, Fourier series are most appropriate for use with y defined by a function rather than the y_j being discrete data.

From discrete transforms to series

The transition from the discrete Fourier transform to the Fourier series is made by letting the number of x values in the DFT, N, become indefinitely large in such a way that

$$\cdot \; \alpha \, x_j = 2\pi \, (j-1)/N \rightarrow x \qquad (9.39)$$

and that the x increments become infinitesimal

$$\alpha(x_{j+1} - x_j) = 2\pi \, /N \rightarrow dx \qquad (9.40)$$

This makes the range of x from 0 to 2π. The Fourier expansion (9.7) thus becomes

$$y(x) = \frac{\displaystyle\sum_{k=-N}^{N} c_k \exp{(ikx)}}{\sqrt{2\pi}} \qquad (9.41)$$

The Fourier series coefficients are obtained by applying the same limiting process to (9.6). Thus, for $k = -N$ to N, the Fourier series coefficients are

$$c_k = \frac{\displaystyle\int_0^{2\pi} dx \; y(x) \exp{(-ikx)}}{\sqrt{2\pi}} \qquad (9.42)$$

Exercise 9.13
Work through in detail the algebraic steps from the discrete Fourier transform to the Fourier series. ∎

The more usual trigonometric form of the Fourier series is obtained by expanding the exponential in the preceding formulas into sines and cosines. Thus, the Fourier series expansion in terms of cosines and sines is written

$$y(x) = \sum_{k=0}^{N} [a_k \cos{(kx)} + b_k \sin{(kx)}] \qquad (9.43)$$

You will find directly that the Fourier coefficients of the cosine terms are

$$a_k = \frac{\displaystyle\int_0^{2\pi} dx \; y(x) \cos{(kx)}}{(1 + \delta_{k,0})\pi} \qquad (9.44)$$

The Fourier coefficients of the sine terms are given by

$$b_k = \frac{\displaystyle\int_0^{2\pi} dx \; y(x) \sin{(kx)}}{\pi} \qquad (9.45)$$

We have made the conventional division of the normalizing factors, putting the π uniformly in the coefficients, (9.44) and (9.45), and therefore omitting it from the expansion (9.43).

Exercise 9.14
Derive (9.44) and (9.45) from the corresponding exponential form of the Fourier series, (9.41) and (9.42), by following the indicated steps. ∎

Now that we have the formulas for the Fourier series amplitudes, how are they to be interpreted?

Interpreting Fourier coefficients

Consider equations (9.44) and (9.45) for the Fourier coefficients. The integrals for a_k and b_k indicate to what degree $y(x)$ varies like the oscillatory functions $\cos(kx)$ and $\sin(kx)$, respectively. Both of these functions have a period of oscillation of $2\pi/k$. Thus, as k increases, the more rapidly varying parts of $y(x)$ are the most important in determining the corresponding a_k and b_k. In musical notation, $k = 1$ gives the *fundamental* frequency, and all the higher frequencies are its *harmonics*. An alternative name for the Fourier series expansion is the "harmonic expansion."

Since, as (9.44) shows, the coefficient a_0 is just the average value of y over the interval 0 to 2π, it is often assumed that this average value has been subtracted out before y is Fourier analyzed. Thus, the formulas for the a_k and b_k become symmetric, and there need be no special treatment for the DC ($k = 0$) amplitude. The treatment of Fourier series then becomes closer to the usual treatment of the discrete Fourier transform.

Exercise 9.15
Show that for the Fourier series (9.43), with the expansion in the range $x = 0$ to $x = 2\pi$, one predicts a periodicity of y of 2π, that is, $y(x + 2\pi) = y(x)$. ∎

This exercise shows that the Fourier series expansion that we have derived above generates a function with the same period as the range of x values used to generate the coefficients. This is the same property as found for the discrete Fourier transform in Section 9.2, where N data values produced N coefficients. The generalization to Fourier series for arbitrary intervals is made in the next subsection.

Fourier series for arbitrary intervals

Suppose that we have a function of x and want to compute its Fourier series by using the range from x_- to x_+. Given the above expansion formulas for the range 0 to 2π, how should they be modified to account for this different range? The answer to this question is to map the range x_- to x_+ by

$$x' = \frac{2\pi (x - x_-)}{x_+ - x_-} \tag{9.46}$$

which ranges over 0 to 2π as x ranges over x_- to x_+. also note that

$$dx' = \frac{2\pi\, dx}{x_+ - x_-} \tag{9.47}$$

We may now use the Fourier expansion formulas (9.41) and (9.42) in terms of the x' variable to get the Fourier series for the arbitrary interval. With a relabeling of x' by x, the expansion formulas for the complex-exponential series are

$$y(x) = \frac{\displaystyle\sum_{k=-N}^{N} c_k \exp\left[\frac{2\pi i k (x - x_-)}{x_+ - x_-}\right]}{\sqrt{2\pi}} \tag{9.48}$$

in which the coefficients are given by

$$c_k = \sqrt{2\pi}\ \frac{\displaystyle\int_0^{2\pi} dx\ y(x) \exp\left[\frac{-2\pi i k (x - x_-)}{x_+ - x_-}\right]}{x_+ - x_-} \tag{9.49}$$

Exercise 9.16
Work through the steps from the change of variables (9.46) through formulas (9.41) and (9.42) to verify the expansions (9.48) and (9.49) for the complex Fourier series for arbitrary intervals. ∎

We have now accomplished a considerable amount of analysis for Fourier series. Let us work out some applications to practical and interesting examples.

9.5   SOME PRACTICAL FOURIER SERIES

To illustrate the delicate superposition of high and low harmonics that is often required to achieve a good Fourier series representation of a function, we now consider periodic functions with rapid variation near some values of x and much slower variation near others. The examples of functions that we consider are the square pulse, the wedge, the window and the sawtooth. Because of the discontinuities of these functions, they cannot be described by the Taylor-expansion methods covered in Chapter 3. Recall from Chapter 6.2 that we are making least-squares fits to these unconventional but useful functions, since the trigonometric functions or complex exponentials form orthogonal functions for the appropriate expansions.

The functions that we consider are real-valued; it is therefore most convenient to use the cosine and sine series expansions (9.44) and (9.45) for their Fourier series rather than the complex-exponential expansion (9.42). Further, various reflection and translation symmetries of the functions (which we will derive) will result in only cosine or only sine terms appearing in the expansions. The x interval on which we define each example function is $0 \le x \le 2\pi$. This is done primarily for convenience and to obtain simple results. If you wish to use a different interval, you may use the results at the end of Section 9.4 to make the conversion.

By sneaking a look ahead, you may have noticed that we use examples of functions that have discontinuous values, and sometimes discontinuous derivatives. Justification for this is provided in Section 9.6 in the diversion on the Wilbraham-Gibbs overshoot phenomenon.

The square-pulse function

A square pulse may be used, for example, to describe voltage pulses as a function of time, if $y = V$, the voltage (suitably scaled), and $x = t$, the time (in suitable units). This square pulse function is described by

$$y(x) = \begin{vmatrix} 1 & 0 < x < \pi \\ -1 & \pi < x < 2\pi \end{vmatrix} \tag{9.50}$$

We can quite directly find the Fourier series coefficients a_k and b_k by substituting into formulas (9.44) and (9.45) and performing the indicated integrals. The results for the coefficients of the cosines are

$$a_k = 0 \tag{9.51}$$

and for the coefficients of the sines we have

$$b_{2k} = 0 \quad b_{2k+1} = \frac{4}{\pi(2k+1)} \tag{9.52}$$

The nonzero Fourier series coefficients, the sine coefficients of the odd harmonics, $k = 1,3,...$, are displayed in Figure 9.8. Note that their envelope is a rectangular hyperbola because of their inverse proportionality to k.

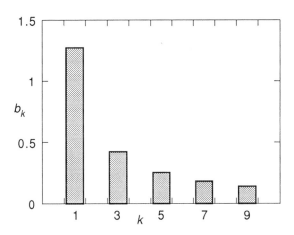

FIGURE 9.8 Fourier series coefficients, b_k, for the square wave as a function of k.

Why do most of the Fourier coefficients vanish? If we consider the square pulse defined by (9.50), we see that it is antisymmetric about $x = \pi$. It must therefore be represented by cosine or sine functions with the same symmetry. But the cosine has reflection symmetry about $x = \pi$, so the cosines cannot appear, thus their coefficients, the a_k, must all be zero, as (9.51) claims. Why do the sine coefficients vanish for k even? Notice that the square pulse changes sign under translation of x by π. The function $\sin(kx)$ has this property only when k is odd, just as we see for the b_k in (9.52).

We conclude that *reflection symmetries* and *translation symmetries* determine to a large degree which Fourier series coefficients will be nonzero. If such symmetry conditions are invoked before the coefficients are evaluated, much of the effort of integration can usually be avoided.

With the expansion coefficients given by (9.51) and (9.52), the approximate Fourier series reconstruction of the square pulse becomes

$$y(x) \approx y_M(x) \tag{9.53}$$

where the sum up to M (taken as odd, for simplicity of expression) is

$$y_M(x) = \frac{4}{\pi}\left[\sin(x) + \frac{\sin(3x)}{3} + \dots + \frac{\sin(Mx)}{M}\right] \tag{9.54}$$

Note that this is the best that one can make the fit, using the least-squares criterion, for a given value of M.

The sine series expansion (9.54) converges as a function of M as indicated in Figure 9.9. From $M = 3$ to $M = 31$ the contribution of the last term should decrease by a factor of more than 10, and the approach to a square pulse shape appears to proceeding smoothly, even if rather slowly.

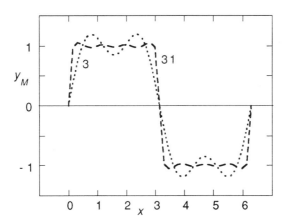

FIGURE 9.9 Fourier series approximation to the square-pulse function up to the Mth harmonic, shown for $M = 3$ and $M = 31$.

Exercise 9.17

(*a*) Carry out all the integrations for the square-pulse Fourier-series expansion coefficients in order to verify formulas (9.51) and (9.52).

(*b*) Check the symmetry conditions discussed above for the square pulse, and verify in detail the claimed reflection and translational symmetry properties of the square pulse and the nonzero sine terms.

(*c*) Verify the algebraic correctness of the expansion (9.54). ∎

Convergence of Fourier series is of practical importance, both for computational software (the time involved and the accuracy of the result) and for practical devices, such as pulse generators based on AC oscillations and their harmonics. Therefore, it is interesting to have programs available for computing the series.

Program for Fourier series

For studying the convergence of Fourier series expansions of interesting functions, it is helpful to have a program. Here is a simple program, Fourier Series, that computes for given M the series expansion (9.54) for $x = 0$ to $x = 2\pi$.

Four cases of function — square, wedge, window, and sawtooth — are included in Fourier Series. Other cases can easily be included by increasing the range of values that choice takes on, and by writing a simple program function for each case. The programming for producing an output file for use by a graphics application is not included, since the programming depends quite strongly on the computer system used. However, you won't find a list of numerical printout very enlightening, so you should add some code for your own graphics output.

PROGRAM 9.2 Fourier series for square, wedge, window, and sawtooth functions.

```
#include <stdio.h>
#include <math.h>

main()
{
/* Fourier Series
   for real functions; data for plots */
double pi,dx,x,y;
int M,nx,choice;
void FSsquare(),FSwedge(),FSwindow(),FSsawtooth();

pi = 4*atan(1.0);
printf("Fourier Series: Input M: ");
scanf("%i",&M);
nx = 2*M+1; /* Number of x steps in 0 - 2pi */
dx = 0.5*pi/M;
```

```
printf("\nChoose function:\n"
  "1 square, 2 wedge, 3 window, 4 sawtooth: ");
scanf("%i",&choice);
for ( x = 0; x <= 2*pi; x = x+dx )
  {
  switch (choice)
    {
    case 1: FSsquare(pi,x,M,&y);   break;
    case 2: FSwedge(pi,x,M,&y);   break;
    case 3: FSwindow(pi,x,M,&y);   break;
    case 4: FSsawtooth(pi,x,M,&y);   break;
    default:
      {
      printf("\n!! No such choice\n\n");
      printf("\nEnd  Fourier Series"); exit(1);
      }
    }
  printf("\n%f %f",x,y);
  }
printf("\nEnd  Fourier Series");   exit(0);
}

void FSsquare(pi,x,M,y)
double pi,x,*y;   /* Square wave series */
int M;
{
double sum;   int k;
sum = 0;
for ( k = 1; k <= M; k = k+2 )
  {
  sum = sum+sin(k*x)/k;
  }
*y = 4*sum/pi;
}

void FSwedge(pi,x,M,y)
double pi,x,*y;   /* Wedge series */
int M;
{
double sum;   int k;
sum = 0;
for ( k = 1; k <= M; k = k+2 )
  {
  sum = sum+cos(k*x)/(k*k);
  }
```

```
*y = 0.5-4*sum/(pi*pi);
}

void FSwindow(pi,x,M,y)
double pi,x,*y;   /* Window series */
int M;
{
double sum;   int k,phase;
sum = 0;
for ( k = 1; k <= M; k = k+2 )
  {
  phase = (k-1)/2;
  if ( phase-2*(phase/2) > 0 )
    sum = sum-cos(k*x)/k;
  else
    sum = sum+cos(k*x)/k;
  }
*y = 0.5-2*sum/pi;
}

void FSsawtooth(pi,x,M,y)
double pi,x,*y;   /* Sawtooth series */
int M;
{
double sum;   int k,phase;
sum = 0;
for ( k = 1; k <= M; k = k+1 )
  {
  phase=k-1;
  if ( phase-2*(phase/2) > 0 )
    sum = sum-sin(k*x)/k;
  else
    sum = sum+sin(k*x)/k;
  }
*y = 2*sum/pi;
}
```

Based on the program, there are several interesting aspects of these series that you may explore.

Exercise 9.18

(*a*) Code the program Fourier Series for your computer system and check it for correctness as follows. When it executes without runtime errors (so that it's satisfying the programming rules), verify the reflection and translation symmetries discussed in the preceding subsection. Then spot check some of the values for the square pulse (choice = 1) for reasonableness against the graphical display in Figure 9.9.

(b)   Run the program for increasingly larger (odd) values of M. Sketch out both how the overall approximation to the square pulse improves, then how the details of the fit near the discontinuity of the square pulse at $x = \pi$ do *not* improve as increases. This "Wilbraham-Gibbs phenomenon" is discussed in detail in Section 9.6.   ■

Now that we have a program to do the busywork of numerics, it is interesting to try the Fourier series for other functions.

The wedge function

The wedge (isosceles triangle) function is of special interest for its application to image enhancement, in which x (or its extension to two dimensions) is spatial position and $y(x)$ becomes image intensity as a function of position. Pratt's text has an extensive discussion of this use of the wedge function.

The wedge function with unity maximum height is defined by

$$y(x) = \begin{vmatrix} x/\pi & 0 < x < \pi \\ 2 - x/\pi & \pi < x < 2\pi \end{vmatrix} \qquad (9.55)$$

As for the square pulse in the preceding subsection, we can just plug this $y(x)$ into each of the integrals for the Fourier coefficients, (9.44) and (9.45).

Exercise 9.19

(a)   Carry out all the integrations for the wedge-function Fourier-series coefficients in order to obtain

$$a_0 = \frac{1}{2} \qquad a_{2k} = 0 \ (k > 0) \qquad a_{2k+1} = -\frac{4}{\pi^2(2k+1)^2} \qquad (9.56)$$

$$b_k = 0 \qquad (9.57)$$

(b)   Use the reflection symmetry of this function about $x = \pi$ in order to explain why sine terms are missing from this expansion.

(c)   Use the translation property of the wedge function

$$1 - y(x + \pi) = y(x) \qquad (9.58)$$

to explain why only odd values of k contribute to the expansion.   ■

The Fourier coefficients in k space is shown in Figure 9.10. They are very similar to those for the seemingly quite different square pulse shown in Figure 9.8.

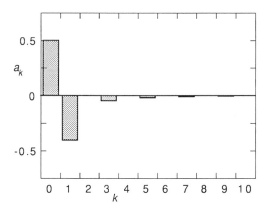

FIGURE 9.10 Fourier series coefficients, a_k, for the wedge function (9.55) as a function of k.

The wedge function may be approximately reconstructed by using the sum of the series up to M terms, namely

$$y(x) \approx y_M(x) \tag{9.59}$$

where the sum up to M (odd, for simplicity) is

$$y_M(x) = \frac{1}{2} - \frac{4}{\pi^2}\left[\cos(x) + \frac{\cos(3x)}{9} + \dots + \frac{\cos(Mx)}{M^2}\right] \tag{9.60}$$

This series converges very rapidly to the wedge function as M increases, as shown in Figure 9.11, where for $M = 3$ we have as close a representation of the wedge as for $M = 31$ for the square pulse shown in Figure 9.9.

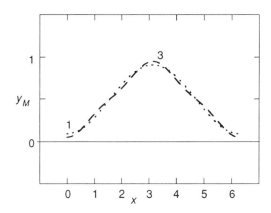

FIGURE 9.11 Fourier series approximation of the wedge function up to the Mth harmonic for $M = 1$ (dotted line) and for $M = 3$ (dashed line).

By running Fourier Series with choice = 2 for the wedge function, you may explore the rate of convergence of the series as a function of M. A clue to the rapid convergence of this Fourier series is the appearance of reciprocal squares of k in the wedge-function Fourier amplitudes rather than just the reciprocal of k, as occurs in (9.52) for the square pulse. This comparison leads to a relation between the wedge and square functions, namely that π times the derivative of the wedge function just produces the square-pulse function, as you may easily verify by comparison of Figures 9.9 and 9.11.

Exercise 9.20

(a) Prove the derivative relation just stated for the relation between wedge and square functions.

(b) Use this relation to obtain the square-pulse Fourier expansion (9.54) by differentiation of the wedge function Fourier expansion (9.60). ∎

Thus we realize that various Fourier expansions may be interrelated through term-by-term comparison of their series. This is both a practical method of generating Fourier series and a good way to check the correctness of expansions derived by other means. Similar connections appear for the window and sawtooth functions which we now consider.

The window function

A window, which allows a signal in a certain range of x, but blocks out the signal outside this range, is a very important function in image and signal processing. For example, if x denotes spatial position and y is the intensity level of an object, then we have literally an optical window. If $x = t$, the time, and $y = V$, the voltage across a circuit element, the window allows the voltage signal through only for a finite time. We consider only a very simple window; more general classes of window functions are considered in the books by Oppenheim and Schafer, and by Pratt. The window function is also called the "boxcar" function, from its resemblance to the silhouette of a railroad boxcar.

In our example of the window function we have the somewhat artificial (but algebraically convenient) window of width π centered on $x = \pi$. For a window of different width or height, one may use the interval scaling in (9.48) and (9.49). Magnitude scaling is done by direct multiplication of each coefficient by the appropriate scale factor because Fourier expansions are linear expansions. Our window function is

$$y(x) = \begin{vmatrix} 0 & 0 < x < \pi/2 \\ 1 & \pi/2 < x < 3\pi/2 \\ 0 & 3\pi/2 < x < 2\pi \end{vmatrix} \tag{9.61}$$

The quickest way to get the Fourier series coefficients for the window function is to relate it to the square pulse (9.50), by evaluating the latter at $x - \pi/2$, dividing it by 2, then adding 1/2. We obtain immediately

$$a_0 = \frac{1}{2} \quad a_{2k} = 0 \ (k > 0) \quad a_{2k+1} = \frac{2\,(-1)^{k-1}}{\pi\,(2k+1)} \tag{9.62}$$

$$b_k = 0 \tag{9.63}$$

This is the first of our examples in which the coefficients alternate in sign.

Exercise 9.21
Use the integral formulas (9.44) and (9.45) directly to verify the above results for the Fourier coefficients of the window function. ■

The window function in k space is shown in Figure 9.12. Because of its simple relation to the square pulse, the magnitudes of these coefficients have the same dependence on k (except for the first) as do the square-wave coefficients in Figure 9.8.

The approximate reconstruction of the window function is very similar to that of the square pulse in (9.54). By using (9.62) and (9.63) for the Fourier amplitudes, we have immediately that

$$y(x) \approx y_M(x) \tag{9.64}$$

where (with M assumed to be odd)

$$y_M(x) = \frac{1}{2} - \frac{2}{\pi}\left[\cos(x) - \frac{\cos(3x)}{3} + \ldots + \frac{(-1)^{(M-1)/2}\cos(Mx)}{M}\right] \tag{9.65}$$

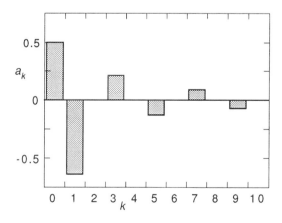

FIGURE 9.12 Fourier series coefficients, a_k, for the window function (9.61).

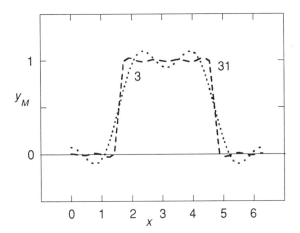

FIGURE 9.13 Fourier series reconstruction of the window function (9.61) up to the M th harmonic for $M = 3$ (dotted) and for $M = 31$ (dashed).

Expression (9.65) has a similar dependence on M to that of the square pulse shown in Figure 9.9. The window is displayed in Figure 9.13 for the same M values as used for the square pulse. By running program Fourier Series with choice = 3, you can discover how the convergence of the series depends on the harmonic number M.

The sawtooth function

Our last example of a Fourier series is that of the sawtooth function

$$y(x) = \begin{vmatrix} x/\pi & 0 < x < \pi \\ -2 + x/\pi & \pi < x < 2\pi \end{vmatrix} \tag{9.66}$$

Such a function is used, for example, in electronics to provide a voltage sweep and a flyback (raster) in video imaging. Here y would be voltage and x time. In such applications, the number of harmonics needed to give an accurate approximation to the sudden drop in y which occurs at $x = \pi$ is important.

The Fourier coefficients may be found either by direct integration using (9.44) and (9.45) or by using the proportionality of the derivative of the sawtooth to the square pulse, namely

$$\pi \frac{d[y(x)]}{dx}_{\text{sawtooth}} = [y(x)]_{\text{squarepulse}} \tag{9.67}$$

Therefore, by integrating the square pulse with respect to x, dividing by π, then including the appropriate constant of integration, you can find the sawtooth Fourier series. It is given by

$$y_M(x) = \frac{2}{\pi}\left[\sin(x) - \frac{\sin(2x)}{2} + \dots + \frac{(-1)^{(M-1)/2}\sin(Mx)}{M}\right] \qquad (9.68)$$

The Fourier amplitudes of the sawtooth can therefore be read off by inspection; namely:

$$a_k = 0 \qquad (9.69)$$

$$b_k = \frac{2(-1)^{k-1}}{\pi k} \qquad (9.70)$$

Exercise 9.22

(a) Make the indicated integration of (9.67), then find the constant of integration (for example by insisting that $y = 0$ at $x = 0$), to derive the sawtooth Fourier series (9.68).

(b) Verify the Fourier amplitude formulas by substituting the sawtooth function (9.68) directly into the integral formulas (9.44) and (9.45). ■

The sawtooth function in k space has the representation shown in Figure 9.14. The envelope of the amplitudes is a rectangular hyperbola. Convergence of the sawtooth series will be fairly slow compared with that for the wedge function investigated above. This is illustrated in Figure 9.15 for $M = 2$ and $M = 20$. Note for the latter the strong oscillations ("ringing") near the flyback at $x = \pi$. This is characteristic of functions that have a sudden change of value, as seen for the square pulse above and discussed more completely in Section 9.6 as the Wilbraham-Gibbs phenomenon.

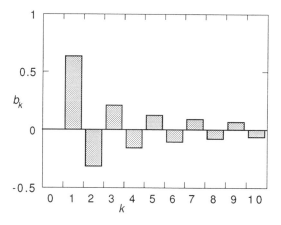

FIGURE 9.14 Fourier series coefficients for the sawtooth function (9.66) as a function of harmonic number k.

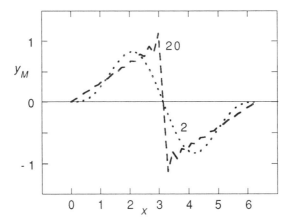

FIGURE 9.15 Approximation of the sawtooth function (9.66) by the first two harmonics (dotted curve) and by the first twenty harmonics (dashed line).

Exercise 9.23

Investigate the convergence of the Fourier series for the sawtooth, (9.66), by running `Fourier Series` for a range of values of M and with `choice = 4` in order to select the sawtooth. Comment both on the convergence as a function of M and on the ringing that occurs near the discontinuity at $x = \pi$. ∎

The sawtooth example concludes our discussion of detailed properties of some Fourier series. In the next section we explore two important general properties of Fourier series that limit the accuracy of such expansions.

9.6 DIVERSION: THE WILBRAHAM-GIBBS OVERSHOOT

We noticed in the preceding section, particularly for the square pulse and the sawtooth, that near the discontinuity there seems to be a persistent oscillation of the Fourier series approximation about the function that it is describing. This is the so-called "Gibbs phenomenon," the persistent discrepancy, or "overshoot," between a discontinuous function and its approximation by a Fourier series as the number of terms in the series becomes indefinitely large. What aspect of this pulse gives rise to the phenomenon, and does it depend upon the function investigated?

Historically, the explanation of this phenomenon is usually attributed to one of the first American theoretical physicists, J. Willard Gibbs, in the two notes published in 1898 and 1899 that are cited in the references. Gibbs was motivated to make an excursion into the theory of Fourier series because of an observation of Albert Michelson that his harmonic analyzer (one of the first mechanical analog computers) produced persistent oscillations near discontinuities of functions that it Fourier analyzed, even up to the maximum harmonic ($M = 80$) the machine could handle. The phenomenon had, however, already been observed numerically and explained fairly completely by the English mathematician Henry Wilbraham 50 years earlier in correcting a remark by Fourier on the convergence of Fourier series. It is therefore more appropriate to call the effect the "Wilbraham-Gibbs phenomenon" than the "Gibbs phenomenon," so that is the name used here.

The first extensive generalization of the phenomenon, including the conditions for its existence, was provided by the mathematician Bôcher in 1906 in a treatise. Both this treatment and those in subsequent mathematical treatises on Fourier series are at an advanced level. A readable discussion is, however, provided in Körner's book. Here we investigate by a rigorous method the problem of Fourier series for functions with discontinuities; we present the essence of the mathematical treatments without their complexity; and we discuss how to estimate the overshoot numerically. I have given a similar treatment elsewhere (Thompson 1992).

We first generalize the sawtooth function to include in a single formula the conventional sawtooth, the square-pulse, and the wedge functions, already considered in Section 9.5. Their Fourier amplitudes can be calculated as special cases of the Fourier series formula that we derive to provide the starting point for understanding the Wilbraham-Gibbs phenomenon. Finally, we give some detail on numerical methods for estimating the overshoot values so that you can readily make calculations yourself.

Fourier series for the generalized sawtooth

The generalized sawtooth function that we introduce is sketched in Figure 9.16 (*a*). It is defined by

$$y(x) = \begin{vmatrix} 1 - (\pi - x)s_L & 0 < x < \pi \\ 1 - D + (x - \pi)s_R & \pi < x < 2\pi \end{vmatrix} \qquad (9.71)$$

in terms of the slopes on the left- and right-hand sides of the discontinuity, s_L and s_R, and the extent of the discontinuity, D. From this definition we can obtain all the functions investigated in Section 9.5; the square pulse, for which $s_L = s_R = 0$, $D = 2$; the wedge ($s_L = -s_R = 1/\pi$, $D = 0$); and the sawtooth ($s_L = s_R = 1/\pi$, $D = 2$).

For our purposes the Fourier amplitudes, a_k and b_k, can most conveniently be obtained from the complex-exponential form (9.42), so that

$$a_k = \frac{\text{Re}[c_k]}{1 + \delta_{k,0}} \qquad b_k = \text{Im}[c_k] \qquad (9.72)$$

Recall that, according to Section 6.2, the Fourier amplitudes, for given M, provide the best fit in the least-squares sense of an expansion of the function y in terms of cosines and sines. Therefore, attempts to smooth out the Wilbraham-Gibbs phenomenon by applying damping factors necessarily worsen the overall fit.

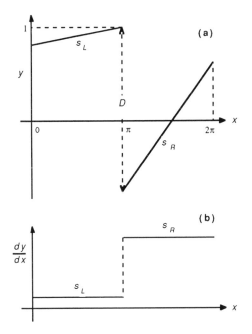

FIGURE 9.16 Generalized sawtooth function for discussing the Wilbraham and Gibbs phenomenon. The solid lines in part (a) show the function and in part (b) they show its derivative with respect to x.

We note immediately, and most importantly for later developments, that a discontinuity in any *derivative* of a function is no stumbling block to calculating its Fourier series. For example, in Figure 9.16 (b) we see the discontinuity in dy/dx that occurs for our generalized sawtooth at $x = \pi$. This discontinuity, of amount $s_R - s_L$, is independent of the discontinuity in y at the same point, namely D. Derivative discontinuities can be removed as follows. If instead of y in (9.42) we had its nth derivative, then integration by parts n times would recover the integral as shown, with some k-dependent factors and additional endpoint values. Therefore, there is no Wilbraham-Gibbs phenomenon arising merely from discontinuities of slope, but only from discontinuities of y itself. We show this explicitly towards the end of this section for the wedge function, for which the slope discontinuity at $x = \pi$ is $2/\pi$, but for which there is no discontinuity in value ($D = 0$).

With these preliminary remarks, we are ready to calculate the c_k in (9.72). For $k = 0$ we obtain, as always, that $a_0 = c_0/2$, the average value of y in the interval of integration, namely

$$a_0 = 1 - \frac{D}{2} + \frac{\pi(s_R - s_L)}{4} \tag{9.73}$$

For $k > 0$ the integrals can be performed directly by the methods in Section 9.5. Upon taking real and imaginary parts of the resulting c_k you will obtain

$$a_k = \frac{(s_R - s_L)(1 - \pi_k)}{\pi\, k^2} \tag{9.74}$$

$$b_k = \frac{D(1 - \pi_k)/\pi - (s_R + s_L)}{k} \tag{9.75}$$

where the phase factor

$$\pi_k = e^{i\pi k} = (-1)^k \tag{9.76}$$

has the property that

$$\pi_k (1 - \pi_k) = -(1 - \pi_k) \tag{9.77}$$

a result that is useful when manipulating expressions subsequently.

Exercise 9.24
Start from the generalized sawtooth (9.71) in the Fourier series formula (9.42) to derive formulas (9.74) and (9.75) for the Fourier coefficients. ∎

The formulas for the Fourier amplitudes of the generalized sawtooth, (9.74) and (9.75), can be used directly to generate the amplitudes for the examples in Section 9.5. The appropriate slope and discontinuity values are given at the beginning of this section, or can be read off Figure 9.16 (a). Note that we have introduced a considerable labor-saving device by allowing all the exercises in Section 9.5 to be composed in a single formula. (The purpose of exercises is not, however, to save labor but, rather, to develop fitness and skill.)

What is the value of the Fourier series right at the discontinuity, $x = \pi$? It is obtained directly by using $x = \pi$ in the Fourier expansion with the a_k values from (9.74) and the sine terms not contributing. The result is simply

$$y_M(\pi) = a_0 - \frac{(s_R - s_L)}{\pi} \sum_{k=1}^{M} \frac{1 - (-1)^k}{k^2} \tag{9.78}$$

There are several interesting conclusions from this result:

(1) The Fourier series prediction at the discontinuity depends on M independent of the extent of the discontinuity, D.

(2) For any number of terms in the sum, M, if there is no change of slope across the discontinuity (as for the square pulse and conventional sawtooth), the series value is just the mean value of the function, a_0. For the square pulse the series value at the discontinuity is just the average of the function values just to the left and right of the discontinuity, namely zero.

(3) The series in (9.78) is just twice the sum over the reciprocal squares of the odd integers not exceeding M. It therefore increases uniformly as M increases. This convergence is illustrated for the wedge function in Figure 9.10. The limiting value

of the series can be obtained in terms of the Riemann zeta function, as $3\zeta(2)/2 = \pi^2/4$ (Abramowitz and Stegun, formulas 23.2.20, 23.2.24). The Fourier series then approaches

$$y_\infty(\pi) = a_0 - \frac{\pi(s_R - s_L)}{4} = 1 - \frac{D}{2} \tag{9.79}$$

Thus, independently of the slopes on each side of the discontinuity, the series tends to the average value across the discontinuity —a commonsense result.

Exercise 9.25

(a) Derive the result for the Fourier series at the discontinuity, (9.79), by using the steps indicated above that equation.

(b) Write a small program that calculates the sum in (9.78) for an input value of M. Check it out for a few small values of M, then take increasingly larger values of M in order to verify the convergence to $\pi^2/4$.

(c) Justify each of the conclusions (1) through (3) above. ∎

Notice that in the above we took the limit in x, then we examined the limit of the series. The result is perfectly reasonable and well-behaved. The surprising fact about the Wilbraham-Gibbs phenomenon, which we now examine, is that taking the limits in the opposite order produces quite a different result.

The Wilbraham-Gibbs phenomenon

Now that we have examined a function without a discontinuity in value but only in slope, we direct our steps to studying the discontinuity. Consider any x not at a point of discontinuity of y. The overshoot function, defined by

$$O_M(x) = y_M(x) - y(x) \tag{9.80}$$

is then well-defined, because both the series representation and the function itself are well-defined. As we derived at the end of the last subsection, if we let x approach the point of discontinuity with M finite, we get a sensible result. Now, however, we stay near (but not at) the discontinuity. We will find that the value of O_M depends quite strongly on both x and M. In order to distinguish between the usual oscillations of the series approximation about the function and a Wilbraham-Gibbs phenomenon, one must consider the behavior of O_M for large M and identify which parts (if any) persist in this limit.

For the generalized sawtooth shown in Figure 9.16 (a) we can substitute for the series the expressions (9.73), (9.74) and (9.75), and for the function (9.71), to derive the explicit formula for the overshoot function

$$O_M(x) = (\pi - x)s_L + \frac{\pi}{4}(s_R - s_L) - \frac{D}{2}$$
$$- \frac{s_R - s_L}{\pi}S_- - \frac{s_R + s_L}{2}S_+ + \frac{D}{\pi}S_D \tag{9.81}$$

where the trigonometric sums are

$$S_- \equiv -\sum_{k=1}^{M} \frac{(1 - \pi_k)\cos(kx)}{k^2} \tag{9.82}$$

$$S_+ \equiv 2\sum_{k=1}^{M} \frac{\sin(kx)}{k} \tag{9.83}$$

$$S_D \equiv \sum_{k=1}^{M} \frac{(1 - \pi_k)\sin(kx)}{k} \tag{9.84}$$

The signs in the definitions of these sums are chosen so that the sums are positive for x close to, but less than, π.

Exercise 9.26
Fill in the steps in deriving the overshoot function equation (9.81), including the trigonometric series (9.82)–(9.84). ∎

You may investigate the overshoot values directly as a function of the maximum harmonic in the Fourier series, M, and as a function of the values of x near the discontinuity at $x = \pi$. We first see what progress we can make analytically. In particular, for what value of x, say x_M, does $O_M(x)$ have a maximum for x near π? To investigate this, let us do the obvious and calculate the derivatives of the terms in (9.81), which requires the derivatives of the trigonometric series, (9.82)–(9.84). We have

$$\frac{dS_-}{dx} = S_D \tag{9.85}$$

$$\frac{dS_+}{dx} = 2\sum_{k=1}^{M} \cos(kx) \tag{9.86}$$

$$\frac{dS_D}{dx} = \sum_{k=1}^{M} (1 - \pi_k)\cos(kx) \tag{9.87}$$

Exercise 9.27
Carry out the indicated derivatives of the trigonometric series in order to verify (9.85) – (9.87). ∎

To evaluate the latter two series in closed form, we write the cosine as the real part of the complex-exponential function, then recognize that one has geometric series in powers of exp (ix), which can be summed by elementary means then converted to sine form by using the formulas in Section 2.3. Thus

$$\frac{dS_+}{dx} = -1 + \frac{\sin\left[(M + 1/2)x\right]}{\sin\left(x/2\right)} \tag{9.88}$$

$$\frac{dS_D}{dx} = \frac{\sin\left[(M + 1)x\right]}{\sin\left(x\right)} \tag{9.89}$$

In the second equation we assume that M is odd, else $M + 1$ is replaced by M. Since S_D is not known in closed form, there is probably no simple way to evaluate the derivative of S_- in closed form. It turns out that we will not need it.

Collecting the pieces together, we finally have the result for the derivative of the overshoot at any $x < \pi$:

$$\frac{dO_M}{dx} = (s_R - s_L)\left(\frac{1}{2} - \frac{1}{\pi}\frac{dS_-}{dx}\right) - \frac{(s_R - s_L)}{2}\frac{\sin\left[(M + 1/2)x\right]}{\sin\left(x/2\right)}$$
$$+\frac{D}{\pi}\frac{\sin\left[(M + 1)x\right]}{\sin\left(x\right)} \tag{9.90}$$

It is worthwhile to check out these derivations yourself.

Exercise 9.28
(a) Sum the series (9.86) and (9.87) as indicated in order to verify (9.88) and (9.89).
(b) Verify equation (9.90) for the derivative of the overshoot function. ∎

The derivative in (9.90) is apparently a function of the independently chosen quantities s_R, s_L, D, and M. Therefore, the position of the overshoot extremum (positive or negative) seems to depend upon all of these.

The wedge was the only example that we considered in Section 9.5 that had $s_R \neq s_L$, and it was well-behaved near $x = \pi$ because $D = 0$. Figure 9.17 shows the overshoot function for the wedge, for three small values of M, namely, 1, 3, and 15. Only the S_- series, (9.82), is operative for O_M, and this series converges as $1/k^2$ rather than as $1/k$ for the other series. Note the very rapid convergence, which improves about an order of magnitude between each choice of M. Clearly, there is no persistent overshoot.

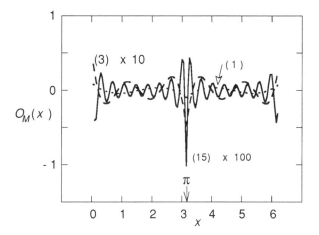

FIGURE 9.17 Oscillation of the Fourier series for the wedge function, (9.80), shown for $M = 1$ (dotted curve), $M = 3$ (dashed curve), and $M = 15$ (solid curve). The overshoot function for $M = 3$ has been multiplied by 10, and that for $M = 15$ has been multiplied by 100. The derivative of the wedge function, but not the wedge function itself, has a discontinuity at $x = \pi$.

Exercise 9.29

(a) Modify the program `Fourier Series` in Section 9.5 so that it can prepare output for a graphics program to make plots like those shown for the wedge series oscillations in Figure 9.17. (I found it most convenient to write a file from the program, then to do the plotting from a separate application program. That way, it was not necessary to recalculate the Fourier series each time I modified the display.)

(b) Calculate $O_M(x)$ for the wedge for a range of maximum k values, M, similarly to Figure 9.17. Note that the convergence goes as $1/k^2$, so you should find a very rapid approach to zero overshoot. Scaling of the function as M increases, as indicated in Figure 9.17, will probably be necessary. ∎

Now that we have experience with the overshoot phenomenon, it is interesting to explore the effects with functions that have discontinuities.

Overshoot for the square pulse and sawtooth

For the square-pulse function, the common example for displaying the Wilbraham-Gibbs phenomenon, $s_R = s_L = 0$, so that in (9.90) the extremum closest to (but not exceeding) π will occur at

$$x_M = \pi - \frac{\pi}{M+1} \tag{9.91}$$

By differentiating the last term in (9.90) once more, you will find that the second derivative is negative at x_M, so this x value produces a maximum of O_M. Indeed, from (9.90) it is straightforward to predict that there are equally spaced maxima below $x = \pi$ with spacing $2\pi/(M+1)$. Thus, the area under each excursion above the line $y =$ must decrease steadily as M increases.

Exercise 9.30

(a) Verify the statement that there are equally spaced maxima below $x = \pi$ by successive differentiation of (9.90).

(b) After $(M+1)$ such derivatives you will have maxima at negative x. Explain this result. ■

The square-pulse overshoot behavior is shown in Figure 9.18, in which we see the positive overshoot position shrinking proportionally closer to π as M increases, but reaching a uniform height that is eventually independent of M. By making numerical calculations, using methods described in the next subsection, you may discover that the extent of the overshoot is also remarkably independent of M for the values shown, namely $O_M(x_M) = 0.179$ to the number of figures given. The fact that O_M converges to a nonzero value for large M identifies this as a genuine overshoot, as discussed below the defining equation, (9.80).

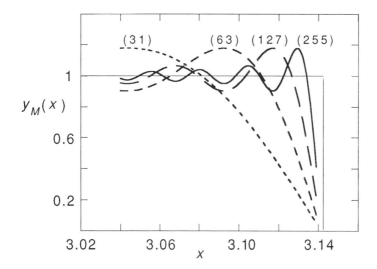

FIGURE 9.18 Wilbraham-Gibbs overshoots for the square pulse of unit amplitude, shown for a range of values of the upper limit of the series sum, $M = 31$, 63, 127, and 255. The function (horizontal segment) has a discontinuity (vertical segment) at $x = \pi$.

The sawtooth function is our final example for the Wilbraham-Gibbs phenomenon. For this function we have $s_R = s_L = 1/\pi$, and $D = 2$. If you insert these values in (9.81) you will find exactly the same position for the location of the maximum overshoot, namely that given by (9.91).

Exercise 9.31
Verify, as indicated, that the sawtooth series maxima are predicted to occur at the same locations as for the square-pulse series maxima. ∎

The sawtooth overshoot also tends to the magic value, 0.179, as you will discover if you investigate the problem numerically, as suggested in Exercise 9.32. Other properties, such as the locations of the zero-overshoot positions on either side of x_M, will be found to depend on the parameters determining the shape of $y(x)$.

Exercise 9.32
Use the modified wedge-function Fourier series program suggested in Exercise 9.29 to investigate the overshoot as a function of s_R, with s_L fixed at zero and $D = 2$. Show numerically and graphically that as s_R becomes very large you begin to get an overshoot type of behavior. ∎

So, what's going on here? According to (9.90), it looks as if x_M should depend on the shape of the function whose Fourier series we determine. But we discovered that if there is a discontinuity ($D \neq 0$), there is no dependence on its position or on the overshoot value, at least for the two such examples that we investigated.

The way to solve this puzzle is to take a different view of the original function $y(x)$, for example as given by (9.71). We may consider this as a linear superposition of a function with no discontinuities in its values, plus a constant (say unity) broken by a drop of amount D at $x = \pi$. Since taking Fourier series is a linear transformation, in that sums of functions have Fourier amplitudes which are the sums of the amplitudes from each series separately, only the discontinuous part gives rise to an overshoot. It is always the same overshoot, except for D as a scale factor, as in (9.81) multiplying S_D, and it is always at the same position, $\pi/(M+1)$ below $x = \pi$, because it arises only from the discontinuity. Because of the symmetry of the discontinuity there will be a mirror undershoot in the opposite direction just above $x = \pi$. The terms in (9.90) that do not involve S_D are just from oscillations of the series about the function. These oscillations damp out for large enough M, as we saw for the wedge in the preceding subsection. A very asymmetric wedge, having $D = 0$, $s_L > 0$, $s_R < 0$, but $|s_R| >> s_L$, may be used to very nearly reproduce the effects of a discontinuity, but this will surely have a very slow convergence.

Thus, we conclude from our analysis and numerical explorations that as soon as we have studied the square-pulse function, we understand all there is to the Wilbraham-Gibbs phenomenon for Fourier series. Since we are learning numerics as well as analysis, we now work out where the magic overshoot number comes from.

Numerical methods for summing trigonometric series

The numerical methods for investigating the Wilbraham-Gibbs phenomenon are straightforward but require care if a large number of terms, M, is used in the series. This is because for x near π, summation over terms containing sine or cosine of kx oscillate in sign very rapidly as k or x changes.

With the precision of my computer I found that the values of S_+ and S_D were calculated accurately for M up to about 300. Past this, even with the use of double-precision arithmetic, unreliable results were produced. In order to make this claim, an alternative estimate of the sum is needed, as follows. The defining (9.84) for S_D can first be replaced by twice the summation over only odd values of k; then this sum can be approximated by an integral if M is large. The result can be converted to the integral

$$S_D(x) \approx \int_0^{(M-1)x'} \frac{\sin t}{t}\, dt \tag{9.92}$$

where $x' = \pi - x$. Since we want S_D evaluated at the peak value nearest to π, we set x to x_M given by (9.91). Then, within a correction of order $1/M$, the resulting integral is to be evaluated with an upper limit of π. The integral in (9.92) can easily be done by expanding the sine function in its Maclaurin series, dividing out the t, then integrating term by term, to obtain

$$S_D(x_M) \approx \int_0^{\pi} \frac{\sin t}{t}\, dt = \pi\left(1 - \frac{\pi^2}{3!\,3} + \frac{\pi^4}{5!\,5} - \frac{\pi^6}{7!\,7} + \cdots\right) \tag{9.93}$$

When the value of this integral, summed out to π^{10}, is substituted in the overshoot formula (9.81) for the square pulse, we find $O_M(x_M) = 0.179$ to three significant figures, in complete agreement with direct calculation of the series overshoot found for both the square pulse and the sawtooth in the preceding subsection.

Exercise 9.33

(a) Develop the integrand of (9.92) as indicated, then integrate term by term (which is permissible for a convergent series) to derive the formula for the integral (9.93).

(b) Numerically evaluate the series in (9.93), then substitute the value in the formula for $O_M(x_M)$ to verify the value of 0.179 given.

(c) Investigate the M dependence of the overshoot by using (9.89) to write $S_D(x_M)$ as an integral from π to x_M, note that $S_D(\pi) = 0$ by (9.84), then transform the variable of integration to t so that the integral is from 0 to π. Now expand the denominator of the integrand in a Taylor series, factor out t, then use the geometric series to get $S_D(x_M)$ in terms of (9.92) and integrals of $t^n \sin t$ from 0 to π. Thus show that, assuming M odd,

$$S_D(x_M) = 1.85194 + \frac{1}{3(M+1)^2} + \frac{7(\pi^2-6)}{180(M+1)^4} \qquad (9.94)$$

where terms in $1/(M+1)^6$ and smaller are neglected.

(*d*) Predict that the *M* dependence is less than 1 part per thousand for $M > 13$. This dependence is surely small enough for any reasonable scientist or engineer.

(*e*) Check (*c*) and (*d*) numerically by using the program modification suggested in Exercise 9.32. ∎

So, that's enough analysis for a while. Let's follow up on discrete Fourier transforms and apply the FFT to a practical example.

9.7   PROJECT 9A: PROGRAM FOR THE FAST FOURIER TRANSFORM

In this section we develop a program that implements the FFT radix-2 algorithm for the discrete Fourier transform, as derived in Section 9.3. Our emphasis is on producing a working program that is similar to the algorithm derived rather than on a program that is very efficient. For the latter, you should use an applications package that is tailored to your computer to produce maximum speed. For example, the book of numerical recipes by Press et al. contains a suite of FFT functions. One way to get speed, at the expense of memory, is to use table lookup, which avoids recomputing cosines and sines.

In our radix-2 FFT the number of data points, *N*, is assumed to be a power of 2, $N = 2^v$, where *v* is input by the user. An option in the program allows you to predict the running time of the FFT program on your computer and thereby to check out the timing-estimate formula (9.37). The driver and testing routine is made to be a program with bare-bones input, execution, and output. For convenience and elegance you should adapt it to your computing environment.

Building and testing the FFT function

The program structure of `Fast Fourier Transform` has the algorithm derived in Section 9.3 coded as function `FFT`. This function computes transforms in subdivided intervals, then reorders these by the bit-reversing function `bitrev`, also programmed according to the algorithm in Section 9.3. The operation of `bitrev` is independent of the computer system used, provided that *N* is not larger than the maximum integer the computer can handle. Program speed could be increased by using machine language to perform the bit reversal.

The `Fast Fourier Transform` program is self-testing by use of the symmetry between the discrete Fourier transform relation (9.6) and the inverse relation (9.7). All that needs to be done is to reverse the sign of the variable `IT`, which has `IT = +1` to obtain the coefficients c_k from the data y_j, and `IT = -1` to invert the c_k

and obtain the y_j. Recall that, according to Sections 9.2 and 9.3, our derivation assumes that the data are real variables.

PROGRAM 9.3   Radix-2 fast Fourier transform checking program.

```
#include <stdio.h>
#include <math.h>
#define MAX 33

main()
{
/*  Fast Fourier Transform;
   radix 2; Checking program */
double yreal[MAX],yimag[MAX];
/* Fortran compatibility; arrays are used from [1] */
double yin,twopi;
int nu,N,MAX1,j,it,iftest;
void FFT();

twopi = 8*atan(1.0);
printf("FFT: Input nu: ");
scanf("%i",&nu);
N = pow(2,nu);   MAX1 = MAX-1;
printf("nu = %i, so # data N = %i\n",nu,N);
if ( N > MAX1 )
   {
   printf("\n!! N=%i > MAX1=%i\n",N,MAX1);
   printf("End FFT");   exit(1);
   }
printf("\nInput %i y data\n",N);
for ( j = 1; j <= N; j++ )
   {
   scanf("%f",&yin);
   yreal[j] = yin;  yimag[j] = 0;
   }

it = 1; /* FFT from y  back into y  */
/* Warning; Original y is destroyed; copy if needed */
FFT(nu,N,it,twopi,yreal,yimag);
printf("\nOutput of FFT; yreal, yimag\n");
for ( j = 1; j <= N; j++ )
   {
   printf("\n%g %g",yreal[j],yimag[j]);
   }
it = -1; /* Inverse FFT from y  back into  y
            should approximately restore  y  */
```

```
FFT(nu,N,it,twopi,yreal,yimag);
printf("\nOutput of FFT; yreal, yimag\n");
printf("{yimag  should be zero to machine tolerance\n}");
for ( j = 1; j <= N; j++ )
  {
  printf("\n%g %g",yreal[j],yimag[j]);
  }
iftest = 1; /* FFT speed test option; manual timing */
while ( iftest == 1 )
  {
  printf("\nInput a  1  for FFT speed test: ");
  scanf("%i",&iftest);
  if ( iftest == 1 )
    {
    printf("Input nu, < 11 : ");   scanf("%i",&nu);
    N = pow(2,nu);
    printf("Start timing now\n");
    for ( j = 1; j <= N; j++ )
      {
      yreal[j] = j;  yimag[j] = 0;
      }
    it = 1; /*  FFT */
    FFT(nu,N,it,twopi,yreal,yimag);
    printf("Stop timing now\n");
    }
  else
    {
    printf("End FFT");  exit(0);
    }
  } /* end  iftest  speed loop */
}

void FFT(nu,N,it,twopi,yreal,yimag)
/* Fast Fourier Transform of N data in yreal+i(yimag) */
double yreal[],yimag[];
double twopi;
int nu,N,it;
{
double tpn,angle,wr,wi,tempr,tempi,norm;
int ii,inu,nu1,nsub,i,j,inup1,in1,jsort;
int bitrev();

tpn = twopi/N;
ii = -it;
inu=0;  nu1=nu-1;  nsub=N/2; /* set up first subinterval */
/* Loop over levels; lengths N/2, N/4, ... */
```

```
for ( i = 1; i <= nu; i++ )
  {
  while ( inu < N )
    {
    for ( j = 1; j <= nsub; j++ )
      { /* Transform subinterval */
      angle = tpn*bitrev((inu/pow(2,nu1)),nu);
      wr = cos(angle);  wi = ii*sin(angle);
      inup1 = inu+1;  in1 = inup1+nsub;
      tempr = yreal[in1]*wr - yimag[in1]*wi;
      tempi = yreal[in1]*wi + yimag[in1]*wr;
      yreal[in1] = yreal[inup1]-tempr;
      yimag[in1] = yimag[inup1] -tempi;
      yreal[inup1] = yreal[inup1]+tempr;
      yimag[inup1] = yimag[inup1]+tempi;
      inu = inu+1;
      }
    inu = inu+nsub;
    }
  inu=0;  nu1=nu-1;  nsub=nsub/2; /* next subinterval */
  }
/* Reverse bit pattern and sort transform */
for ( j = 1; j <= N; j++ )
  {
  jsort = bitrev(j-1,nu)+ 1;
  if ( jsort > j )
    {
    tempr = yreal[j]; /* swap values */
    yreal[j] = yreal[jsort];  yreal[jsort] = tempr;
    tempi = yimag[j];
    yimag[j] = yimag[jsort];  yimag[jsort] = tempi;
    }
  }
norm = 1/sqrt((double) N); /* Normalize */
for ( j = 1; j <= N; j++ )
  {
  yreal[j] = norm*yreal[j];
  yimag[j] = norm*yimag[j];
  }
}

int bitrev(j,nu)
/* Bit reverse the  nu  bits of  j  into jr */
int j,nu;
{
```

```
int jt,i,jd,jr;

jt = j;  jr = 0;
for ( i = 1; i <= nu; i++ )
  {
  jd = jt/2;  jr = 2*jr + jt-2*jd;
  jt = jd;
  }
return jr;
}
```

The following exercise uses the analytically-calculated discrete transform for the harmonic oscillator (single frequency) derived in Section 9.2. It illustrates the usefulness of analytical special cases for checking numerics and programs.

Exercise 9.34

(a) Run Program 9.1, Discrete Fourier Transform for Oscillator, from Section 9.2 for $N = 16$, 64, and 512, with your choice of the oscillator angular frequency k_0, in order to generate the discrete Fourier transform coefficients c_k for $k = 0$ to $N - 1$. Save these coefficients in a file.

(b) Run Fast Fourier Transform to compute the inverse Fourier transform of the array of c_k values generated in (a). That is, input IT = -1 and get the data from the file. You should recover a complex exponential having angular frequency k_0. Check the accuracy of this inversion. ∎

Another way to check the FFT program is to generate the discrete Fourier coefficients for the decaying exponential (Section 9.2), which is quite different from that in Exercise 9.34, thus providing an alternative check of the FFT program.

Exercise 9.35

(a) Convert (9.20) into expressions for the real and imaginary parts of the c_k for exponential decay. Then code and spot check the program. For example, you can check the symmetry condition (9.10). Use $N = 16$, 64, and 512, with your choice of the stepsize h, in order to generate the discrete Fourier transform coefficients c_k for $k = 0$ to $N - 1$. Save these coefficients in a file.

(b) Run the program Fast Fourier Transform to compute the inverse Fourier transform of the array of c_k values generated in (a). That is, input IT = -1 and get the data from the file created by the program for exponential decay. You should recover a decaying exponential at the stepsize h. By taking the natural logarithm of the resulting coefficients you can check how well the inverse transform performs. ∎

Speed testing the FFT algorithm

Speed testing of the FFT algorithm in order to verify the scaling rule indicated in (9.37) is given as a worked example in Exercise 9.12. The program Fast Four ier Transform has timing as an option in the main program after the self test. You should now run the timing tests yourself.

Exercise 9.36

Use the timing (done manually) of the FFT algorithm coded in the program Fast Fourier Transform. Use $N = 128$, 256, and 1024 if you are using a personal computer. Values about factors of 4 or 8 larger than these may be appropriate for a workstation or for a desktop computer with an accelerator board. (If necessary, increase the maximum array size, MAX, to accommodate the larger values of N.) From your timing results estimate a range for T_a in (9.37) and thus estimate the FFT time for a data sample of $N = 8192 = 2^{13}$ points. ■

With the experience gained from running Fast Fourier Transform you should be ready to analyze real data that do not have an obvious pattern of frequencies. Such are the data from electroencephalograms.

9.8   PROJECT 9B: FOURIER ANALYSIS OF AN ELECTROENCEPHALOGRAM

This project provides an introduction to practical Fourier analysis and it illustrates how the discrete Fourier transform theory in Section 9.2 is applied to the analysis of real data. The electroencephalogram (EEG, or "brainwave") data are analyzed by the FFT algorithm from Section 9.3 with the program provided in Section 9.7.

To the untrained observer the output voltage of the EEG as a function of time indicates very little pattern. The analysis of the EEG data transforms it from the time domain to the frequency domain, in which the dominance of a few frequencies becomes evident, as you will soon discover. Recall that, as shown in Sections 6.2 and 9.1, these transformed amplitudes also provide the best fit to the data (in the least-squares sense) that can be obtained by using a linear superposition of cosines and sines.

Data from many other physiological rhythms may also be analyzed by the Fourier expansion methods in this chapter and Chapter 10. Among the several examples discussed in Cameron and Skofronick's book on medical physics are magnetoencephalograms and magnetocardiograms, which use ultrasensitive magnetometers to measure the tiny magnetic fields from the head and heart.

Overview of EEGs and the clinical record

Since the 1920s the variation of potential differences between points on the scalp as a function of time have been associated with electrical activity of the brain. The early measurements were subject to much controversy, partly because of the small voltages involved, in the microvolt (μV) range, and particularly because direct interpretation was so subjective. The advent of computers in the 1950s enabled objective analysis of the frequency components of brain waves, even if not objective interpretation. A typical EEG of an adult human is shown in Figure 9.19.

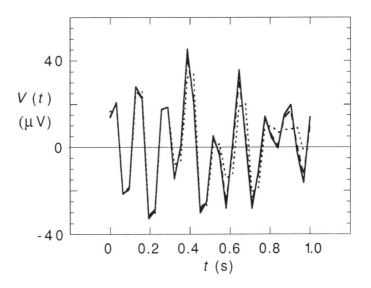

FIGURE 9.19 Electroencephalogram (EEG) voltages over a 1-second interval. The data are connected by solid lines, and the approximation of the data by the first sixteen harmonics is shown dotted. The effects of Lanczos filtering are not visible on this scale.

The main frequencies in an adult human EEG are 0.3 to 3.5 Hz (called δ waves, predominant in sleeping adults and awake young children), 8 to 13 Hz (α waves, predominant in awake, relaxed adults), and 18 to 30 Hz (β waves, appearing sporadically during sleep). The interpretation of such patterns, especially for diagnosis of neural dysfunction, is discussed at an introductory level by Cameron and Skofronick. Instrumentation for EEG data acquisition is described in the book by Cromwell, Weibell, and Pfeiffer. The monograph by Spehlmann provides many examples of EEG traces and their clinical interpretation.

For our analysis of EEG data, three EEG patterns of patient MAC (Mental Aptitude Confidential) are provided in Table 9.1. For illustration purposes, we use only a very short 1-second interval, conveniently taken over 32 data points each separated by 1/31 of a second. Therefore, we may use the radix-2 FFT program developed in Section 9.7, choosing the index $\nu = 5$, since $2^5 = 32$. In order to achieve a periodic function, the first and last points of the data have been forced to be the same. The dominant frequencies are characteristic of those of the EEGs of adult humans, as discussed above. Indeed, you can see a dominant rhythm in Figure 9.19 by noting a fairly regular crossing of the axis that occurs 16 times in 1 sec, so there has to be a strong amplitude at about 8 Hz, in the α-rhythm section.

Three data sets are given in Table 9.1 and displayed in Figure 9.19. As you can see, they are nearly the same, because they differ only in the amount of noise present in the data. I inserted the noise voltages artificially, in such a way that V_1 has the least noise and V_3 has the most noise.

TABLE 9.1 Data sets for the EEG analysis.

IT	t (s)	$V_1(t)$ (μV)	$V_2(t)$ (μV)	$V_3(t)$ (μV)
1	0.0000	13.93	12.07	13.00
2	0.0322	20.77	19.56	19.64
3	0.0645	-21.62	-12.14	-17.18
4	0.0967	-19.28	-11.45	-14.66
5	0.1290	28.15	29.75	29.42
6	0.1612	23.04	22.18	21.61
7	0.1935	-32.87	-32.32	-33.07
8	0.2258	-29.30	-38.27	-32.49
9	0.2580	17.61	18.82	18.51
10	0.2903	18.63	26.64	21.16
11	0.3225	-14.58	-17.42	-16.00
12	0.3548	0.38	-8.74	-2.71
13	0.3871	45.21	35.97	40.30
14	0.4193	20.91	27.56	22.94
15	0.4516	-30.27	-27.19	-28.26
16	0.4838	-25.33	-26.71	-25.02
17	0.5161	5.17	9.14	6.68
18	0.5483	-3.66	-7.36	-6.22
19	0.5806	-28.17	-26.13	-26.86
20	0.6129	0.97	6.47	4.25
21	0.6451	35.69	30.31	33.00
22	0.6774	3.56	1.79	2.15
23	0.7096	-28.18	-30.39	-29.58
24	0.7419	-11.30	-8.54	-9.21
25	0.7741	14.25	10.63	12.94
26	0.8064	4.56	-1.78	0.39
27	0.8387	-0.48	-6.38	-3.90
28	0.8709	14.99	16.67	17.13
29	0.9032	19.68	22.84	21.56
30	0.9354	-2.17	1.69	-1.71
31	0.9677	-16.35	-20.65	-18.00
32	1.0000	13.93	12.07	13.00

A characteristic of noise in data is that, because it is random from one data point to the next, it appears to have a frequency about equal to the reciprocal of the sampling interval; thus, here the noise frequency should be about 30 Hz. You may test this by seeing whether the voltages reconstructed from the Fourier amplitudes, but omitting the high-frequency components, are in essential agreement with each other. In these EEG data, the average value (the DC level of the EEG voltage) has been removed. Thus, the a_0 coefficient in the frequency spectrum should be zero. Now that we have the EEG data, we are ready to make Fourier analyses.

Program for the EEG analysis

In this subsection we construct a program for analyzing the EEG data in Table 9.1 by modifying the FFT program developed in Section 9.7. Because there are several data sets and filtering options to choose from, I have included file input and analysis selections within the program.

The program EEG/FFT consists of three main sections:

(*i*)   Optional preparation of EEG data input to file EEGVn, where n = 1, 2, or 3.

(*ii*)   Analysis of one of these EEG data sets, including the one just just prepared, to obtain the Fourier amplitudes, c_k. These amplitudes are written to file EEGoutN where n = 1, 2, or 3.

(*iii*)   Remaking of the EEG with three options for filtering:
  1. No filtering.
  2. Lanczos filtering, as explained in the following subsection on filtering the EEG.
  3. Truncated filtering, with equal weight up to maxf, then zero weight.

For each filtering option, the remade EEG is written to file FilterEEGn, where n = 1, 2, or 3. The file may either be rewritten or added to. These options allow you to explore the EEG analysis and to save the results for input to a graphics application, as shown in Figure 9.19. The source program is as follows, with the FFT and bitrev functions being just those given in Section 9.7, so they are omitted from the program listing.

PROGRAM 9.4 EEG FFT analysis, with variable data and filtering options.

```
#include <stdio.h>
#include <math.h>
#define MAX 33

main()
{
/* EEG FFT analysis;
   Prepare input files of V1, V2, or V3;
   Analyze from files EEGV1, EEGV2, or EEGV3;
   Outfile file for plots */
```

```
FILE *fin,*fout;
FILE *fopen();
double twopi,yin,sigma,arg;
double Vreal[MAX],Vimag[MAX],VFiltr[MAX],VFilti[MAX];
/* Fortran compatibility; arrays are used from [1] */
int nu,N,MAX1,j,it,choice,k,filter,maxf,maxk;
char yn,yesno,wa,waFilter;
void FFT();

twopi = 8*atan(1.0);
nu = 5; /* so */ N = 32; /* Number of time steps */
printf("EEG FFT input & analysis:\n"
     "Prepare input file? (y or n):\n");
scanf("%s",&yn);
if ( yn == 'y' )
  {
  printf("Input EEG data set (1,2,3): ");
  scanf("%i",&choice);
  switch (choice)
    {
    case 1: fout = fopen ("EEGV1","w"); break;
    case 2: fout = fopen ("EEGV2","w"); break;
    case 3: fout = fopen ("EEGV3","w"); break;
    default:
    printf("\n!! No such input file\n\n");exit(1);
    }
  printf("\nInput EEG data:\n");
  for ( j = 1; j <= N; j++ )
    {
    printf("\n%i:  ",j);
    scanf("%lf",&yin);
    Vreal[j] = yin;
    fprintf(fout,"%lf\n",yin);
    }
  fclose(fout); rewind(fout); /* Ready for reuse */
  }
printf("\nAnalyze which EEG data set (1,2,3)? ");
scanf("%i",&choice);
printf("\nWrite over (w) or Add on (a): ");
scanf("%s",&wa);
switch (choice) /* Open input & output files */
  {
  case 1: fin = fopen("EEGV1","r");
          fout = fopen ("EEGout1",&wa); break;
  case 2: fin = fopen("EEGV2","r");
```

```
            fout = fopen ("EEGout2",&wa); break;
   case 3: fin = fopen("EEGV3","r");
            fout = fopen ("EEGout3",&wa); break;
   default:
   printf("\n!! No such input file\n\n");exit(1);
   }
for ( j = 1; j <= N; j++ ) /* Data input */
   {
   fscanf(fin,"%lf\n",&yin);
   Vreal[j] = yin;  Vimag[j] = 0;
   }
it = 1; /* FFT from  V  back into  V  */
/* Original V is destroyed */
FFT(nu,N,it,twopi,Vreal,Vimag);
printf("\nOutput of FFT; Vreal, Vimag\n");
for ( k = 1; k <= N; k++ ) /* V has c-sub-k */
   {
   fprintf(fout,"%lf %lf\n",Vreal[k],Vimag[k]);
   printf("\n%6.2lf %8.2lf",Vreal[k],Vimag[k]);
   }
/* Remaking EEG with filter options */
yesno = 'y';
while ( yesno == 'y' )
   {
   printf("\n\nFilter EEG (y or n)? ");
   scanf("%s",&yn); yesno = yn;
   if ( yesno == 'y' )
     {
     printf("Choose filter type:\n");
     printf("        1, no filter\n");
     printf("        2, Lanczos filter\n");
     printf("        3, truncated filter\n");
     scanf("%i",&filter);
     if ( abs(filter-2) > 1 )
       {
       printf("\n!! No such filter\n\n");exit(1);
       }
     printf
     ("Write (w) or Add (a) FilterEEG%i\n",choice);
     scanf("%s",&waFilter);
     switch (choice) /* Open file for filtered EEG */
     {
     case 1:fout=fopen("FilterEEG1",&waFilter);break;
     case 2:fout=fopen("FilterEEG2",&waFilter);break;
     case 3:fout=fopen("FilterEEG3",&waFilter);break;
```

```
      }
    if ( filter == 3 )  /* Truncated filter */
      {
      printf("Maximum frequency (0 to %i)? ",MAX-2);
      scanf("%i",&maxf);   maxk = maxf+1;
      if ( maxk > MAX-1 )   maxk = MAX-1;
      }
    for ( k = 1; k <= N; k++ )
      {
      switch (filter)
        {
        case 1: sigma = 1; break;/* No filter */
        case 2: /* Lanczos filter */
          {
          if ( k == 1 )  { sigma = 1; break; }
          else
            {
            arg = twopi*(k-1)/(4*N);
            sigma = sin(arg)/arg; break;
            }
          }
        case 3: /* Truncated filter */
          {
          if ( k <= maxk )  sigma = 1;
          else  sigma = 0;  break;
          }
        }
      VFiltr[k] = sigma*Vreal[k];
      VFilti[k] = 0;
      }
    it = -1; /* Remake voltages using FFT  */
    FFT(nu,N,it,twopi,VFiltr,VFilti);
    printf("\nFiltered FFT; VFiltr,VFilti\n");
    for ( j = 1; j <= N; j++ )
      {
      fprintf(fout,"%lf %lf\n",VFiltr[j],VFilti[j]);
      printf("\n%6.2lf %8.2lf",VFiltr[j],VFilti[j]);
      }
    }/* ends  if yesno=='y' */
  }/* ends  while yesno=='y' */
printf("\nGoodbye from this brain wave");
}/* ends  main */

FFT(nu,N,it,twopi,Vreal,Vimag) /* See Program 3, Section 9.7 */
bitrev(j,nu) /* See Program 3, Section 9.7 */
```

Exercise 9.37

(a) Code program EEG FFT Analysis, borrowing the functions FFT and bit-rev from Program 9.3. They are not listed in Program 9.4. Test the various options to verify that the whole program is working correctly. For debugging, it is simplest to use a very small data set, nu = 2, so that the number of data N = 4. This will minimize output and will obviate much retyping if you accidentally destroy your EEG data files.

(b) Make another test of the FFT by using constant data, so that the Fourier coefficients for $k > 0$ should be zero. Also check that the inverse transform recovers the original data. ■

Frequency spectrum analysis of the EEG

Since the EEG data in Table 9.1 are discrete and we are using the Fast Fourier transform (FFT) method of analyzing the data, we will obtain from the program the discrete Fourier transform (the DFT, Section 9.2) of the data from the time domain to the frequency domain. Therefore, in our formalism in Section 9.2 the variable x is to be interpreted as the time variable t (in seconds), so that the variable k is interpreted as the frequency in hertz (Hz), that is, per second.

The appropriate formulas for the DFT are those for the amplitudes c_k, (9.6). The square of the modulus of each c_k is proportional to the power contained in that frequency, as discussed under (9.8) for Parseval's theorem. Therefore, we define for each frequency, k, the power, P_k, by

$$P_k = |c_k|^2 \qquad (9.95)$$

The program outputs the Fourier amplitudes c_k, but you may easily also adapt it to produce the power. In Figure 9.20 you can see the power spectrum from the FFT analysis of data set V_1 in Table 9.1.

Now that you have seen a sample of the EEG analysis results, try your own brain waves on the data in Table 9.1.

Exercise 9.38

(a) Select an EEG-analysis output file EEGoutchoice, where choice = 1, 2, or 3, and use it as input to a graphics application program in order to display the power spectrum as a function of k, similarly to Figure 9.20.

(b) Is the pattern of Fourier amplitudes that you obtained as a function of frequency, k, consistent with that of a normal adult human, as discussed in the introduction to the EEG? If you run more than one data set, compare them and discuss for what frequency range the analysis is relatively insensitive to noise. Recall that the average noise level increases from V_1 to V_2 to V_3. ■

Now that we have an understanding of the basic Fourier analysis of data, it is time to introduce some data massaging techniques.

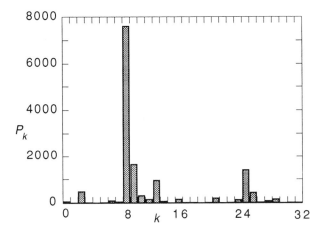

FIGURE 9.20   Power spectrum, P_k, as a function of frequency, k, for the EEG voltage in Table 9.1 having the lowest noise, namely $V_1(t)$.

Filtering the EEG data: The Lanczos filter

Here we explore how the reconstructed FFT can be modified by selectively down-weighting various frequencies, usually the highest ones, that are usually associated with noise rather than with the desired signal.

We use the code options for reconstructing $V(t)$ by the inverse FFT. Thus, in the coding you see that the parameter passed to function FFT is it = -1, rather than it = 1, as used for the frequency analysis. It's as simple as that for the discrete Fourier transform. Truncation-type filtering is a simple choice.

Exercise 9.39

This exercise consists of using the program segment in EEG FFT Analysis to reconstruct the voltage from the Fourier amplitudes computed in Exercise 9.38, but using a choice of filtering options, either none or truncation.

(a) Use the filter type 1 (no filtering) to verify that the program reproduces the original data, within the roundoff error of your computer and its arithmetic routines for cosines and sines.

(b) Next, choose a maximum frequency, maxf, in EEG FFT Analysis. This frequency may not be larger than the number of data minus one, but the program will force this requirement. Run the program, then graph the output to compare it with the input data. (An example is shown in Figure 9.19.) Describe qualitatively how and why the agreement between filtered and original data changes with maxf. For what range of maxf values do you judge that you are probably reproducing the EEG signal, as opposed to the EEG data, which contain some noise? ■

If you followed the discussion in Section 9.6, you will object to this kind of filtering, because we have introduced a discontinuity into the Fourier amplitudes, c_k, by suddenly (at maxf) effectively turning them to zeros. When we Fourier transform these modified amplitudes we will get a Wilbraham-Gibbs overshoot introduced into the reconstructed EEG. With such a bumpy function as our data represent and with such coarse time steps, this will not be obvious by looking at the results, but we know that the overshoot has to be there.

The *Lanczos damping factors* provide a smooth transition from complete inclusion of the lowest frequencies in the filtered reconstruction to suppression (by a factor of about 0.64) of the highest frequency.

Exercise 9.40
Show analytically that integration of a continuous function over the range $\pm\pi/(2N)$ about each point x of the function is equivalent to multiplying its k th frequency component amplitude, c_k, by the damping factor

$$\sigma_k = \frac{\sin[\pi k/(2N)]}{\pi k/(2N)} \tag{9.96}$$

These factors are called *Lanczos damping factors*. ∎

The Lanczos damping factors, σ_k, are shown in Figure 9.21. Note that they are independent of the data being analyzed. In our analysis, since the major amplitudes are concentrated around 8 Hz (Figure 9.19), suppressing higher-frequency amplitudes has relatively little effect on the reconstructed amplitudes. That is why in Figure 9.19 the Lanczos-filtered and original EEG data nearly coincide. Try for yourself the effects of Lanczos damping on the reconstruction of the EEG.

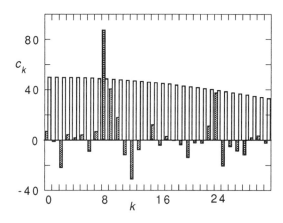

FIGURE 9.21 EEG amplitudes (with sign) as a function of frequency, k, for the EEG data set V_1 (from Table 9.1) are shown by shaded bars. The Lanczos-filter function, σ_k, from (9.96) is shown by the hollow bars.

Exercise 9.41

Explore the effects of Lanczos damping in filtering Fourier expansions by using the program `EEG FFT Analysis` as follows:

(*a*) Use filtering option 2 (Lanczos filtering) applied to one of the EEG data sets given in Table 9.1. (By now, you probably have typed in these three data sets.) Make a plot comparing the original and filtered EEG. In this case, because filtering effects are likely to be small, it is a good idea to plot the *difference* (including sign) between the original and filtered values.

(*b*) Modify the program slightly so that you can input your own choice of Fourier amplitudes, c_k. Then make a data set such that the highest frequencies are emphasized. From this set of amplitudes reconstruct the original function by using the inverse FFT after the program has Lanczos-filtered the amplitudes. Now you should notice a much larger effect from Lanczos filtering than in the EEG data. ∎

Filtering is further considered in the texts on signal processing by Hamming, by Oppenheim and Schafer, and by Embree and Kimble. The last book has many algorithms, plus programs written in C, for digital signal processing. A suite of programs, primarily for FFT calculations, is provided in Chapter 12 of the numerical recipes book by Press et al. Bracewell's text on Fourier transforms presents many engineering examples.

REFERENCES ON FOURIER EXPANSIONS

Abramowitz, M., and I. A. Stegun, *Handbook of Mathematical Functions*, Dover, New York, 1964.

Bôcher, M., "Introduction to the Theory of Fourier's Series," Annals of Mathematics, 7, 81, Sect. 9 (1906).

Bracewell, R. N., *The Fourier Transform and Its Applications*, McGraw-Hill, New York, second edition, 1986.

Bracewell, R. N., *The Hartley Transform*, Oxford University Press, Oxford, England, 1986.

Brigham, E. O., *The Fast Fourier Transform and Its Applications,* Prentice Hall, Englewood Cliffs, New Jersey, 1988.

Cameron, J. R., and J. G. Skofronick, *Medical Physics*, Wiley, New York, 1978.

Champeney, D. C., *A Handbook of Fourier Theorems*, Cambridge University Press, Cambridge, England, 1987.

Cromwell, L., F. J. Weibell, and E. A. Pfeiffer, *Biomedical Instrumentation and Measurements*, Prentice Hall, Englewood Cliffs, New Jersey, 1980.

Embree, P. M., and B. Kimble, *C Language Algorithms for Digital Signal Processing*, Prentice Hall, Englewood Cliffs, New Jersey, 1991.

Gibbs, J. W., "Fourier's Series," Nature, **59**, 200 (1898); *erratum*, Nature, **59**, 606 (1899): reprinted in *The Collected Works of J. Willard Gibbs*, Laymans Green, New York, Vol. II, Part 2, p. 258, 1931.

Hamming, R. W., *Digital Filters*, Prentice Hall, Englewood Cliffs, New Jersey, third edition, 1989.

Körner, T. W., *Fourier Analysis*, Cambridge University Press, Cambridge, England, 1988.

Oppenheim, A. V., and R. W. Schafer, *Discrete-Time Signal Processing*, Prentice Hall, Englewood Cliffs, New Jersey, 1989.

Peters, R. D., "Fourier Transform Construction by Vector Graphics," American Journal of Physics, **60**, 439 (1992).

Pratt, W. K., *Digital Image Processing*, Wiley, New York, 1978.

Press, W. H., B. P. Flannery, S. A. Teukolsky, and W. T. Vetterling, *Numerical Recipes in C,* Cambridge University Press, New York, 1988.

Protter, M. H., and C. B. Morrey, *Intermediate Calculus*, Springer-Verlag, New York, second edition, 1985.

Roberts, R. A., and C. T. Mullis, *Digital Signal Processing*, Addison-Wesley, Reading, Massachusetts, 1987.

Spehlmann, R., *EEG Primer*, Elsevier, New York, 1981.

Thompson, W. J., "Fourier Series and the Gibbs Phenomenon," American Journal of Physics, **60**, 425 (1992).

Weaver, H. J., *Applications of Discrete and Continuous Fourier Analysis*, Wiley-Interscience, New York, 1983.

Wilbraham, H., "On a Certain Periodic Function," Cambridge and Dublin Mathematics Journal, **3**, 198 (1848).

Chapter 10

FOURIER INTEGRAL TRANSFORMS

In this chapter we continue the saga of Fourier expansions that we began in Chapter 9 by exploring the discrete Fourier transform, the Fourier series, and the Fast Fourier Transform (FFT) algorithm. A major goal in this chapter is to extend the Fourier series to the Fourier integral transform, thus completing our treatment of Fourier expansions that was outlined in Section 9.1.

The outline of this chapter is as follows. In Section 10.1 we make the transition from Fourier series to Fourier integrals, then in Section 10.2 we give several examples of these transforms that are interesting for practical applications, especially their application to Lorentzian and Gaussian functions. By Section 10.3 we have enough analytical preparation that we can start to emphasis applications of Fourier integral transforms and their numerical approximations, so we investigate convolutions calculated from Fourier integral transforms, including the Voigt function that is used extensively in analyzing optical spectra. In Project 10 (Section 10.4) we develop a program for calculating convolutions by using the FFT, then apply it to calculate the line profile of a stellar spectrum. References on Fourier integral transforms round out the chapter.

10.1 FROM FOURIER SERIES TO FOURIER INTEGRALS

In the discussion of discrete Fourier transforms and Fourier series in Chapter 9, we found that they predict a periodicity of any function expressed in terms of these two expansions. Suppose, however, that we want to describe an impressed force, a voltage pattern, or an image, that is not periodic. Can the Fourier series be adapted to this use? It can, as we show in the following.

The transition from series to integrals

One way to make the transition from Fourier series to Fourier integrals is to allow the upper and lower limits of the interval for a Fourier series to lie far outside the range of x values for which we will use the series. The periodicity of the Fourier series will then be inconsequential. This idea also leads to the discrete harmonics, k, in the Fourier series becoming continuous variables, thus establishing symmetry of treatment with the x variables, as outlined in Section 9.1.

Our method of deriving the Fourier integral transform is to set $x_- = -L$ and $x_+ = L$, so that the interval $x_+ - x_- = 2L$ in the Fourier series for arbitrary intervals (Section 9.4). Eventually we let $L \to \infty$ to produce the integral transform from the series. In more detail, we temporarily set $\omega_k = \pi k /L$ in (9.46), so that unit step of k produces the change $\Delta \omega_k = \pi/L$. The Fourier series expansion of a function $y(x)$, equation (9.48), then becomes

$$y(x) = \frac{\sum_{k=-N}^{N} Y(\omega_k)\, e^{i \omega_k x}\, \Delta \omega_k}{\sqrt{2\pi}} \tag{10.1}$$

in which the Fourier coefficients $Y(\omega_k)$ are given by

$$Y(\omega_k) = \frac{\int_{-L}^{L} y(x)\, e^{i \omega_k x}\, dx}{\sqrt{2\pi}} \tag{10.2}$$

In these two expressions, in order to achieve symmetry we have split the factors in the denominators as shown. Now one lets L and N tend to infinity carefully (for mathematical rigor, see Churchill's text), so that the summation in (10.1) becomes an integral and the variable ω_k becomes continuous. Also, we can revert to the notation of k for the variable complementary to x. Now, however, k varies continuously rather than discretely. Finally, we have the result for the Fourier integral transform pair

$$y(x) = \frac{\int_{-\infty}^{\infty} Y(k)\, e^{ikx}\, dk}{\sqrt{2\pi}} \tag{10.3}$$

and the inverse transformation

$$Y(k) = \frac{\int_{-\infty}^{\infty} (x)\, e^{-ikx}\, dx}{\sqrt{2\pi}} \tag{10.4}$$

Both integrals in this transform pair are on an equal footing, and there is no mathematical distinction between the variables x and k. For this reason, one will often see the expressions with the signs of the exponents reversed. As long as one maintains complete consistency of sign throughout the use of a Fourier integral transform pair, such a sign change is not significant.

Exercise 10.1
Verify all the steps between (10.1) and (10.4) for the derivation of the Fourier integral transform. ■

For the scientist and engineer, x and k describe physical measurements, so that their dimensions must be reciprocal in order that the argument of the complex exponential be dimensionless. If x is position, then k is the wavenumber, given in terms of the wavelength, λ, by $k = 2\pi/\lambda$, and having dimensions of reciprocal length. When x represents time, t, then k is replaced by the complementary variable the angular frequency, $\omega = 2\pi/T = 2\pi f$, where T is the period and f is the frequency. The dimensions of ω are radian per unit time.

Waves and Fourier transforms

Since, as just described, we usually have the transform combination kx or ωt, it is interesting to consider briefly the extension of the Fourier integral transform to two variables, namely

$$y(x, t) = \frac{\int_{-\infty}^{\infty} \int_{-\infty}^{\infty} Y(k, \omega) \, e^{-i(kx - \omega t)} \, dk \, d\omega}{2\pi} \qquad (10.5)$$

Here the complex exponential represents a monochromatic (single wavelength and frequency) plane wave travelling along the x direction with wavenumber k at angular frequency ω, as we discussed in Section 2.4. Its associated amplitude is $Y(k, \omega)$. Thus, the double integral over wavenumbers and frequencies, $y(x,t)$, represents a physical wave, called a wave packet. Suppose that the speed of the wavelet of given frequency, given by $c(\omega) = \omega/k = f\lambda$ is fixed, as it is for a light wave in vacuum, then the double integral must collapse to a single integral, because when one of ω or k is given, the other is determined. This is called a nondispersive wave. How to handle the mathematics of this is considered below in discussing the Dirac delta distribution.

Dirac delta distributions

A rather strange integral property can be deduced from the Fourier integral transform pair (10.3) and (10.4). Indeed, when first used by the physicist Dirac it disquieted mathematicians so much that it took a few years to provide a sound basis for its use. Suppose that we substitute (10.3) into (10.4), taking care to use a different variable under the integral sign in (10.3), say k', since it is a dummy variable of integration. One then finds, by rearranging the integrals, that

$$Y(k) = \int_{-\infty}^{\infty} Y(k')\, \delta(k - k')\, dk' \tag{10.6}$$

where the *Dirac delta distribution* is

$$\delta(k - k') \equiv \int_{-\infty}^{\infty} e^{i(k - k')x}\, dx/2\pi \tag{10.7}$$

Exercise 10.2
Work through the steps leading from (10.3) and (10.4) to the last two equations.
■

To the extent that in (10.6) Y is any reasonably well-behaved function that allows construction of its Fourier integral transform and rearrangement of the integrals that lead to (10.7), we have a remarkable identity. The Dirac delta distribution must be such that it can reproduce a function right at the point k, even though its values at many other points k' appear under the integral. Clearly, δ must be distributed with respect to $k - k'$ such that it is concentrated near $k' = k$.

Recollection of a property of the discrete Fourier transform is illustrative at this point, and it makes an analogy between the Kronecker delta and the Dirac delta. Referring to (9.9), we see that summation over the discrete data corresponds to integration over the continuous variable in (10.7). The summation leads to selection by the Kronecker delta of *discrete* matching k and l values, corresponding to matching of k and k' values that is implied for the *continuous* variables in the Dirac delta. Our examples of Fourier transforms in Section 10.2, especially the integral transform of the Gaussian function, provide examples of Dirac delta distributions.

10.2 EXAMPLES OF FOURIER TRANSFORMS

We now explore several examples of Fourier integral transforms, initially to understand their relations to other Fourier expansions and to be able to use them in applications such as convolutions and calculation of spectral profiles.

Exponential decay and harmonic oscillation

We discussed in Section 9.2 the discrete Fourier transforms for exponential decay and harmonic oscillation, and we obtained analytic expressions for both of them. Suppose that the function to be transformed is

$$y(x) = e^{-\alpha x} \qquad \text{Re } \alpha > 0 \tag{10.8}$$

The transition to the Fourier integral transform can be made from the expression for the general complex exponential discrete transform (9.18) by letting $Nh \to \infty$ as

$h \to 0$ and assuming that Re $\alpha > 0$ to guarantee convergence. By expanding the denominator to lowest order in αh, then letting $h \to 0$, we readily find that

$$c_k \to Y(k) = \frac{1}{\alpha + ik} \tag{10.9}$$

where k is now a continuous variable. The transition from the sum in the discrete transform to the integral in the integral transform is a good example of the limit processes involved in the Riemann definition of an integral.

Exercise 10.3
Show all the steps between (9.18) and (10.9), paying particular attention to the order in which the limits on N and h are taken. ∎

This integral transform of the complex exponential can now be applied to the two limiting behaviors for α. If α is purely real and positive, then we have exponential decay, while for α purely imaginary we have harmonic oscillation. These are just the analytical examples considered in Section 9.2 for the discrete transform.

Exponential decay integral transforms can be scaled, as we discussed in Section 9.2, so that $\alpha = 1$ and thus k is measured in units of α. To understand the transition from discrete to integral transform it is interesting to consider the case of N large and h small but finite. You can easily calculate from (9.18) and (10.9) the ratio of the midpoint discrete transform to the integral transform, where the midpoint frequency is calculated from (9.15) as π/h. As $h \to 0$ we find that

$$\frac{c_{N/2}}{Y(\pi/h)} \approx \left(\frac{\pi}{2}\right)^2 + \left(\frac{h}{2}\right)^2 \to 2.46740 \tag{10.10}$$

For zero frequency ($k = 0$) and in the limit of large N, one can show simply from (9.18) and (10.9) with $\alpha = 1$ that the DFT overpredicts the transform compared with the FIT by an amount that increases as h increases.

Exercise 10.4
(a) Make graphs of the real and imaginary parts of the integral transform (10.9) and compare them with calculations of the discrete transform for large N, say $N = 128$ for $h = 1$. A small computer program, such as you may have written for Exercise 9.6 (b), would be helpful for this.

(b) For the midpoint ratio of discrete to integral transforms fill in the steps in the derivation of the result (10.10).

(c) For the zero-frequency difference between discrete and integral transforms show that for $Nh \gg 1$ the difference is given by

$$c_0 - 1 \approx \frac{h}{2}\frac{1 + e^{-h}}{1 - e^{-h}} - 1 \approx \frac{h^2}{12} \tag{10.11}$$

where the last approximation is from a Maclaurin expansion in h. ∎

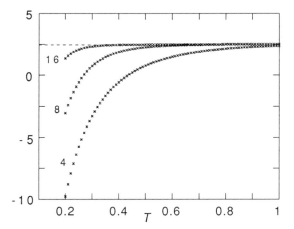

FIGURE 10.1 Ratios of the discrete Fourier transform (9.20) at the midpoint frequency $k = N/2$ to the Fourier integral transform at the corresponding frequency p/h, shown for discrete transforms of lengths $N = 4, 8$, and 16. The dashed line is the asymptotic ratio given by (10.10).

Numerical comparisons between discrete and integral transforms of the damped exponential at the midpoint value $k = N/2$ are shown in Figure 10.1 for small values of N. The dashed line in Figure 10.1 is the result (10.10). Notice that for N and h both small the discrete transform actually gives a negative real transform at $k = N/2$ that is quite different from the real part of the integral transform, which is everywhere positive. This emphasizes the need for care in the limit process of deriving the integral transform from the discrete transform, as described above (10.3) and worked out by the conscientious reader in Exercise 10.3.

Harmonic oscillation integral transforms are included in (10.9), as in Section 9.2 for the discrete transform, by setting $\alpha = -ik_0$, a pure oscillator with a single frequency k_0. In order to consider this rigorously, a small positive real part, ε, has to be included in α to produce convergence of the geometric series for N large, as discussed above (9.21). Then one may let $\varepsilon \to 0$ to obtain the following result for the Fourier integral transform of the harmonic oscillator:

$$Y(k) = i \, \frac{1}{k - k_0} \qquad (10.12)$$

This shows a simple pole at the resonance frequency k_0, whereas the discrete transform has a distribution of values that are peaked in magnitude near k_0, as shown in Figure 9.4.

In the next examples we do not have nice functions that readily allow the discrete and integral Fourier transforms to be compared directly and analytically. For most of these examples the discrete transform can be computed numerically, for example by using the FFT algorithm described in Sections 9.3 and 9.7, whereas it is the integral transform that can usually be expressed in analytical form.

The square-pulse function

For the first example of a Fourier integral transform we choose the first example from the Fourier series in Section 9.5, the square pulse. It has the same definition as in (9.50), namely

$$y(x) = \begin{vmatrix} 1 & 0 < x < \pi \\ -1 & \pi < x < 2\pi \end{vmatrix} \qquad (10.13)$$

In order to compute the integral transform in k space you need only plug this expression into (10.4). Because the integration is over all x, we have to supplement the definition (10.13) with the constraint that

$$y(x) = 0 \quad \text{elsewhere} \qquad (10.14)$$

This function can now be used in the defining equation for the Fourier integral transform, (10.4).

Exercise 10.5
Perform the indicated integration in (10.4) to show that the Fourier integral transform of the square pulse is given by

$$Y(0) = 0 \qquad (10.15)$$

and

$$Y(k) = \sqrt{2\pi} \, \frac{\sin{(\pi k/2)}}{\pi k/2} \qquad (10.16)$$

for $k \neq 0$. ∎

The correspondence of these results with those for the Fourier series coefficients of the square pulse can be clarified as follows. For the series we used an expansion in terms of cosines and sines, and we found in Section 9.5 that only the sine terms are nonzero. Equation (9.52) and Figure 9.8 show that the envelope of the series coefficients, $b_k \propto 1/k$, and that the coefficients vanish for k an even integer. Do these properties also hold for the Fourier integral transform, in which k varies continuously?

Exercise 10.6
(a) Use (10.16) to show that when k is an even integer, including zero, $Y(k) = 0$, just as for the series.
(b) Show that the modulus of the Fourier transform, $|Y(k)| \propto 1/k$ when k is an odd integer, just as for the Fourier series. ∎

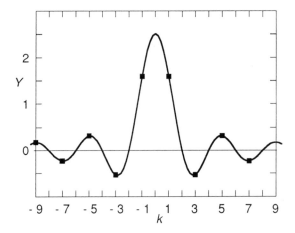

FIGURE 10.2 Fourier integral transform of the square pulse, as given by (10.16), but plotted with the phase factor omitted. The solid points are the discrete Fourier transform values at integer frequencies.

You will have noticed something peculiar about the k-space Fourier amplitude, $Y(k)$, for k near zero. As long as k is infinitesimally removed from zero (10.16) applies and $Y(k)$ varies smoothly with k. But right at $k = 0$ the amplitude must be zero. It is, however, usual to ignore this fact and to extend the definition of $Y(k)$ given by (10.16) into the origin. A Maclaurin expansion of the sine function followed by division by the denominator then produces

$$Y(k) = \sqrt{2\pi} \qquad k \approx 0 \qquad\qquad (10.17)$$

On this basis, we may now graph the Fourier integral transform, apart from the complex factor in (10.16), as shown in Figure 10.2. The values that correspond in magnitude to the Fourier series coefficients, b_k, for the square pulse are also indicated for the odd integers.

If you have some experience with diffraction of waves, you will notice that $Y(k)$ varies with k just like the amplitude of a wave diffracted through an aperture. Indeed, detailed study (presented, for example, in optics texts) will show you that this is exactly so.

Now that we have seen the similarities and differences of Fourier series and integral transforms for the square pulse, it is interesting to consider another example to reveal other facets of integral transforms.

Fourier transform of the wedge function

The wedge function, considered for the discrete transform in Section 9.5, illustrates how the behavior of the integral transform varies as the input function is scaled in shape. We therefore have a wedge whose shape is adjustable, namely

$$y(x) = \begin{vmatrix} (L - |x|)/L^2 & |x| < L \\ 0 & |x| > L \end{vmatrix} \tag{10.18}$$

Although the wedge always has unit area, its width (and therefore its height) is controlled by the value of L. This property makes it convenient for later investigation.

Exercise 10.7
Derive the Fourier integral transform of the wedge function by inserting the definition of the wedge function, (10.18), in the formula (10.4) for the transform, then carrying out the indicated integration. This is straightforward but tedious. Show, after several carefully executed steps, that

$$Y(k) = \frac{1}{\sqrt{2\pi}} \left[\frac{\sin{(kL/2)}}{kL/2} \right]^2 \tag{10.19}$$

Unlike the square-pulse example above, no special treatment is necessary for $k = 0$, except that the limit as k approaches zero must be taken carefully. ∎

You may surmise, correctly, that the behavior of the Fourier integral transform near $k = 0$ is related to the Wilbraham-Gibbs phenomenon discussed in Section 9.6, in that functions that have discontinuities in their values need special treatment.
Note that there is dimensional consistency between the arguments in the original function and the arguments in the transform, as follows. Suppose, for example, that x is a distance, then so must be L. Then in the transform (10.19) k has dimensions of reciprocal distance, but the product kL, being the argument of the sine function, must be dimensionless, as required for any mathematical function.

Exercise 10.8
Suppose that $x = t$, where t represents time, and that $L \to T$, also a time. Show that the Fourier transform (10.19) becomes

$$Y(\omega) = \frac{1}{\sqrt{2\pi}} \left[\frac{\sin{(\omega/2)}}{\omega T/2} \right]^2 \tag{10.20}$$

where ω represents angular frequency. Is the product ωT dimension-free, as required? ∎

The Fourier integral transform of the wedge function is shown in Figure 10.3 for two values of L, namely $L = 1$ (in the same units as x) and $L = 2$.

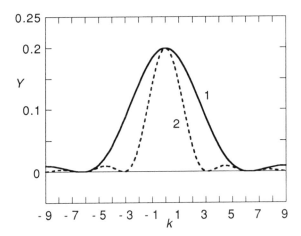

FIGURE 10.3 Fourier integral transform of the wedge function (10.18), as given by (10.19). Shown for a wedge of width $L = 1$ (solid curve) and for a wider wedge ($L = 2$, dashed curve) which produces a narrower transform.

Notice immediately in Figure 10.3 that as the wedge function becomes broader in the x space (although always keeping unit area) it becomes proportionally narrower in the k space. Indeed, the Fourier integral transform as a function of L is obtained merely by scaling the k values so that the product kL is constant, because (10.19) shows that $Y(k)$ depends only upon this product. For example, in Figure 10.3 the zeros of the transform occur with twice the frequency for $L = 2$ as for $L = 1$.

Exercise 10.9
Verify the following properties of the Fourier integral transform of the wedge function:
(a) $Y(k)$ is symmetric about $k = 0$.
(b) $Y(k)$ falls to half its maximum value at $k = 2.78/L$.
(c) The zeros of $Y(k)$ occur at $k = 2n\pi/L$, with n a nonzero integer.
(d) A secondary maximum of $Y(k)$ occurs near $k = 3\pi/L$, where its height is about 0.045 that of the first maximum. ∎

The results from this exercise are important in the theory of wave diffraction from apertures. See, for example, the books on vibrations and waves by Ingard and by Pippard.

Gaussian functions and Fourier transforms

We investigated the Gaussian function in the context of maximum likelihood and least squares in Section 6.1. We repeat its definition here as

$$g(x, \sigma_x) = \frac{1}{\sigma_x \sqrt{2\pi}} \exp\left[-\frac{x^2}{2\sigma_x^2}\right] \tag{10.21}$$

We have renamed the function, commemorating Gauss, and we have included the standard deviation, σ_x, as a parameter in the function definition to emphasize its importance. Also, this Gaussian is centered on $x = 0$, rather than on an average value as in Section 6.1. Such a shift of origin is not important for our current use and interpretation of the Gaussian. You probably already derived some properties of the Gaussian function by working Exercise 6.2. The Gaussian is of interest in Fourier transforms because of its simple behavior under the integral transform. Also, it is a smoothly varying function that is often suitable for use as a windowing or damping function, of the kind discussed in Section 9.8.

The process of deriving the Fourier integral transform of a Gaussian is quite tricky. It is therefore interesting to begin with some numerical explorations by Fourier transforming some discretized Gaussian data using the FFT program developed in Section 9.7.

Exercise 10.10

(a) Write a simple main program that generates an array of data from the Gaussian function (10.21), given the user's choice of σ_x, x range, and the index ν that determines the number of points, N, in the FFT through $N = 2^\nu$. Include an option that makes a file to output the FFT results or plotting. The program should then use the FFT for these data.

(b) Run the program for a range of parameter values. Suitable values are $\sigma_x = 1/2$, 1, 2, and $-2 < x < 2$, and $\nu = 7$ so that $N = 128$.

(c) Plot the FFT output from each run. By plotting $\ln(\text{FFT})$ versus k^2, show that, through about $k \approx 1$ such a plot is linear and its slope is about $-\sigma_x^2/2$. ∎

You will notice after the calculations in this exercise that the function obtained in k space is not very smooth when k increases beyond unity. The limiting process that we carried out analytically to obtain the Fourier integral from the Fourier series is therefore not so easy numerically.

This exercise has probably suggested to you that the Gaussian and its integral transform have the same functional dependence on their arguments, x or k, and that their width (standard-deviation) parameters are the inverse of each other. Let us demonstrate this analytically.

Exercise 10.11

(a) Insert the Gaussian function (10.21), which is normalized to unity when integrated over all x space, into formula (10.4) for its Fourier integral transform, which we will call $G(k, \sigma_k)$. Change variables to $y = x/(\sigma_x\sqrt{2})$ in order to show that

$$G(k, \sigma_k) = \frac{1}{\pi\sqrt{2}} \int_{-\infty}^{\infty} e^{-(y+\alpha)^2} dy \, e^{-\sigma_x^2 k^2/2} \tag{10.22}$$

Here the variable α is given by

$$\alpha = i\sigma_x k/\sqrt{2} \qquad (10.23)$$

(*b*) Use the value for the integral over y in (10.22) from Abramowitz and Stegun, equations (7.1.1) and (7.1.16), namely $\sqrt{\pi}$, to show that

$$G(k, \sigma_k) = \frac{1}{\sqrt{2\pi}} \, e^{-k^2/2\sigma_k^2} \qquad (10.24)$$

where in k space

$$\sigma_k = \frac{1}{\sigma_x} \qquad (10.25)$$

is the standard deviation of the Gaussian in k space. ∎

This result confirms the numerical surmise made in Exercise 10.10.

Now that we have the Gaussian in both x and k spaces, we can examine some of their properties. First, we can display both on the same graph, which is very economical for the author and very enlightening for the student. Figure 10.4 shows three general-purpose Gaussians that may be used for both x and k spaces. Because of the reciprocal relations between the standard deviations, the plot for $\sigma = 1$ is the same for both. Further, if in x space we choose $\sigma = \sigma_x = 2$, the broadest of the three curves, then $\sigma = \sigma_k = 1/2$ in k space, and vice versa.

The reciprocity between widths in x and k spaces has important interpretations in optics and quantum physics, since a narrow pulse in x space (well-localized) is necessarily broad in k space (delocalized). When appropriately interpreted in quantum mechanics, this provides an example of the Heisenberg Uncertainty Relations.

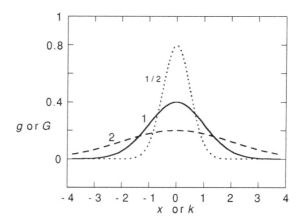

FIGURE 10.4 Gaussian distributions and their Fourier integral transforms. The distribution with magnitude $\sigma = 1/2$ in x space transforms into that with magnitude $\sigma = 2$ in k space. For $\sigma = 1$ the transforms are identical.

A further property of the Gaussian function may be introduced, namely its relation to Dirac delta distributions discussed at the end of Section 10.1. If the Gaussian width σ is steadily decreased, the function steadily becomes more and more concentrated towards the origin. Thus you can imagine that the limiting behavior as $\sigma \to 0$ of the Gaussian is a Dirac delta distribution, because if it occurs inside an integral with another function only the value of that function near the origin will survive in the integration, so that eventually $\delta(0)$ will be obtained.

Lorentzian functions and their properties

We previously encountered the Lorentzian function in Section 8.1 when investigating forced motion and resonances. Because the Lorentzian's properties are almost exclusively used in connection with resonances, we express it in terms of the angular frequency, ω. This is the variable most often encountered in acoustics, optics, and engineering, whereas the energy, E, usually occurs as the variable of interest in atomic and subatomic physics.

We define the Lorentzian function by

$$L(\omega) = \frac{(\gamma/2)^2}{(\omega - \omega_0)^2 + (\gamma/2)^2} \qquad (10.26)$$

where ω_0 is called the resonance frequency and γ is called the Full Width at Half Maximum (FWHM) of the resonance. You will understand the use of these terms by looking at Figure 10.5 and working Exercise 10.12.

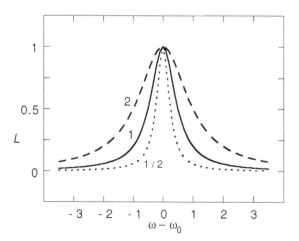

FIGURE 10.5 The Lorentzian function (10.26) as a function of frequency from the central frequency, measured in units of the FWHM γ. Shown for $\gamma = 1/2$ (dotted curve), $\gamma = 1$ (solid curve), and $\gamma = 2$ (dashed curve).

Exercise 10.12

Show that the Lorentzian decreases to half its maximum value of unity at points $\gamma/2$ above and below the resonance frequency ω_0, as seen in Figure 10.5. ∎

Both Gaussians and Lorentzians have convenient properties for describing data, so they are occasionally used as alternative line shapes. They also appear commingled in the Voigt function — the convolution of a Lorentzian with a Gaussian — whose properties are discussed in Section 10.4. It is therefore interesting to compare and contrast the Gaussian and Lorentzian. In order to do this we need a common unit for the variables x and ω. We choose the Lorentzian measured in units of FWHM, γ, relative to ω_0 and a Gaussian with the same FWHM as the Lorentzian, which therefore sets the value of σ_x. You may wish to derive the following comparative properties.

Exercise 10.13

(*a*) Use the definition of the Gaussian, (10.21), with the definition of its FWHM to show that if the Gaussian FWHM is to be unity, then its standard deviation $\sigma_x = 1/(2\sqrt{2\ln 2}) = 0.4246$.

(*b*) Show that if a Gaussian and Lorentzian are to have the same FWHM, then the appropriate scale for the Gaussian relative to the Lorentzian scale chosen above is $2\sqrt{\ln 2}\, x = 1.665\, x$.

(*c*) Show that when its argument is two half widths on either side of the maximum the Gaussian has decreased to 1/16 of its peak value, but the Lorentzian has decreased to only 1/5 of its peak value.

(*d*) Compare the second derivatives of Gaussian and Lorentzian at the peak to their values at the peak to show that the Gaussian with unit FWHM is decreasing as $-16\ln(2) = -11.09$, somewhat faster than for the Lorentzian, for which the ratio is decreasing as -8. ∎

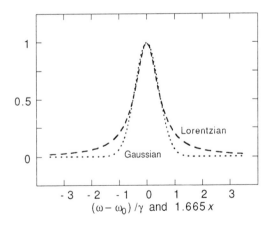

FIGURE 10.6 Comparison of Lorentzian (dashed) with Gaussian (dotted) for the same FWHM.

These analytic properties are illustrated in Figure 10.6, in which the Lorentzian and Gaussian are plotted on a common scale in units of the FWHM. The most noticeable difference between them is the broader wings on the Lorentzian. When the two functions are convolved into the Voigt function (Section 10.4) a behavior intermediate between the Lorentzian and Gaussian is obtained when their arguments are measured in the units of this figure.

Now that we have compared Gaussians and Lorentzians, is the Fourier integral transform of the Lorentzian as easy to derive as for the Gaussian?

Fourier integral transform of a Lorentzian

We have now understand some properties of the Lorentzian function (10.26), so it is time to determine its Fourier transform. Because we are using the variable ω (for reasons described at the start of the preceding subsection), we denote the conjugate variable in its Fourier transform by t, reminiscent of time.

The Fourier integral transform of the Lorentzian, (10.26), we denote by $P(t)$. From the definition (10.4) of the Fourier transform and by making some changes of variables, we can write compactly

$$P(t) = \frac{e^{-i\omega_0 t}\,\gamma}{\sqrt{8\pi}}\,I(s) \tag{10.27}$$

where the integral

$$I(s) = \int_{-\infty}^{\infty} \frac{e^{-isv}}{1+v^2}\,dv \tag{10.28}$$

in terms of the variables

$$v = 2(\omega - \omega_0)/\gamma \tag{10.29}$$

and the parameter

$$s = \gamma t/2 \tag{10.30}$$

Although the integral in (10.28) can readily be evaluated by the method of contour integration in the complex plane, it is interesting and instructive to use another method.

Exercise 10.14

(a) Verify the substitutions of variables made to express $P(t)$.

(b) By differentiating I in (10.28) under the integral sign twice with respect to s, then rearranging the integrand, show that I satisfies the differential equation

$$\frac{d^2I(s)}{ds^2} = I(s) - \int_{-\infty}^{\infty} e^{-isv}\, dv \tag{10.31}$$

(c) Assume that the integral in this expression can be set to its average value, namely zero, then verify that this differential equation is satisfied by

$$I(s) = I_- e^{-s} + I_+ e^{+s} \tag{10.32}$$

(d) Show that for solutions that don't diverge as $|t| \to \infty$, only solutions with exponent $-s$ | are acceptable.

(e) Determine the pre-exponentials in (10.32) by evaluating the integral (10.28) explicitly for $s = 0$ by making use of the substitution $v = \tan\theta$ to show that $I(0) = \pi$. ■

So, after all this exercising we have the Fourier integral transform of the Lorentzian function (10.26) as

$$P(t) = \sqrt{\frac{\pi}{8}}\, \gamma\, e^{-i\omega_0 t}\, e^{-\gamma|t|/2} \tag{10.33}$$

In order to emphasize the complementarity between the widths of functions and the widths of their Fourier transforms, we have in Figure 10.7 the exponential part of $P(t)$ for the three values of the Lorentzian FWHM γ used in Figure 10.6. Notice that if the Lorentzian, L, is broad in the ω space, then its Fourier transform, P, is narrow in t space, and vice versa. This is similar to the behavior of the wedge and Gaussian functions and their transforms discussed above.

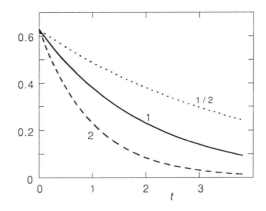

FIGURE 10.7 Magnitude of the Fourier integral transform of the Lorentzian, $|P(t)|$, expressed in units of γ, as a function of t. The exponential-decay curves are for $\gamma = 1/2$ (dotted), $\gamma = 1$ (solid), and $\gamma = 2$ (dashed). Notice the inverse relation between the Lorentzians, Figure 10.5, and their transforms shown here.

A further interesting property of the Lorentzian and its transform is that although $L(\omega)$ has a very nonlinear dependence on the resonance frequency ω_0 and on the FWHM γ, $P(t)$ can be simply transformed to produce a linear dependence on γ.

Exercise 10.15

(a) Show that ω_0 can be determined from the oscillations in $P(t)$.

(b) Show that the graph of $\ln P(t)$ against t is a straight line with slope given by $-\gamma/2$. ■

10.3 CONVOLUTIONS AND FOURIER TRANSFORMS

In this section we introduce the convolution operation, which is of broad applicability in both mathematical analysis and numerical applications. Convolutions become particularly simple in the context of Fourier integral transforms, in that in the transform space, convolution is replaced by multiplication. We show how this result greatly simplifies the mathematics of computing transforms, and we culminate our analysis with the practical example, much used in optics and in astrophysics, of the Voigt profile, which is the convolution of a Gaussian with a Lorentzian. The numerics and a program for computing the Voigt profile are presented in Project 10 in Section 10.4.

Convolutions: Definition and interpretation

Suppose that we have two functions, y_1 and y_2, of the same variable x. Let us construct the following integral of their product:

$$y_1 * y_2(x) \equiv \int_{-\infty}^{\infty} y_1(x') \, y_2(x - x') \, dx' \qquad (10.34)$$

We use the special "star" notation to denote convolution. After the convolution the result is a function of the variable x, which is kept fixed during the integration over x' under the integral. Some books show on the left-hand side an x argument after both y_1 and y_2. This is misleading, because it is the result of the convolution that is a function of x, rather than y_1 and y_2 separately.

What is the interpretation of the convolution? Consider the integral of the product of two functions evaluated at the same argument. This is just the overlap between the two functions, including their relative signs. The convolution, on the other hand, has the arguments of the two functions shifted by an amount depending on the variable x. The particular choice of arguments is made clear by considering y_1 as a function peaked around $x' = 0$. Then the integral samples only values of y_2 near the particular choice of argument x.

In spite of this viewpoint, the convolution is independent of the order in which the functions are written down. That is,

$$y_1 * y_2(x) = y_2 * y_1(x) \tag{10.35}$$

Exercise 10.16
In the definition of the convolution, (10.34), change the variable of integration x' by setting $x' = x - x''$. Thus show that the convolution definition is then satisfied if the two functions are interchanged, that is, (10.35) holds. ■

This commutation property of convolutions is useful in practical calculations.

Convoluting a boxcar with a Lorentzian

As an illustration of convolution in the context of resolution of a measurement, consider the "boxcar-averaging" function with unit area

$$y_1(x) = \begin{vmatrix} \dfrac{1}{2L} & |x| < L \\ 0 & |x| > L \end{vmatrix} \tag{10.36}$$

Three examples of this boxcar are shown in Figure 10.8. (For $L = 0.1$ this is, to continue our agricultural analogy, a grain silo rather than a boxcar.)

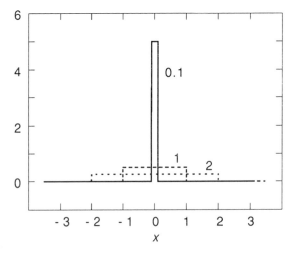

FIGURE 10.8 The boxcar (or "window") function for convolutions, shown for widths of $L = 0.1$, 1, and 2.

The Lorentzian is given by, as discussed at the end of Section 10.2,

$$y_2(x) = \frac{(\gamma/2)^2}{x^2 + (\gamma/2)^2} \tag{10.37}$$

Lorentzians with various widths are shown in Figure 10.5. If the boxcar and Lorentzian functions are inserted in the convolution definition (10.34) and the change of variable to $x - x' = (\gamma/2)\tan\theta$ is made, then the integration can be readily performed to produce the convolution of a boxcar with a Lorentzian

$$y_1 * y_2(x) = \frac{\gamma}{4L} \tan^{-1}\left[\frac{\gamma L}{x^2 + (\gamma/2)^2 - L^2}\right] \tag{10.38}$$

If $L < \gamma/2$, the inverse tangent is to be taken in the first quadrant, but otherwise multiples of π will need to be added to the values returned for the angle by your calculator, which is in radians in (10.38).

Exercise 10.17

(a) Perform the integration with the change of variables suggested in order to verify (10.38).

(b) Show that if $L \ll \gamma$, that is, if the width of the averaging function is much less than the width of the Lorentzian, then the convolution just collapses to the original Lorentzian, y_2. ∎

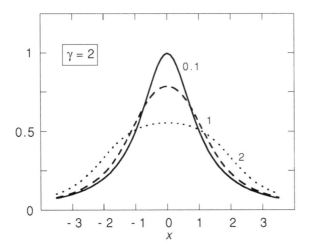

FIGURE 10.9 Convolution of a boxcar (Figure 10.8) with a Lorentzian (Figure 10.5) for a Lorentzian FWHM $\gamma = 2$, shown for the three boxcar widths in Figure 10.8, namely $L = 0.1$, 1, and 2.

The property worked out in Exercise 10.17 (*b*) is shown in Figure 10.9 (which has $\gamma = 2$), in which the convolution for $L = 0.1$ is essentially the original Lorentzian, as one can see because its maximum value is essentially unity and its FWHM is nearly 2 units of x, that is, $\gamma = 2$. You can also see from this figure that the convolution becomes broader and squatter as L increases. The decrease of peak resolution as the boxcar width, L, increases can readily be seen by considering the peak height as a function of L.

Exercise 10.18

(*a*) Show that the peak value of the convoluted distribution (10.38) is at $x = 0$, just as for the Lorentzian.

(*b*) Use the formula relating tangents of doubled angles to the tangents of the angles to show that the peak value is at the origin, where

$$y_1 * y_2 (0) = \frac{\tan^{-1} (2L/\gamma)}{2L/\gamma} \qquad (10.39)$$

(*c*) From the first two terms of the Taylor expansion of the arctangent function, show that the peak value given by (10.39) can be approximated by

$$y_1 * y_2 (0) = 1 - \frac{4}{3}\left(\frac{L}{\gamma}\right)^2 \qquad (10.40)$$

in which the leading term is just the peak value of the original Lorentzian (10.37). ∎

You may wish to investigate in more detail the effect of binning on a Lorentzian spectrum.

Exercise 10.19

Consider the effect of binning a Lorentzian-shaped spectrum extending from $x = -2\gamma$ to $x = 2\gamma$ into N bins, which are sometimes called channels in the parlance of multichannel analyzers.

(*a*) Show that the fractional error from binning, $E(x)$, defined as (convoluted – Lorentzian) / (Lorentzian) is given by

$$E(x) = \frac{[(2x/\gamma)^2 + 1]N}{16}\left\{\frac{\pi}{2} - \tan^{-1}\left[\frac{[(2x/\gamma)^2 + 1]N}{16} - \frac{4}{N}\right]\right\} - 1 \quad (10.41)$$

(*b*) Use the Taylor expansion of the arctangent function to approximate this as

$$E(x) \approx \frac{(8/N)^2}{1 + (2x/\gamma)^2} \qquad (10.42)$$

(*c*) Assume that the binning is done into $N = 128$ bins, about the lowest resolution that one would typically use. Show that the fractional error at the peak of the Lorentzian is only -1.3×10^{-3}. ■

The increasing width with decreasing height of all the convolutions that we have made explicitly suggests that convolution preserves area. This is indeed so, as can be shown from the definition by integrating both sides with respect to x, interchanging the order of integrations on the right-hand side, then changing the variable of integration for y_2, keeping x' fixed, to prove the area-conserving property of a convolution, namely

$$\int_{-\infty}^{\infty} y_1 * y_2 (x) dx = \int_{-\infty}^{\infty} y_1 (x) \, dx \int_{-\infty}^{\infty} y_2 (x) \, dx \qquad (10.43)$$

Exercise 10.20
Show in detail the steps indicated for proving this area-conserving property of convolutions. ■

Another observation that you may have made is that, starting with two functions that are symmetric about the origin, such as the Lorentzian (Figure 10.5) and the boxcar (Figure 10.8), their convolution is also symmetric about the origin. Is this a general property for convolutions? Try it and see.

Exercise 10.21
Suppose that y_1 and y_2 are symmetric about the origin, $y_i (-x) = y_i (x)$ for $i = 1,2$. Use this property in the convolution definition (10.34), together with a change of variable of integration from x' to $-x'$, to show that their convolution is also symmetric about the origin. ■

The filtering and windowing of Fourier integral transforms, of which the boxcar convolution provides an example, proceeds similarly to that discussed in Section 9.8. The techniques are especially important in the reconstruction of three-dimensional images by tomography, as discussed at an introductory level by Solomon in his teaching module. The subject of filtering is covered in, for example, Hamming's book on digital filters. Bracewell's text on Fourier transforms and their applications makes extensive use of convolutions.

We now have some insight into the properties of convolutions and into possible applications of them. There are few interesting examples of analytical functions whose convolutions are straightforward to calculate from the definition (10.34). Therefore, we proceed to two alternatives: first, numerical estimation of convolutions of discretized functions, then convolutions performed analytically by using Fourier integral transforms.

Program for convoluting discretized functions

In most practical applications the functions that one wants to convolute are experimental data, necessarily at discrete points. Therefore, the integral convolutions that we have discussed so far are idealizations when applied to data. It is therefore useful, and even interesting, to construct the analogous discretized convolution. This will bear a similar relation to the integral definition of the convolution, (10.34), as the discrete Fourier transform in Section 9.2 bears to the Fourier integral transform in this chapter.

We write the discretized convolution as

$$y_1 * y_2[i] = \sum_{j=1}^{N} y_1[j]\, y_2[i-j] \qquad (10.44)$$

where array notation is used both for the two inputs and for the output convolution. There is an ambiguity in such a definition, because array elements less than zero (in C, or less than unity in Fortran) are often problematic. In the program below, I made the simplest, and often realistic, choice of ignoring terms which have the index value $[i-j] < 1$. This is referred to as "truncation." An alternative choice, called "wraparound," is to associate with zero or negative index values the values read down from the highest index in the array of y_2 values.

The discrete convolution function `convolve_arrays` given below implements (10.44) assuming truncation. In order to test the function, the main program was written to allow preparation of input arrays, and also output to a file for graphing the results. The file CONARRAY1 contains the N values of y_1, and CONARRAY2 contains the values of y_2. The discrete convolution is output to the console and to the file CONV. The range of i values for which the convolution in (10.44) is computed is from `imin` to `imax`, inclusive. No checking is done to ensure that this range is sensible.

PROGRAM 10.1 Direct discretized convolution of two arrays according to (10.44).

```c
#include <stdio.h>
#include <math.h>
#define MAX 513

main()
{
/* Convolute Arrays */
/* Input arrays then convolute them */
FILE *fin,*fout,*fin1,*fin2;
FILE *fopen();
double y1[MAX],y2[MAX],y12[MAX],yin,yin1,yin2;
int  N,i,j,imin,imax;
```

```
char yn,ynfile,wa;
void convolve_arrays();

printf("Convolute Arrays: Input # of data, N:\n");
scanf("%i",&N);
if ( N > MAX-1 )
  {
  printf("!! N=%i exceeds array sizes %i\n",N,MAX-1);
  exit(1);
  }
printf("Prepare input file(s)? (y or n): ");
scanf("%s",&yn);
if ( yn == 'y' )
  {
  for ( i = 1; i <= 2; i++ )
    {
    printf("\nPrepare y%i file? (y or n): ",i);
    scanf("%s",&ynfile);
    if ( ynfile == 'y' )
      {
      switch (i) /* Open file for write */
        {
        case 1: fout = fopen("CONARRAY1","w"); break;
        case 2: fout = fopen("CONARRAY2","w"); break;
        }
      printf("Input %i data for file %i:\n",N,i);
      for ( j = 1; j <= N; j++ )
        {
        printf("%i:   ",j);
        scanf("%lf",&yin);
        switch (i)
          {
          case 1: y1[j] = yin;
          case 2: y2[j] = yin;
          }
        fprintf(fout,"%lf\n",yin);
        }     /* Ready to reuse */
      fclose(fout); rewind(fout);
      }
    }
  } /* end prepare input files */
printf("\nConvolution calculation\n");
printf("Write over output (w) or Add on (a): ");
scanf("%s",&wa);
fout =  fopen("CONV",&wa);
```

```
fin1 =fopen("CONARRAY1","r");
fin2 = fopen("CONARRAY2","r");
for ( j = 1; j <= N; j++ )
   {              /* Data input from arrays */
   fscanf(fin1,"%lf\n",&yin1);  y1[j] = yin1;
   fscanf(fin2,"%lf\n",&yin2);  y2[j] = yin2;
   }
printf("Choose min (>=1) & max (>=min,<=N) index:\n");
scanf("%i%i",&imin,&imax);
/* Convolution */
convolve_arrays(N,imin,imax,y1,y2,y12);
for ( i = imin; i <= imax; i++ )
   {     /* Output convolution */
   printf("\n%i %8.4lf",i,y12[i]);
   fprintf(fout,"%i %lf\n",i,y12[i]);
   }
printf("\n\nConvolution of arrays ends");
}


void convolve_arrays(N,imin,imax,y1,y2,y12)
/* Convolute arrays y1 & y2 for index imin to imax */
/*   Array indices  <= 0  do not contribute */
double y1[],y2[],y12[];
int N,imin,imax;
{
int i,j;

for ( i = imin; i <= imax; i++ )
  {
  y12[i] = 0;
  for ( j = 1; j <= N; j++ ) /* Accumulate convolution */
    {
    if ( i > j )  y12[i] = y12[i]+y1[j]*y2[i-j];
    }
  }
}
```

This convolution program, Convolute Arrays, appears to be mostly a main program for file handling. The program will, however, be useful for performing discretized convolutions of Gaussians and Lorentzians (whose integral transforms we consider below) and for convoluting data that you may have.

Exercise 10.22
In order to test convolve_arrays and the main program, Convolute Arrays, one may check that the dependence of an integral convolution, (10.34), on x

and of a discretized convolution, (10.44), on i are similar when the steps in x are small compared to the rate of variation of the functions y_1 and y_2.

To do this, for y_2 use the Lorentzian function with $\gamma = 2$ and x in the range from -3.5 to $+3.5$ by steps of $\Delta x = 0.2$; this has $N = 36$. For y_1 use the box-car function, Figure 10.8, with $L = 1$. After making use of the correspondence $x = -3.5 + (i - 1)\, \Delta x$, verify that the shape of this curve (but not the vertical scale) is the same as the curve for $L = 1$ in Figure 10.9 for the integral convolution, except near the endpoints where the truncation assumption spoils the agreement. ∎

An advanced treatment of convolution algorithms that is suitable for use with the FFT is given in Nussbaumer's book.

Now that we have some experience with numerical convolutions and insight into their properties, let us return to the analytical developments.

Fourier integral transforms and convolutions

We now derive analytical results for convolutions that depend on properties of Fourier integral transforms. These results will provide a powerful means of calculating convolutions, as we will show for Gaussian and Lorentzian functions.

The general result is the *Fourier convolution theorem*. This theorem states that the Fourier integral transform of the convolution of two functions is proportional to the product of their Fourier transforms. In symbols the convolution theorem reads

$$Y_{12}(k) = \sqrt{2\pi}\; Y_1(k)\, Y_2(k) \tag{10.45}$$

where the integral

$$Y_{12}(k) = \frac{1}{\sqrt{2\pi}} \int_{-\infty}^{\infty} y_1 * y_2(x)\, e^{-ikx}\, dx \tag{10.46}$$

denotes the Fourier integral transform of the convolution of the two functions, whose Fourier transforms are $Y_1(k)$ and $Y_2(k)$.

Exercise 10.23
To prove the convolution theorem, first write down $Y_1(k)$ and $Y_2(k)$ in terms of their definition, (10.4), using integration variables x' and x''. Then change the x'' variable to $x - x'$ and perform the x' integration first, which produces the convolution of y_1 with y_2. Show that the integral that remains is just proportional to the Fourier transform of this convolution, as (10.46) claims. ∎

This remarkable theorem has an immediate use for calculating convolutions. By expressing the convolution in terms of its Fourier transform using (10.45), then using the convolution theorem (10.46) for the transform you will find immediately that

$$y_1 \star y_2\,(x) = \int_{-\infty}^{\infty} Y_1(k)\,Y_2(k)\,e^{+\,ikx}\,dk \qquad (10.47)$$

Exercise 10.24
Find this result for a convolution in terms of the Fourier integral transforms of the two functions by carrying through the indicated steps. ■

We immediately put this convolution formula to use in computing the convolution of two Gaussians or of two Lorentzians.

Convolutions of Gaussians and of Lorentzians

You probably wondered why, when working examples of convolutions at the beginning of this section, we did not use the common Gaussian and Lorentzian functions, of which I seem to be so fond. Although convolution of Gaussians can be done directly from the definition (10.34), the derivation is messy. Convolution of two Lorentzians can be carried out directly, but it is also tedious unless complex-variable methods are used.

Convolution of two *Gaussians* by use of the convolution theorem, (10.46), is very direct because the Fourier transform of a Gaussian is just a Gaussian, as we showed in Section 10.2. The proof is easier done than said, so why not try it?

Exercise 10.25
(a) Use (10.46) to express the Fourier transform of the Gaussian convolution, $G_{12}(k)$, in terms of the product of the two transforms, G_1 and G_2, where (10.24) is used for each of these. Thus show directly that

$$G_{12}\,(k) = \frac{1}{\sqrt{2\pi}}\,e^{-\sigma_{x12}^2 k^2/2} \qquad (10.48)$$

where the convoluted width is obtained from

$$\sigma_{x12}^2 = \sigma_{x1}^2 + \sigma_{x2}^2 \qquad (10.49)$$

in terms of the standard deviations σ_{x1} and σ_{x2} of the two Gaussians in x space.
(b) From (10.48) show immediately, by using the inverse Fourier transform result (10.47), that

$$g_1 \star g_2(x) = g(x,\,\sigma_{x\,12}) \qquad (10.50)$$

where g is the Gaussian defined by (10.21). ■

Therefore, two Gaussians convolute to a single Gaussian whose standard deviation is given by the sum-of-squares formula (10.49). This result can readily be generalized, as follows:

Exercise 10.26
Use the method of mathematical induction to show that successive convolutions of K Gaussians always produces Gaussians with total standard deviation, σ_K, obtained from the sum-of-squares formula

$$\sigma_K^2 = \sum_{i=1}^{K} \sigma_i^2 \tag{10.51}$$

a result which holds for widths in either x space or in k space, depending on which space is used to perform the convolution. ■

This formula is the same as that used in statistical error analysis for K independent and normally-distributed variables. The Gaussian (normal) distribution was introduced in Section 6.1 in the context of maximum likelihood and least squares. But what is the connection of error analysis with convolutions? If the error distribution for the ith source of errors is Gaussian with standard deviation σ_i, and if the errors are all independent, then each point on a given error distribution has to be convoluted with those due to all the other sources of error in order to estimate the total error distribution. The standard deviation of this distribution is just given by (10.51).

We derived the convolution theorem and its application to convolution of Gaussians by formal analysis. How can you be convinced that the result (10.49) is plausible? Just use the program `Convolute Arrays` that we developed in the preceding subsection.

Exercise 10.27
(*a*) Modify program `Convolute Arrays`, given above, so that `CONARRAY1` and `CONARRAY2` are filled with Gaussians extending well into their tails for input choices of the two standard deviations. Both Gaussians should have the same stepsize, chosen small compared with the smaller of the two standard deviations. Then run the program to estimate the discretized convolution (10.44).

(*b*) Use the trapezoidal rule to estimate the area under the resulting distribution, then divide each point of the distribution by this area, in order to produce a function with unit area.

(*c*) Add some code to compare how well the discretized convolution approximates the Gaussian with standard deviation given by (10.49). Discuss the reasons for discrepancies as a function of position along the distribution and the stepsize. ■

Convolution of two *Lorentzians* is also most easily accomplished by using the convolution theorem (10.46) applied to the exponentials, (10.33), that result from

Fourier transforming Lorentzians. Several changes of notation are required. We write the convolution theorem as

$$L_1 * L_2(\omega) = \int_{-\infty}^{\infty} P_1(t) P_2(t) e^{+it\omega} dt \qquad (10.52)$$

Now there are several straightforward analytical steps to obtain the convolution formulas.

Exercise 10.28
Substitute the expressions for the Fourier transforms P_1 and P_2, obtained using (10.33), into the integral (10.46), then regroup real and imaginary exponents in the two exponentials. Next, identify the integral as the Fourier transform of a Lorentzian whose parameters are additive in the parameters of the original Lorentzians, namely

$$L_1 * L_2(\omega) = \frac{\pi}{2} \frac{\gamma_1 \gamma_2}{\gamma_{12}} L_{12}(\omega) \qquad (10.53)$$

where the convolved Lorentzian is

$$L_{12}(\omega) = \frac{(\gamma_{12}/2)^2}{(\omega - \omega_{12_0})^2 + (\gamma_{12}/2)^2} \qquad (10.54)$$

in which the total FWHM is additive in the input FWHM values

$$\gamma_{12} = \gamma_1 + \gamma_2 \qquad (10.55)$$

as is the resonance frequency

$$\omega_{12_0} = \omega_{1_0} + \omega_{2_0} \qquad (10.56)$$

Thus, Lorentzians convolve into Lorentzians with additive parameters. ∎

The additivity of Lorentzian widths expressed by (10.55) has a consequence for observed spectral line widths of radiating systems, such as atoms or molecules emitting (or absorbing) light or other electromagnetic radiation, as discussed at the end of Section 10.4 and in Dodd's book. The width that one observes is the sum of the widths of the initial and final states between which the transition occurs. Therefore, the lifetime for decay (which is proportional to reciprocal of the width) depends on the widths of both initial and final states. This is discussed in the context of stellar spectroscopy in Section 9-1 of the book by Mihalas.

Note in the convolution result (10.54) that extra factors are obtained because we chose to be conventional and scaled the initial Lorentzians to unity peak values.

Since convolutions preserve areas, as shown by (10.43), one then cannot force the peak value of the convolved Lorentzian, L_{12}, to be unity. In order to understand this normalization problem for Lorentzians, it is instructive to define the *normalized Lorentzian* by

$$L_N(\omega) \equiv \frac{N}{(\omega - \omega_0)^2 + (\gamma/2)^2} \tag{10.57}$$

where the normalization factor, N, is to be chosen so that the integral over ω, negative as well as positive values, is unity.

Exercise 10.29

Apply a tangent transformation to the variable of integration ω in (10.57), then integrate $L_N(\omega)$ to show that N must be given by

$$N = \gamma/2\pi \tag{10.58}$$

if the integral is to be unity. ■

Thus, the normalized Lorentzian is given by

$$L_N(\omega) \equiv \frac{\gamma/2\pi}{(\omega - \omega_0)^2 + (\gamma/2)^2} \tag{10.59}$$

The convolution result for the normalized Lorentzian now reads

$$L_{N_1} * L_{N_2}(\omega) = L_{N_{12}}(\omega) \tag{10.60}$$

where the FWHM and resonance frequency relations are given by (10.55) and (10.56). Convolution of such normalized Lorentzians conserves the area under each Lorentzian.

Exercise 10.30

Use mathematical induction to prove that successive convolutions of any number of normalized Lorentzians produces a normalized Lorentzian whose FWHM is the sum of its component Lorentzians and whose resonance frequency is the sum of its component resonance frequencies. ■

For Lorentzians, as for the Gaussians, it is interesting to see convolution results occurring in practice, and therefore to try a numerical example with a discretized convolution (10.44).

Exercise 10.31

(*a*) Modify program `Convolute Arrays` so that the files `CONARRAY1` and `CONARRAY2` are filled with Lorentzians extending well into their tails for input choices of the two standard deviations. As we discussed in Section 10.2 when comparing Gaussians and Lorentzians (Figure 10.6), this will require a wider range of x values than for a Gaussian of the same FWHM. Both Lorentzians should use the same stepsize, which should be chosen small compared with the smaller FWHM. Run the program to estimate the discretized convolution (10.44).

(*b*) Add some code to compare how well the discretized convolution approximates the Lorentzian with FWHM given by (10.55). Discuss the reasons for discrepancies as a function of position along the distribution and of stepsize. ∎

We now have a good appreciation of the analysis, numerics, and applications of convolving two Gaussians or two Lorentzians. It might appear that it will be as simple to convolute a Gaussian with a Lorentzian, which is a very common example in applications. This is not so, however, and extensive analysis and numerics are required, as we now develop.

Convoluting Gaussians with Lorentzians: Voigt profile

As we derived in Section 8.1, a Lorentzian as a function of the frequency ω often arises in the description of resonance processes in mechanical, optical, or quantal systems. The Gaussian usually arises from random processes, such as random vibration of a mechanical system or of molecules in a gas. Therefore, convolution of Gaussians with Lorentzians will be required whenever a resonance process occurs in a system which is also subject to random processes, such as when atoms in a very hot gas emit light. The emission spectra of light from stars is an example we develop in Section 10.4.

We now derive formulas for the convolution of a Gaussian with a normalized Lorentzian, which is called the Voigt profile. (Woldemar Voigt, 1850 – 1919, was a German physicist who developed the science of crystallography, discovered the Lorentz transformations, and coined the term "tensor.") In Project 10 (Section 10.4) we present the numerics and applications of the formulas for the Voigt profile. The method we develop is most suitable for preparation of tables (in computer memory or on paper) that have to be frequently used. If only a few values of the Gaussian and Lorentzian parameters are to be used, then direct numerical integration of the convolution integral may be more appropriate. The derivations that follow also require several mathematical analysis skills that are commonly needed in numerical and scientific applications.

To perform the convolution, it is most convenient (and also instructive) to use the convolution theorem, since we know the Fourier transforms of Gaussians, $G(k, \sigma_k)$, and of Lorentzians, $P(k)$. If their formulas, (10.24) and (10.33), are inserted in the convolution formula (10.34), and if ω is made the variable of the convolution, then some redefinition of variables leads to a simple-looking expression for the Voigt profile.

Exercise 10.32

Substitute the Gaussian and normalized Lorentzian Fourier transforms in the convolution formula (10.34) as indicated, then make the following substitution of variable

$$x = \sqrt{2}\, \sigma_x k \qquad (10.61)$$

and introduce the scaled parameters

$$a \equiv \frac{\gamma}{2\sqrt{2}\, \sigma_x} \qquad (10.62)$$

$$v \equiv \frac{\omega - \omega_0}{\sqrt{2}\, \sigma_x} \qquad (10.63)$$

to show that the convolution of a Gaussian with a normalized Lorentzian is given by

$$g \star L_N(\omega) \equiv \frac{1}{\sqrt{\pi}} H(a,v) \qquad (10.64)$$

where $H(a,v)$ is the *Voigt function*, defined by

$$H(a,v) \equiv \frac{1}{\sqrt{\pi}} \int_0^\infty e^{-\left(ax + x^2/4\right)} \cos(vx)\, dx \qquad (10.65)$$

in which all variables are dimensionless. ∎

The Voigt function is often also called the Voigt profile. Note that $H(a,v)$ is an even function of v, since it is the convolution of two functions that are even functions of their arguments. This result is useful in numerical work, both in programming and in debugging code.

It is interesting and relevant that the results about Gaussians together convoluting to Gaussians and Lorentzians with Lorentzians producing Lorentzians implies that Voigt functions have the same property, so that these convolutions all stay in the same families.

Exercise 10.33

(a) Prove that the convolution of two Voigt functions is also a Voigt function. In order to do this use the results from Exercise 10.26 that the convolution of two Gaussians produces a Gaussian and from Exercise 10.30 that two Lorentzians when convolved produce a Lorentzian. In addition, you need the results that the process of convolution is associative and commutative.

(b) Demonstrate this analytical result, at least approximately, by using Program 10.1, Convolute Arrays, to convolute two Voigt-profile arrays that are the outputs from Program 10.2, Voigt Profile. Compare the convolution that results with that obtained by this program, using as input an appropriate Gaussian of total width given by (10.49) and a Lorentzian of total width from (10.56), in terms of the widths input to form the two Voigt profiles.

(c) Generalize the result in (a) to prove that convolution of any number of Voigt profiles produces a Voigt profile, similarly to (and dependent on) the results derived in Exercises 10.26 and 10.30 for Gaussians and Lorentzians. ∎

These results have a very practical consequence, as follows. Suppose that there are several sources of Gaussian-shape line broadening, and possibly several contributions to Lorentzian-type line widths, all modifying the same spectral line, with other conditions being constant. The resulting spectral profile will be a Voigt profile that can readily be obtained by first combining all the Gaussian widths quadratically by (10.51) and all the Lorentzian widths additively by the generalization of (10.55), then computing a single Voigt profile.

Computationally, the result in Exercise 10.33 is a convenience, but scientifically it is a hindrance because it prevents one from recognizing the individual line-broadening contributions, and thereby having some possibility of determining their physical origin. Sources of such line broadening in stellar environments and models for estimating the broadening are discussed in Chapter 9 of the book by Mihalas.

The integral expression for the Voigt function $H(a,v)$ in (10.65) generally cannot be simplified. For any given a and v it may be estimated by numerical integration, which is one option considered in Project 10 below. A very common situation is that the resonance, which has FWHM γ, is very much broadened by convolution with the Gaussian, that is, one has $a \ll 1$. It is then practical to expand $H(a,v)$ in a Taylor series in powers of a by

$$H(a,v) \equiv \sum_{n=0}^{\infty} a^n H_n(v) \tag{10.66}$$

If this series converges rapidly, then it is a practical series, especially since the auxiliary functions $H_n(v)$ may be computed independently of the choice of a.

Exercise 10.34

(a) In the integrand of the Voigt function (10.65) make a Maclaurin expansion of the exponential, then interchange integration and summation in order to show that

$$H_n(v) = \frac{(-1)^n}{n! \sqrt{\pi}} \int_0^{\infty} e^{-x^2/4} x^n \cos(vx) \, dx \tag{10.67}$$

(b) Differentiate this expression twice with respect to variable v, which may be taken under the integral sign because the only dependence of the integral on v is in the cosine function. Identify the result as proportional to the function with the n value two larger. Thus derive the recurrence relation

$$H_{n+2}(v) = -\frac{1}{(n+1)(n+2)} \frac{d^2 H_n(v)}{dv^2} \tag{10.68}$$

which enables all the $H_n(v)$ to be calculated by recurrence from $H_0(v)$ and $H_1(v)$. ■

Our computation is now reduced to finding appropriate formulas for the two lowest H functions, then being able to differentiate these with respect to v.

The function $H_0(v)$ may be derived either by some laborious mathematics or by noting that it is proportional to the value of the convolution when $a = 0$, which is just the Gaussian function with altered normalization. Thus we can show that

$$H_0(v) = e^{-v^2} \tag{10.69}$$

Exercise 10.35

For those who are skeptical of the above reasoning, derive $H_0(v)$ from the definition (10.64). To do this, you will need to change variables to a complex variable z, then to use the normalization integral for the Gaussian function. After some tedious and error-prone steps you should obtain (10.69). ■

Now that we have $H_0(v)$, it is straightforward to generate by differentiation the successive $H_n(v)$ with n even. For example, the second derivative of $H_0(v)$ leads immediately to

$$H_2(v) = (1 - 2v^2) e^{-v^2} \tag{10.70}$$

Exercise 10.36

Carry out the indicated differentiation and normalization by using (10.69) in (10.68) with $n = 0$ in order to derive (10.70). ■

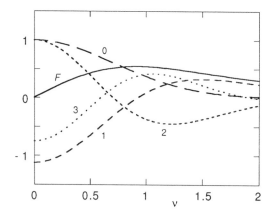

FIGURE 10.10 The H_n functions in (10.66) for $n = 0, 1, 2, 3$, and the Dawson integral F in (10.72), each as a function of v.

For usual values of a in the expansion (10.66), $H_n(v)$ with $n > 3$ is not required, but you can produce H_4, etc., if you are willing to keep on differentiating. In
Figure 10.10 you can see the exponentially damped behavior of H_0 and H_2.

The values of $H_n(v)$ for n odd require more mathematical analysis, and eventually more computation, than for n even. Although one may calculate H_1 directly
from the definition (10.67), some manipulation of the integral puts it into a form that
is more suitable for differentiating to produce the $H_n(v)$ for larger odd-n values.
As usual, you can exercise your analysis skills by completing the intermediate steps.

Exercise 10.37

(a) Use integration by parts in (10.67) for $n = 1$, with $\cos(vx)$ being the quantity that is differentiated. Transform variables to $z = x^2$ on the part that is integrated. Thus show that

$$H_1(v) = \frac{2}{\sqrt{\pi}} [v F(v) - 1] \tag{10.71}$$

where $F(v)$ is called *Dawson's integral*, defined by

$$F(v) \equiv \int_0^\infty e^{-x^2/4} \sin(vx)\, dx \tag{10.72}$$

(b) From the definition, show that Dawson's integral is an odd function of v,
and therefore that $H_1(v)$ is an even function of v, as required.

(c) By expressing the sine function in (10.72) in terms of the complex exponential of vx by Euler's theorem, then introducing the variable $u = -ix/2 - v$ and
manipulating the limits of integration before using the normalization integral for
the Gaussian, show that Dawson's integral can be written as

$$F(v) = e^{-v^2} \int_0^v e^{u^2}\, du \tag{10.73}$$

This form is more suitable than (10.72) for analysis and numerical work. ∎

It appears that we are sinking deeper and deeper into the quicksands of mathematical complexity; fear not, we will drag ourselves out by our bootstraps. We now
have H_1 in terms of a reasonable-looking integral, (10.73). For H_3 we will need derivatives of $F(v)$, which are straightforward to obtain from this second form of
Dawson's integral.

Exercise 10.38

(a) Differentiate $F(v)$ given by (10.73) with respect to v, noting that the derivative of an integral with respect to its upper limit is just the integrand evaluated at
this limit, in order to show that

$$F'(v) = 1 - 2v F(v) \tag{10.74}$$

(*b*) By reusing this result, show that the second derivative of Dawson's integral is given by

$$F''(v) = -2\left[\left(1 - 2v^2\right)F(v) + v\right] \qquad (10.75)$$

Note that the first derivative is an even function of v and that the second derivative is odd in v, as required from the evenness of F under reflection of v. ∎

This completes our analysis of Dawson's integral. In Project 10 we address the practical concerns of computing this function, making use of the results that we have just derived.

We now return to the job of obtaining formulas for the H_n. The only one that we will require but do not yet have is H_3. This can be obtained from H_1 by using the derivative relation (10.68).

Exercise 10.39

Differentiate H_1 from (10.71) twice with respect to v, substitute the derivatives of F according to (10.74) and (10.75), then use the recurrence relation (10.68) in order to show that

$$H_3(v) = \frac{4}{3\sqrt{\pi}}\left[\left(3 - 2v^2\right)v\,F(v) + v^2 - 1\right] \qquad (10.76)$$

which is again explicitly even in v. ∎

The functions H_1 and H_3, given by (10.71) and (10.76), are shown in Figure 10.10. They slowly damp out as v increases because they are basically components of a Gaussian. The computation of $F(v)$ that is needed for these functions is described in Section 10.4.

We have finally produced an expression for the Voigt profile through terms in the ratio of Lorentzian to Gaussian widths, a, up to cubic. By continuing to higher derivatives the function $H(a, v)$ may be expressed in terms of exponential functions and Dawson's integral function $F(v)$, (10.73).

10.4 PROJECT 10: COMPUTING AND APPLYING THE VOIGT PROFILE

In this section we develop the numerics of computing the Voigt profile, starting from the analysis in the preceding subsection. Then we apply this profile to relate temperatures to the widths of atomic emission lines from stellar sources. This project also illustrates how to make the step-by-step development and programming of algorithms for numerical computation that results in effective and tested programs that can be adapted to various uses in science and engineering.

The numerics of Dawson's integral

Dawson's integral, the function $F(v)$ defined as (10.72) or (10.73), arose as an intermediate step in our calculation of the functions $H_n(v)$ for the series expansion in powers of a for the Voigt profile, formula (10.66). Although we display $F(v)$ in Figure 10.10, we have not yet shown how to approximate it numerically.

Series expansion is the most obvious way to express $F(v)$, since inspection of Figure 10.10 or of tables (such as Abramowitz and Stegun, Table 7.5) shows that it is a smooth function of small range that varies approximately proportional to v when v is small. Further, for large v, F appears to be about $1/(2v)$. Appropriate series can be found as follows.

Exercise 10.40

(*a*) Consider the first-order nonlinear differential equation for F, (10.75). Assume a series solution of odd powers of v, since we know that $F(v)$ is an odd function of v. Thus show that

$$F(v) = v - \frac{2v^3}{3} + \frac{2v^5}{5} - \cdots \tag{10.77}$$

(*b*) Express the first derivative of F with respect to v as a first derivative with respect to $1/v$, then use (10.74) to show that this quantity is finite as $v \to \infty$ only if

$$F(v) \to \frac{1}{2v} \qquad v \to \infty \tag{10.78}$$

which provides the asymptotic value of Dawson's integral. ∎

Although the series in v, (10.77), is appropriate for very small v, its convergence is too slow for much practical use, since $v \le 2$ is a fairly common argument.

Numerical integration from the formula (10.73) is a practical method for v values that are not large enough for (10.78) to be accurate. Some care needs to be taken because both large and small exponentials appear in the integration. Computer underflow and overflow will therefore be a problem for large enough v. One way to reduce this problem is to rewrite (10.73) as

$$F(v) = e^{-v^2/2} \int_0^v e^{u^2 - v^2/2} \, du \tag{10.79}$$

The range of exponentials that appear is then $\exp(-v^2/2)$ to $\exp(+v^2/2)$.

Exercise 10.41

Show that if the maximum power of 10 that the computer can handle is PMAX, and that a safety factor is 10 is allowed for, then the maximum value of v that can be used with the form (10.79) is VLIM = $\sqrt{[2*\ln(10)*(\text{PMAX}-1)]}$. ∎

This condition is checked in the function FDawson given in Program 10.2.

Another consideration in the numerical integration is that if $v \leq 0$, the integration can either not be started ($v = 0$) or should range over decreasing values of u. To avoid the awkward coding involved for the latter case, one may use the fact that F is an odd function of v, so that $F(v) = (v/|v|)F(|v|)$. When v is nearly zero, the integral also needs to be set to zero in order to avoid underflow in the exponential function.

The program for Dawson's integral can therefore be written compactly as shown in function FDawson in the complete Voigt Profile program in the next subsection. For simplicitly, the trapezoidal rule, (4.43), is used for numerical integration.

Exercise 10.42

(a) Code and run as a stand-alone program FDawson, with PMAX adjusted for your computer. Check your programming against numerical values for both small and large v from the series expansions (10.77) and (10.78), respectively. Table 7.5 in Abramowitz and Stegun may also be used for this purpose.

(b) Plot Dawson's integral, $F(v)$, against v for, say, $0 \leq v \leq 2$. Check the appearance of your graph against the curve for $F(v)$ in Figure 10.10. ∎

Now that the numerics of Dawson's integral has been covered, the $H_n(v)$ functions are relatively straightforward to compute.

Program for series expansion of profile

The series expansion of the Voigt profile $H(a,v)$ in powers of the ratio of Lorentzian to Gaussian widths, a, is given by (10.66). The functions $H_n(v)$ for $n = 0 - 3$ are specified by (10.69), (10.71), (10.70), and (10.76), and for $n = 1$ and 3 require Dawson's F function that we have just considered.

Exercise 10.43

(a) Program the formulas for the $H_n(v)$, using the functions given in Voigt Profile below. Check values for even n against hand calculations and those for odd n against spot checks for values of $F(v)$ from Exercise 10.42.

(b) Plot the $H_n(v)$ functions against v for say $0 \leq v \leq 2$. Compare your curves with those in Figure 10.10. ∎

Now that we are convinced that the component functions for the Voigt function are correct, it is time to assemble them to form the series expansion (10.66). Given that you are using the divide-and-conquer approach to program writing and verification, this is now straightforward (especially if you copy out the `Voigt Profile` code that we provide). Note that for `FDawson`, since each value is used twice, we significantly improve the program efficiency by invoking this function only once, then we save the value. The programming for H_1 and H_3 must be slightly different than in Exercise 10.42, in order to accommodate this sharing of F. We also include the code section for computing the Voigt function by direct numerical integration of its Fourier transform as described in the next subsection.

The program `Voigt Profile` includes writing to an output file, "`VOIGT`", in order to save the Voigt function values for graphing. Also included are comparison values of the appropriate Gaussian and Lorentzian functions that are also output to the file. Our choice of units, which also reduces the complexity of coding and input, is to set the Gaussian standard deviation $\sigma_x = 1/\sqrt{2}$.

PROGRAM 10.2 Voigt profiles by series and by direct integration.

```
#include <stdio.h>
#include <math.h>
#define MAX 102

main()
{
/* Voigt Profile; Series,Integral,Gaussian,Lorentzian */
/*   Gaussian standard deviation set to 1/sqrt(2)   */
FILE *fout;
FILE *fopen();
double gLseries[MAX],gLtrap[MAX],g[MAX],L[MAX];
double pi,sp,gma,vmax,dv,dx,vmin,a,v,even,odd,FD;
int Nx,Niv,iv;
char wa;
double FDawson(),Hzero(),Hone(),Htwo(),Hthree(),trap();

pi = 4.0*atan(1.0);
sp = sqrt(pi);
printf("Voigt Profiles\n");
gma = 1.0;  /* gma is Lorentzian FWHM */
printf("Write over output (w) or Add on (a): ");
scanf("%s",&wa);   fout = fopen("VOIGT",&wa);
while ( gma > 0 )
  {
  printf("Input gma,vmax,dv,dx,Nx (gma<=0 to end):\n");
  scanf("%le%lf%le%le%i",&gma,&vmax,&dv,&dx,&Nx);
  if ( gma > 0 )
    {
    vmin = -vmax;
```

```
    Niv = (vmax-vmin)/dv+1.1;
    if ( Niv > MAX+1 )
      {
      printf("!! Niv=%i > array sizes %i\n",Niv,MAX-1);
      exit(1);
      }
    a = gma/2;
    v = vmin;
    for ( iv = 1; iv <= Niv; iv++ )
      {  /* loop over  v  values */
      even = Hzero(v)+a*a*Htwo(v);/* even series */
      FD = FDawson(v);/* odd series has Dawson */
      odd = a*(Hone(sp,FD,v)+a*a*Hthree(sp,FD,v));
      gLseries[iv] = (even+odd)/sp; /* series value */
      /* Trapezoid integral */
      gLtrap[iv]=trap(a,v,dx,Nx)/pi;
      g[iv]=exp(-v*v)/sp;/* Gaussian; sigma=1/sqrt(2) */
      L[iv]=(a/pi)/(v*v+a*a);/* normed Lorentzian;a=gma/2 */

      printf("%6.2lf %10.6lf %10.6lf %10.6lf %10.6lf\n",
              v,gLseries[iv],gLtrap[iv],g[iv],L[iv]);
      fprintf(fout,"%6.2lf %10.6lf %10.6lf %10.6lf %10.6lf\n",
              v,gLseries[iv],gLtrap[iv],g[iv],L[iv]);
      v = v+dv;
      } /* end  v  loop */
    }
  } /* end  gma  loop */
printf("\nEnd  Voigt  Profile");
}

double FDawson(v)
/* Dawson's integral approximated by trapezoid rule */
double v;
{
double PMAX,VLIM,absv,vsq2,expv,trap,u,du,signv;
int Nu,iu;

absv = fabs(v);
du = 0.005; /* gives 5 sig.fig. accuracy for v <= 2 */
Nu = absv/du;   du = absv/Nu; /* more-exact  du  */
PMAX = 10; /* maximum power of 10 for computer */
VLIM = sqrt(2*log(10)*(PMAX-1)); /* max v */
if ( absv < pow(10,-PMAX) )
  {
  trap = 0;  return trap;
```

```
    }
else
  {
  signv = v/fabs(v);
  if ( absv > VLIM )
    {
    printf("!! |v|=%8.2lf set to VLIM=%8.2lf",absv,VLIM);
      absv = VLIM;
  }
  vsq2 = absv*absv/2;   expv = exp(-vsq2);
  trap = expv/2;
  u = du;
  for ( iu = 1; iu < Nu; iu++ )
    {
    trap = trap+exp(u*u-vsq2);   u = u+du;
    }
  trap = signv*du*(expv*trap+0.5);
  return trap;
  }
}

double Hzero(v) /* Zeroth-order H function */
double v;
{
double h;
h = exp(-v*v);
return h;
}

double Hone(sp,FD,v) /* First-order H function */
double sp,FD,v;
{
double h;
h = 2*(2*v*FD-1)/sp;
return h;
}

double Htwo(v) /* Second-order H function */
double v;
{
double h;
h = (1-2*v*v)*exp(-v*v);
return h;
}
```

```
double Hthree(sp,FD,v) /* Third-order H function */
double sp,FD,v;
{
double h;
h = 4*((3-2*v*v)*v*FD+v*v-1)/(3*sp);
return h;
}

double trap(a,v,dx,Nx)
/* Trapezoid rule direct integral for Voigt profile */
double a,v,dx;
int Nx;
{
double x,sum;
int ix;

sum = 0.5;    x = 0;
for ( ix = 1; ix <= Nx; ix++ )
  {
  x = x+dx;
  sum = sum+cos(v*x)/exp(x*(a+x/4));
  }
sum = dx*sum;
return sum;
}
```

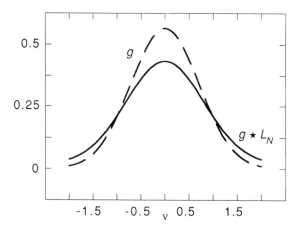

FIGURE 10.11 The Voigt profile, (10.65), with ratio-of-widths parameter $a = 0.25$.

Figure 10.11 illustrates the Voigt profile with a normalized Lorentzian for the parameter $a = 0.25$, corresponding to a ratio of Lorentzian to Gaussian widths for which convergence of the series expansion to about 1% accuracy is expected. As anticipated from our discussion of the convolution of a Gaussian with a Lorentzian and from Figure 10.11, the Voigt profile resembles a Gaussian with wings broadened by the Lorentzian tails. Since the convolution preserves area, according to (10.43), the peak value must be slightly reduced by convolution, as you see.

Now that you are convinced that the analysis and program are correct, it is a good time for you to develop a complete program for the Voigt profile.

Exercise 10.44

(a) Adapt Voigt Profile for your computer system. Note that the parameter PMAX in function FDawson is computer-dependent, as discussed in Exercise 10.41. For check values of the functions H_n consult Exercise 10.43.

(b) Plot the output function $H(a,v)$ against v for interesting values of a that are somewhat less than unity, in order to assure reasonable convergence of the series expansion (10.66). Compare with the curves given in Figure 10.11. ∎

The series expansion method is practical if many small values of a are to be used but the same range of v is to be covered. Then it is practical to save these values for reuse when a is changed. For general-purpose use, direct integration of the function in (10.65) is more appropriate.

Program for direct integration of profile

The final computing part of our Voigt profile project is to integrate the Fourier transform expression directly, thereby being able to check the convergence of the expansion in powers of the width-ratio parameter a, (10.62), as it appears in the series (10.65) for $H(a, v)$.

Exercise 10.45

Code and execute the trapezoid-rule program trap with the integrand from (10.65). Check the convergence of the program as a function of the input step size dx and number of steps Nx. ∎

For dx = 0.05 and Nx = 100, I found agreement to four decimal places between the series expansion and the numerical integration for a = 0.25 out to $|v| = 2$.

A major lesson from our analysis of convolution of a Gaussian with a Lorentzian (calculation of the Voigt profile) is that the complexity of mathematical and numerical analysis, and consequently of programming, can increase remarkably when the functions involved are changed in fairly simple ways. It is therefore also worthwhile to expend a reasonable effort in analysis of a problem before rushing to "give it to the computer." Even if the final results are primarily numerical, as for the Voigt function, the initial analysis may lead to improved understanding of the mathematics and science.

Application to stellar spectra

The spectra of light emitted by stars and observed at the Earth may be used to characterize the plasma environment from which the light was emitted. Allen's book on atoms, stars, and nebulae, provides the astronomy and astrophysics background. The details are described in, for example, Chapter 9 of the book by Mihalas on stellar atmospheres, which has a comprehensive introduction to the subject of spectral-line absorption profiles and line broadening. Here is a schematic summary of the main ideas, sufficient to indicate the order of magnitude of the quantities used to calculate Voigt profiles for stellar spectra.

The simplest example is that of atoms radiating in a gas at temperature T. A spectral "line" of width γ is then Doppler-broadened because of the thermal agitation of the atoms, with some atoms moving towards the observer and some moving away at the moment they emit light. For example, suppose that a spectral line such as that for hydrogen (the predominant element in stars) is of Lorentzian line shape with a FWHM of 0.01 nm at a wavelength of 500 nm. At $T = 10^4$ K for hydrogen the width of the Maxwell-Boltzmann distribution describing the spread of atom velocities in the plasma is about 4.3×10^{-5} times the speed of light.

Under these conditions, the parameter for the Voigt profile, a, given by (10.62), is then about 0.25. The resulting Voigt profile is then that shown in Figure 10.11. It is approximately of Gaussian shape for v in the range –0.5 to 0.5, and therefore the shape of this region of the spectral distribution is dominated by the temperature distribution in the stellar atmosphere. In the wings of the distribution the profile becomes Lorentzian in shape and is affected by the natural line width of the emitting atoms plus effects such as collisions with electrons in the stellar plasma.

For laboratory plasmas, such as in fusion-energy devices, measurements of spectral-line broadening may be used to characterize properties such as the temperature of the major regions of the plasma that are emitting the detected light and other electromagnetic radiation. In such applications, as also in stellar atmospheres, the very strong electric and magnetic fields also produce significant broadening of the spectral lines.

REFERENCES ON FOURIER INTEGRAL TRANSFORMS

Abramowitz, M., and I. A. Stegun, *Handbook of Mathematical Functions*, Dover, New York, 1964.

Allen, L. H., *Atoms, Stars, and Nebulae*, Cambridge University Press, Cambridge, England, third edition, 1991.

Bracewell, R. N., *The Fourier Transform and Its Applications*, McGraw-Hill, New York, second edition, 1986.

Churchill, R. V., *Fourier Series and Boundary Value Problems*, McGraw-Hill, New York, 1963.

Dodd, J. N., *Atoms and Light: Interactions*, Plenum, New York, 1991.

Hamming, R. W., *Digital Filters*, Prentice Hall, Englewood Cliffs, New Jersey, third edition, 1989.

Ingard, K. U., *Fundamentals of Waves and Oscillations*, Cambridge University Press, Cambridge, England, 1988.

Mihalas, D., *Stellar Atmospheres*, W. H. Freeman, San Francisco, second edition, 1978.

Nussbaumer, H. J., *Fast Fourier Transform and Convolution Algorithms*, Springer-Verlag, New York, second edition, 1982.

Pippard, A. B., *The Physics of Vibration*, Cambridge University Press, Cambridge, England, 1988.

Solomon, F., "Tomography: Three-Dimensional Image Reconstruction," UMAP Module 318, in *UMAP Modules Tools for Teaching*, COMAP, Arlington, Massachusetts, 1987, pp. 1 – 20.

EPILOGUE

Now that you have used this guidebook to learn and practise mathematical analysis, numerics, and their applications to the sciences and engineering, I hope that you will be eager to explore further afield, applying your new understanding to pose and solve problems in your own field of endeavor.

As I emphasized in Chapter 1, comprehension of the basics of analytical and numerical techniques is necessary before one travels the smooth highways provided by programming systems. Now that you have such comprehension, many of your computing and graphics tasks can be handled by systems such as Wolfram's *Mathematica*. The book of *Mathematica* examples by Abell and Braselton will also help you acquire proficiency in using this system. If you are using workstations, the book by Landau and Fink should provide useful guidance.

If in your computing you wish to work at the programming level (to make a distinction that was emphasized in Section 1.2), the books of numerical recipes by Press et al. are convenient. They are available with C code and also in Fortran or Pascal. For the programming of mathematical functions of the kind described in the compendium by Abramowitz and Stegun, the handbook and programs provided by Baker are suitable.

Tomorrow to fresh fields and pastures new.

References

Abell, M. L., and J. P. Braselton, *Mathematica by Example*, Academic, New York, 1992.

Abramowitz, M. and I. A. Stegun, *Handbook of Mathematical Functions*, Dover, New York, 1964.

Baker, L., *C Mathematical Function Handbook*, McGraw-Hill, New York, 1992.

Landau, R. H., and P. J. Fink, *A Scientist's and Engineer's Guide to Workstations and Supercomputers*, Wiley-Interscience, New York, 1992.

Press, W. H., B. P. Flannery, S. A. Teukolsky, and W. T. Vetterling, *Numerical Recipes in C*, Cambridge University Press, New York, 1988.

Press, W. H., B. P. Flannery, S. A. Teukolsky, and W. T. Vetterling, *Numerical Recipes: The Art of Scientific Computing (FORTRAN Version)*, Cambridge University Press, New York, 1989.

Press, W. H., B. P. Flannery, S. A. Teukolsky, and W. T. Vetterling, *Numerical Recipes in Pascal: The Art of Scientific Computing*, Cambridge University Press, New York, 1989.

Wolfram, S., *Mathematica: A System for Doing Mathematics by Computer*, Addison-Wesley, Redwood City, California, second edition, 1991.

APPENDIX

TRANSLATING BETWEEN C, FORTRAN, AND PASCAL LANGUAGES

Here, in tabular and example form, is an outline of the main differences between C, Fortran, and Pascal. It is intended to help you overcome the barrier to translating between these three languages, rather than to substitute for learning thoroughly the language of your choice. In particular (as discussed in Section 1.2), the sample programs in this book use only the parts of C relating to numerical applications, which is a limited subset of the full language.

If you wish to become a competent programmer in the C language, the References on Learning and Using C at the end of Chapter 1 should be helpful. The book by Gehani and that by Müldner and Steele, are suitable if you are experienced in other computer languages. Kerrigan's book is specifically aimed to help the transition from Fortran to C, while the book by Shammas introduces C to Pascal programmers.

The material following summarizes the aspects of C language that are used in this book, together with workable counterparts in Fortran and C. The C examples of code segments are drawn from the sample programs, and I indicate in which program a similar code segment can be found. The Fortran and Pascal code segments are likely to be correct, but they have not been tested by compiling them within complete programs. I have not attempted to indicate correspondences between the three languages for topics that are highly dependent on the computing environment, such as file declarations.

In the following table the items in italics are generic names that are substituted by actual names, as shown in the examples.

C	Fortran	Pascal

Overall Structure and Procedure Declarations

C	Fortran	Pascal
main ()	program *name*	program *name* ;
.	.	.
function ()	*function* ()	*function* ();
.	.	.
void *function* ()	subroutine *name* ()	procedure ();
.	.	.
		main *program*

Example; See Program 2.1

Data Type Declarations

C	Fortran	Pascal
double *name* ;	real *name*	var *name* : real;
int *name* ;	integer *name*	var *name* : integer;
double *array* [*SIZE*];	real *array* (*SIZE*)	var array [1..*SIZE*] of real;
int *array* [*SIZE*];	int *array* (*SIZE*)	var array [1..*SIZE*] of integer;
char *name* ;	character *name*	var *name* : char;

Example; Program 3.1

C	Fortran	Pascal
double rmin,dr;	real rmin,dr	var rmin,dr: real;
int nterms,kmax;	integer nterms,kmax	var nterms,kmax: integer;

Input and Output

Console:

C	Fortran	Pascal
scanf(*input list*);	read(*,*format*) *input list*	read(*input list*);
printf(*format*, *output list*);	write(*,*format*) *output list*	write(*output list*);

Example: Program 2.1

C	Fortran	Pascal
printf("Input x1 ");	write(*,*)'Input x1'	write('Input x1');
scanf("%lf",x1);	read(*,*) x1	read(x1);

C	Fortran	Pascal

Files:

```
FILE *file name ;
FILE *fopen();
```

fscanf(*fin, format,*   read(*unit, format*)   get(*pointer to*

    input list);   *input list*   *file item*);

fprintf(*fout, format,*   write(*unit, format*)   put(*pointer to*

    output list);   *output list*   *file item*);

Example; Program 7.1

```
FILE *fin,fout;        (Require system-dependent file allocation)
FILE *fopen();
fprintf(fout,"%lf",    write(10,F10.2) Din    put(Din);
        Din);
```

Control Structures

while

```
while (condition)                      while (condition ) do
  {action }                            begin action   end;
```

Example: Program 2.1

```
while (x1 != 0 )                       while (x1 <> 0 ) do
  {                                    begin
printf("Input x1:");                   writeln('Input x1');
  }                                    end;
```

if

```
if (condition)       if (condition ) then    if (condition ) then
  {action }            action                begin action end;
```

Example: Program 2.1

```
if ( x1 == 0 )       if ( x1.EQ.0 ) then    if ( x1 = 0 )
  { printf ("End");    write (*,*) 'End'    begin
  }                                           writeln ("End");
                                            end;
```

C	Fortran	Pascal

if .. else

C	Fortran	Pascal

if (*condition*)
  {*action* 1}
else
  {*action* 2}
	if (*condition*) then
	  action 1
	else
	  action 2
		if (*condition*) then
		begin
		  action 1;
		end;
		else
		begin
		  action 2;
		end;

Example: Program 2.1

C	Fortran	Pascal

if (den == 0)
 { x1d2 = 0;
 }
else
 { x1d2 =
(x1*x2+y1*y2)/den;
 }
	if (den.EQ.0) then
	  x1d2 = 0;
	else
	  x1d2 =
	  (x1*x2+y1*y2)/den
		if (den = 0)
		begin
		 x1d2 := 0;
		end;
		else
		begin
		 x1d2 :=
		 (x1*x2+y1*y2)/den;
		end;

for

C	Fortran	Pascal
for (*loop condition*) { *action*}	(Sometimes DO loop)	for *loopcondition* do begin *action* end;

Example: Program 3.1

C	Fortran	Pascal

for(k=1;k<=kmax;k++)
 {
 term = r*term;
 sum = sum+term;
 }
	do 2 k=1,kmax,1
	 term = r*term
	 sum = sum+term
	2 continue
		for k:=1 to kmax do
		begin
		 term := r*term;
		 sum := sum+term;
		end;

C	Fortran	Pascal

switch

C	Fortran	Pascal

switch (*expression*)
{
 case *constant* 1:
  {*action*1}; break;
 case *constant* 2:
  {*action*2}; break;
 .
 default: {*action*};
}
	goto(*st* 1,*st* 2,.),
	  expression
	st 1 *action*1
	st 2 *action*2
	.
	C *st* 1 & *st* 2 are
	  statement numbers
		case *condition* of
		 1: *action* 1;
		 2: *action* 2;
		 .
		end;

Example: Program 3.4

C	Fortran	Pascal

switch (choice)
 {
  case 1:
   series=PSexp;
   break;
  case 2:
   series=PScos;
   break;
 }
	goto (1,2),choice
	1 series=PSexp
	2 series=PScos
		case choice of
		 1: series :=PSexp;
		 2: series :=PScos;

Program Operators

Arithmetic Power

C	Fortran	Pascal
pow (x,y)	$x**y$	(Requires function)

Assignment

C	Fortran	Pascal
variable 1=*expression* ;	*variable* 1=*expression*	*variable* 1:=*expression* ;

Increment & Decrement

++   increment by 1
−−   decrement by 1

C	Fortran	Pascal

Arithmetic Compare

Equal to;	==	.EQ.	=
Not equal to;	!=	.NE.	<>
Less than;	<	.LT.	<
Less than or equal;	<=	.LE.	<=
Greater than;	>	.GT.	>
Greater than or equal;	>=	.GE.	>=

Logical Operators

And;	&&	.AND.	AND
Or;	||	.OR.	OR
Not;	!	.NOT.	NOT

INDEX TO COMPUTER PROGRAMS

429

INDEX

First authors are referenced in full on the pages with numbers printed in italics.